Introduction to Micrometeorology

Second Edition

This is Volume 79 in the
INTERNATIONAL GEOPHYSICS SERIES
A series of monographs and textbooks
Edited by RENATA DMOWSKA, JAMES R. HOLTON and H. THOMAS ROSSBY

A complete list of books in this series appears at the end of this volume.

Introduction to Micrometeorology

Second Edition

S. Pal Arya
DEPARTMENT OF MARINE, EARTH AND ATMOSPHERIC SCIENCES
NORTH CAROLINA STATE UNIVERSITY
RALEIGH, NORTH CAROLINA, USA

An Imprint of Elsevier

San Diego San Francisco New York
Boston London Sydney Tokyo

This book is printed on acid-free paper.

Academic Press
An Imprint of Elsevier
525 B Street, Suite 1900, San Diego, California 92101-4495, USA
http://www.academicpress.com

Academic Press
An Imprint of Elsevier
Harcourt Place, 32 Jamestown Road, London NW1 7BY, UK
http://www.academicpress.com

ISBN-13: 978-0-12-059354-5
ISBN-10: 0-12-059354-8

Transferred to Digital printing 2009

To my wife, Nirmal, and
my children, Niki, Sumi and Vishal

Contents

Preface to the Second Edition

It has been nearly twelve years since the publication of the first edition of this book. During this time, many journal papers have been published on various aspects of micrometeorology and the related areas of atmospheric boundary layer and turbulence. Several new books and monographs have also been published (Stull, 1988; Sorbjan, 1989; Monteith and Unsworth, 1990; Garratt 1992; Kaimal and Finnigan, 1994). The primary motivation for preparing this second edition came from the perceived need to update the material, incorporate some worked-out example problems in each chapter, and include a comprehensive treatment of observations, theories and models of the atmospheric boundary layer. I hope that I have accomplished these tasks, while retaining the original scope and objectives given in my Preface to the First Edition. Updating has resulted in an expanded list of references. Worked-out example problems should help the reader in solving other problems and exercises given at the end of each chapter. The new Chapter 13 provides a comprehensive review of stratified atmospheric boundary layers, while much of the material of the old chapter on the marine atmospheric boundary layer has been incorporated in other chapters. Other chapters have also been revised and updated, wherever appropriate.

I would like to thank the reviewers of my proposal for the second edition for their excellent suggestions. I would also like to thank the publishers and authors for their permission to reproduce many figures and tables in the book. The original sources are acknowledged in the legends of figures and tables, whenever applicable. Finally, I wish to acknowledge the help of Ms Mel Defeo and Ms Michele Marcusky for word-processing of the manuscript and in performing the painstaking tasks of spellchecking, proofreading, and correcting several times during the last one and a half years.

Preface to the First Edition

Many university departments of meteorology or atmospheric sciences offer a course in micrometeorology as a part of their undergraduate curricula. Some graduate programs also include an introductory course in micrometeorology or environmental fluid mechanics, in addition to more advanced courses on planetary boundary layer (PBL) and turbulence. While there are a number of excellent textbooks and monographs available for advanced graduate courses, there is no suitable text for an introductory undergraduate or graduate course in micrometeorology. My colleagues and I have faced this problem for more than ten years, and we suspect that other instructors of introductory micrometeorology courses have faced the same problem – lack of a suitable textbook. At first, we tried to use R.E. Munn's 'Descriptive Micrometeorology,' but soon found it to be much out-of-date. There has been almost an explosive development of the field during the past two decades, as a result of increased interest in micrometeorological problems and advances in computational and observational facilities. When Academic Press approached me to write a textbook, I immediately saw an opportunity for filling a void and satisfying an acutely felt need for an introductory text, particularly for undergraduate students and instructors. I also had in mind some incoming graduate students having no background in micrometeorology or fluid mechanics. Instructors may also find the latter half of the book suitable for an introductory or first graduate course in atmospheric boundary layer and turbulence. Finally, it may serve as an information source for boundary-layer meteorologists, air pollution meteorologists, agricultural and forest micrometeorologists, and environmental scientists and engineers.

Keeping in mind my primary readership, I have tried to introduce the various topics at a sufficiently elementary level, starting from the basic thermodynamic and fluid dynamic laws and concepts. I have also given qualitative descriptions based on observations before introducing more complex theoretical concepts and quantitative relations. Mathematical treatment is deliberately kept simple, presuming only a minimal mathematical background of upper-division science majors. Uniform notation and symbols are used throughout the text, although certain symbols do represent different things in different contexts. The list of symbols should be helpful to the reader. Sample problems and exercises given at the end of each chapter should be useful to students, as well as to instructors. Many of these were given in homework assignments, tests, and examinations for our undergraduate course in micrometeorology.

The book is organized in the form of fifteen chapters, arranged in order of increasing complexity and what I considered to be a natural order of the topics covered in the book. The scope and importance of micrometeorology and turbulent exchange processes in the PBL are introduced in Chapter 1. The next three chapters describe the energy budget near the surface and its components, such as radiative, conductive, and convective heat fluxes. Chapter 5 reviews basic thermodynamic relations and presents typical temperature and humidity distributions in the PBL. Wind distribution in the PBL and simple dynamics, including the balance of forces on an air parcel in the PBL, are discussed in Chapter 6. The emphasis, thus far, is on observations, and very little theory is used in these early chapters. The viscous flow theory, fundamentals of turbulence, and classical semiempirical theories of turbulence, which are widely used in micrometeorology and fluid mechanics, are introduced in Chapters 7, 8, and 9. Instructors of undergraduate courses may like to skip some of the material presented here, particularly the Reynolds-averaged equations for turbulent motion. Chapters 10–12 present the surface-layer similarity theory and micrometeorological methods and observations using the similarity framework. These three chapters would constitute the core of any course in micrometeorology. The last three chapters cover the more specialized topics of the marine atmospheric boundary layer, nonhomogeneous boundary layers, and micrometeorology of vegetated surfaces. Since these topics are still hotly pursued by researchers, the reader will find much new information taken from recent journal articles. The instructor should be selective in what might be included in an undergraduate course. Chapters 7–15 should also be suitable for a first graduate course in micrometeorology or PBL.

Finally, I would like to acknowledge the contribution of my colleagues, Jerry Davis, Al Riordan, and Sethu Raman, and of my students for reviewing certain parts of the manuscript and pointing out many errors and omissions. I am most grateful to Professor Robert Fleagle for his thorough and timely review of each chapter, without the benefit of the figures and illustrations. His comments and suggestions were extremely helpful in preparing the final draft. I would also like to thank the publishers and authors of many journal articles, books, and monographs for their permission to reproduce many figures and tables in this book. The original sources are acknowledged in figure legends, whenever applicable. Still, many ideas, explanations, and discussions in the text are presented without specific acknowledgments of the original sources, because, in many cases, the original sources were not known and also because, I thought, giving too many references in the text would actually distract the reader. In the end, I wish to acknowledge the help of Brenda Batts and Dava House for typing the manuscript and Pat Bowers and LuAnn Salzillo for drafting figures and illustrations.

Symbols

A	Amplitude of thermal wave in the subsurface medium (Chapter 4)
A	Aerodynamic surface area of vegetation per unit volume (Chapter 15)
A	Similarity constant or function (Chapters 10, 13)
A_s	Amplitude of thermal wave at the surface
A_θ	Mean advection terms in the thermodynamic energy equation
a	Albedo or shortwave reflectivity of the surface (Chapter 3)
a	Inverse length scale (Chapter 7)
a	An empirical constant (Chapter 10)
a_b	Acceleration due to buoyancy
a_i	Empirical constant
a_θ	Instantaneous advection terms in the thermodynamic energy equation (Chapter 9)
a_1	Empirical constant
a_2	Empirical constant
B	Bowen ratio
B	Buoyancy production term in the turbulent kinetic energy equation (Chapter 8)
B	Similarity constant or function (Chapters 10, 13)
b	Boltzmann constant
C	Heat capacity
C	Empirical constant (Chapter 9)
$\mathbf{C}$	Coriolis force (Chapter 6)
C_D	Surface drag coefficient
C_d	Drag coefficient (Chapter 15)
C_{DN}	Drag coefficient in neutral stability
C_g	Heat capacity of the near-ground soil layer (Chapter 4)
C_H	Heat transfer coefficient
C_{HN}	Heat transfer coefficient in neutral stability
C_S	Smagorinski constant (Chapter 13)
C_s	Volumetric heat capacity of soil (Chapter 4)
C_W	Water vapor transfer coefficient

C_w	A constant (Chapter 9)
C_θ	An empirical constant
C_0	Phase speed of dominant waves
c	Specific heat
c	Speed of light (electromagnetic waves) in vacuum (Chapter 3)
c	Empirical constant (Chapter 10)
c_g	Geostrophic drag coefficient
c_m	Electromagnetic wave speed in a medium (Chapter 3)
c_m	A constant (Chapter 14)
c_p	Specific heat at constant pressure
c_v	Specific heat at constant volume
D	Reference depth in the sub-medium (Chapter 4)
D	Viscous dissipation term in the turbulent kinetic energy equation (Chapter 8)
D/Dt	Total derivative
d	Distance of the earth from the sun (Chapter 3)
d	Damping depth of thermal wave in the submedium (Chapter 4)
d	Dimensionless constant (Chapter 13)
dH	Heat added to a parcel
d_m	Mean distance between the earth and the sun
dP	Change in the air pressure
dT	Change in the air temperature
d_0	Zero-plane displacement
E	Rate of evaporation or condensation
E	Average turbulence kinetic energy
E_a	Drying power of air
E_p	Potential evaporation or evapotranspiration
E_0	Rate of evaporation or condensation at the surface
e	Water vapor pressure (Chapters 5 and 12)
e	Fluctuating turbulence kinetic energy (Chapters 9 and 13)
e_r^*	Saturated vapor pressure at reference height (Chapter 12)
e_s	Water vapor pressure at saturation (Chapter 5)
e_0	Water vapor pressure at the surface or at $z = z_0$
F	Function of certain variables (Chapter 9)
F	Froude number (Chapters 13 and 14)
$\mathbf{F}$	Friction force (Chapter 6)
F_c	Critical Froude number
F_L	Froude number based on length of the hill
F_u	A similarity function
F_v	A similarity function
f	Coriolis parameter
f	Function of variables (Chapter 9)

f'	Normalized velocity as a function of η
G	Magnitude of geostrophic wind
$\mathbf{G}$	Geostrophic wind vector
Gr	Grashof number
G_0	Magnitude of surface geostrophic wind
$\mathbf{G}_0$	Surface geostrophic wind vector
g	Acceleration due to gravity
H	Sensible (direct) heat flux
H	Building height (Chapter 14)
H_1	Soil heat flux at depth z_1
H_F	Anthropogenic heat flux in an urban area
H_G	Ground heat flux to or from the subsurface medium
H_g	Soil heat flux at depth d_g
H_i	Heat flux at the inversion base
H_in	Energy coming in
H_L	Latent heat flux
H_out	Energy going out
H_s	Height of the dividing streamline
H_0	Sensible heat flux at the surface
h	Boundary layer thickness
h	Depth of channel flow or distance between two parallel planes (Chapter 7)
h_E	Ekman depth
h_i	Height of surface inversion (Chapter 5)
h_i	Internal boundary layer thickness (Chapter 14)
h_p	Planck's constant
h_0	Average height of roughness elements or plant canopy
I	Insolation at the surface
I_0	Insolation at the top of the atmosphere
i	Imaginary number $\sqrt{-1}$
i_u	Longitudinal turbulence intensity
i_v	Lateral turbulence intensity
i_w	Vertical turbulence intensity
K	Effective viscosity (Chapter 7)
K	Subgrid-scale eddy viscosity (Chapter 13)
K_E	Diffusivity of turbulence kinetic energy
K_h	Eddy diffusivity of heat
K_m	Eddy viscosity or diffusivity of momentum
K_mr	Eddy viscosity at the reference height
K_w	Eddy diffusivity of water vapor
K_*	Dimensionless eddy viscosity (Chapter 13)
k	Von Karman constant

k	Thermal conductivity (Chapter 4)
k_m	Hydraulic conductivity of the subsurface medium
L	Obukhov length
L	A characteristic length scale (Chapter 9)
L_{AI}	Leaf area index
L_e	Latent heat of evaporation/condensation
l	Large eddy length scale (Chapter 8)
l	Fluctuating mixing length (Chapter 9)
l_b	Buoyancy-limited mixing length (Chapter 13)
l_h	Mean mixing length of heat
l_m	Mean mixing length of momentum
l_o	Mixing length in the outer layer (Chapter 13)
ℓ	Large-eddy length scale (Chapter 13)
ℓ_ε	Dissipation length scale (Chapter 13)
M	Soil moisture flow rate
M_b	Soil moisture flow rate at the bottom of layer
M_d	Mass of dry air
M_w	Mass of water vapor
M_0	Soil moisture flow rate at the surface
m	Mean molecular mass of air (Chapter 5)
m	Exponent in the power-law wind profile equation (Chapter 10)
m	A coefficient (Chapter 15)
m_d	Mean molecular mass of dry air
m_w	Mean molecular mass of water vapor
N	Brunt–Vaisala frequency
Nu	Nusselt number
n	Exponent in the power-law eddy viscosity profile (Chapter 10)
n	A coefficient (Chapter 15)
P	Mean air pressure
P	Period of the wave (Chapter 4)
$\mathbf{P}$	Pressure-gradient force
P_{AI}	Plant area index
Pe	Peclet number
Pr	Prandtl number
P_0	Pressure of the reference atmosphere
p	Instantaneous pressure
p_0	Pressure of the reference state
p_1	Deviation of pressure from the reference state
Q	Mean specific humidity
Q_r^*	Saturation specific humidity at reference height (Chapter 12)
Q_s	Mean specific humidity at saturation (Chapter 5)
Q_0	Specific humidity at $z = z_0$

q	Specific humidity fluctuation
q_*	Specific humidity scale for the surface layer
R	Radiative energy flux or radiation (Chapter 3)
R	Specific gas constant (Chapter 5)
Ra	Rayleigh number
R_D	Diffuse radiation
R_d	Specific gas constant for dry air
Re	Reynolds number
Rf	Flux Richardson number
Ri	Gradient Richardson number
Ri_B	Bulk Richardson number
Ri_c	Critical Richardson number
R_L	Longwave radiation (flux)
$R_{L\downarrow}$	Incoming (downward) longwave radiation
$R_{L\uparrow}$	Outgoing (upward) longwave radiation
R_N	Net radiation
R_O	Surface Rossby number
R_S	Shortwave radiation (flux)
$R_{S\downarrow}$	Incoming (downward) shortwave radiation
$R_{S\uparrow}$	Outgoing (upward) shortwave radiation
R_s	Solar radiation flux at the surface
R_*	Absolute gas constant
R_λ	Radiative energy flux density per unit wavelength
R_0	Solar radiation flux at the top of the atmosphere
r	Correlation coefficient
r_H	Aerial resistance to heat transfer
r_M	Resistance to momentum transfer
r_w	Resistance to water vapor transfer
S	Shear production term in the turbulent kinetic energy equation (Chapter 8)
S	Soil moisture content (Chapter 12)
S	Skewness (Chapter 13)
S	Mean value of a passive scalar
S_F	Speed-up factor
S_H	Source or sink of heat
S_0	Solar constant
s	Static stability
s	Fluctuating passive scalar
T	Mean air temperature
T	Temperature of the subsurface medium (Chapter 4)
T_c	Complex stress variable
T_{eb}	Equivalent blackbody temperature

T_g	Temperature of the near-ground soil layer
T_m	Mean temperature of the surface and submedium
T_m	Mean temperature of the lower soil layer
T_r	Turbulent transport of turbulent kinetic energy
T_s	Surface temperature (Chapter 4)
T_s	Sea-surface temperature (Chapter 12)
T_u	Urban air temperature
T_v	Mean virtual temperature of moist air
T_{vp}	Mean virtual temperature of the parcel
T_{v0}	Virtual temperature at the reference state
T_*	Convective temperature scale
T_x	Normalized shear stress in x direction
T_y	Normalized shear stress in y direction
T_0	Air temperature at the reference state
T_1	Deviation in the temperature from the reference state
T_{10}	Air temperature at 10 m above the surface
t	Time
U	Mean velocity component in x direction
U_g	Geostrophic wind component in x direction
U_h	Velocity at $z = h$
U_h	Velocity of the moving surface (Chapter 7)
U_m	Mixed-layer averaged wind component in x direction
U_r	Wind speed at the reference height
U_∞	Ambient wind speed
U_0	Mean velocity in the approach flow
U_{10}	Wind speed at a reference height of 10 m
u	Fluctuating velocity component in x direction
$\tilde{u}$	Instantaneous velocity in x direction
u_f	Local free-convective velocity scale
u_ℓ	Characteristic velocity scale of turbulence (Chapter 8)
u_*	Friction velocity
V	Mean velocity component in y direction
V	Volume (Chapter 2)
V	Characteristic velocity scale (Chapter 9)
$\mathbf{V}$	Wind vector
V_g	Geostrophic wind component in y direction
V_m	Layer-averaged wind speed in the PBL
V_m	Mixed-layer averaged wind component in y direction
v	Fluctuating velocity component in y direction
$\tilde{v}$	Instantaneous velocity in y direction
W	Mean velocity component in z (vertical) direction
W	Building width (Chapter 14)

W_{AI}	Woody-element area index
W_h	Mean vertical motion at the top of the PBL
W_*	Convective velocity scale
w	Fluctuating velocity component in z (vertical) direction
$\tilde{w}$	Instantaneous velocity in z direction
w_e	Entrainment velocity
z	Height above or depth below the surface or an appropriate reference plane
z_i	Height of inversion base above the surface
z_m	Geometric mean of the two heights z_1 and z_2
z_r	Reference height
z'	Height above the ground level
z_p	Height of a constant pressure surface
z_0	Roughness length parameter
z_{01}	Roughness parameter of the upstream surface
z_{02}	Roughness parameter of the downstream surface
α	Cross-isobar angle (Chapter 6)
α	Empirical constant (Chapter 15)
α_h	Molecular diffusivity of heat or thermal diffusivity
α_m	Soil moisture diffusivity
α_w	Molecular diffusivity of water vapor in air
α_λ	Absorptivity at wavelength λ
α_θ	Dimensionless coefficient (Chapter 12)
α_0	Cross-isobar angle at the surface
β	Coefficient of thermal expansion
β	Slope angle of an inclined plane (Chapter 7)
Γ	Adiabatic lapse rate
Γ_s	Saturated adiabatic lapse rate
γ	Solar zenith angle (Chapter 3)
γ	Psychrometer constant (Chapter 12)
γ	Potential temperature gradient above the PBL (Chapter 13)
Δ	Gradient operator
Δ	Slope of the saturation vapor pressure versus temperature curve (Chapter 12)
ΔA	Elemental area
ΔH_s	Rate of energy storage
ΔM	Rate of storage of soil moisture
ΔT_s	Difference between the maximum and minimum surface temperatures
ΔT_{u-r}	Difference between the urban and rural air temperatures
ΔU	Difference in mean velocities at two heights
ΔU	Velocity deficit in the wake

ΔU_g	Change in U_g over a layer
ΔV_g	Change in V_g over a layer
$\Delta\Theta$	Difference between potential temperatures at two heights
Δx	Small increment in x
Δy	Small increment in y
Δz	Small increment in z
Δz	$z_2 - z_1$
∇^2	Laplacian operator
ε	Overall infrared emissivity (Chapter 3)
ε	Rate of energy dissipation (Chapters 8, 9, 13)
ε_λ	Emissivity at wavelength λ
ζ	Monin–Obukhov stability parameter
η	Normalized distance from the surface (Chapter 7)
η	Sea-surface elevation (Chapter 13)
η	Kolmogorov's microscale of length (Chapter 8)
Θ	Mean potential temperature
Θ_m	Mixed-layer averaged potential temperature
Θ_r	Rural air potential temperature
Θ_u	Urban air potential temperature
Θ_v	Mean virtual potential temperature
Θ_0	Potential temperature at $z = z_0$
Θ_{01}	Temperature of the upstream surface
Θ_{02}	Temperature of the downstream surface
θ	Fluctuating potential temperature
$\tilde{\theta}$	Instantaneous potential temperature
θ_f	Local free convective temperature scale
θ_*	Friction temperature scale
κ	Wave number
λ	Wavelength (Chapter 3)
λ	Frontal area density of roughness elements (Chapter 10)
λ_{max}	Wavelength corresponding to spectral maximum
μ	Dynamic viscosity
μ_g	Heat transfer coefficient
μ_*	Dimensionless stability parameter (Chapter 13)
ν	Kinematic viscosity
ξ	Dimensionless height parameter (Chapters 10, 13)
ξ_τ	Dimensionless PBL height
Π	Dimensionless group
π	The ratio of circumference to diameter of a circle
ρ	Mass density of air or other medium
ρ_p	Mass density of the air parcel
ρ_0	Density of the reference state

ρ_1	Deviation in the density from the reference state
σ	Stefan–Boltzmann constant
σ_q	Standard deviation of specific humidity fluctuations
σ_u	Standard deviation of velocity fluctuations in x direction
σ_v	Standard deviation of velocity fluctuations in y direction
σ_w	Standard deviation of velocity fluctuations in z direction
σ_θ	Standard deviation of temperature fluctuations
τ	Shearing stress
$\boldsymbol{\tau}$	Shear stress vector
τ_0	Surface shear stress or drag
ϕ_h	Dimensionless potential temperature gradient
ϕ_m	Dimensionless wind shear
ϕ_w	Dimensionless specific humidity gradient
ψ_h	M–O similarity function for normalized potential temperature
ψ_m	M–O similarity function for normalized velocity
ψ_w	M–O similarity function for normalized specific humidity
Ω	Rotational speed of the earth
$\boldsymbol{\Omega}$	Earth's rotational velocity vector
ω	Wave frequency

Chapter 1 Introduction

1.1 Scope of Micrometeorology

Atmospheric motions are characterized by a variety of scales ranging from the order of a millimeter to as large as the circumference of the earth in the horizontal direction and the entire depth of the atmosphere in the vertical direction. The corresponding time scales range from a tiny fraction of a second to several months or years. These scales of motions are generally classified into three broad categories, namely, micro-, meso-, and macroscales. Sometimes, terms such as local, regional, and global are used to characterize the atmospheric scales and the phenomena associated with them.

Micrometeorology is a branch of meteorology which deals with the atmospheric phenomena and processes at the lower end of the spectrum of atmospheric scales, which are variously characterized as microscale, small-scale, or local-scale processes. The scope of micrometeorology is further limited to only those phenomena which originate in and are dominated by the shallow layer of frictional influence adjoining the earth's surface, commonly known as the atmospheric boundary layer (ABL) or the planetary boundary layer (PBL). Thus, some of the small-scale phenomena, such as convective clouds and tornadoes, are considered outside the scope of micrometeorology, because their dynamics is largely governed by mesoscale and macroscale weather systems.

In particular, micrometeorology deals with the exchanges of heat (energy), mass, and momentum occurring continuously between the atmosphere and the earth's surface, including the subsurface medium. The energy budget at or near the surface on a short-term (say, hourly) basis is an important aspect of the different types of energy exchanges involved in the earth–atmosphere–sun system. Vertical distributions of meteorological variables such as wind, temperature, and humidity, as well as trace gas concentrations and their role in the energy balance near the surface, also come under the scope of micrometeorology. In addition to the short-term averaged values of meteorological variables, more or less random fluctuations of the same in time and space around their respective average values are of considerable interest. The statistics of these so-called turbulent fluctuations are intimately related to the above-mentioned exchange processes and, hence, constitute an integral part of micrometeorology or boundary-layer meteorology.

1.1.1 Atmospheric boundary layer

A boundary layer is defined as the layer of a fluid (liquid or gas) in the immediate vicinity of a material surface in which significant exchange of momentum, heat, or mass takes place between the surface and the fluid. Sharp variations in the properties of the flow, such as velocity, temperature, and mass concentration, also occur in the boundary layer.

The atmospheric boundary layer is formed as a consequence of the interactions between the atmosphere and the underlying surface (land or water) over time scales of a few hours to about 1 day. Over longer periods the earth–atmosphere interactions may span the whole depth of the troposphere, typically 10 km, although the PBL still plays an important part in these interactions. The influence of surface friction, heating, etc., is quickly and efficiently transmitted to the entire PBL through the mechanism of turbulent transfer or mixing. Momentum, heat, and mass can also be transferred downward through the PBL to the surface by the same mechanism. A schematic of the PBL, as the lower part of the troposphere, over an underlying rough surface is given in Figure 1.1. Also depicted in the same figure is the frequently used division of the atmospheric boundary layer into a surface layer and an outer or upper layer. The vertical dimensions (heights) given in Figure 1.1 are more typical of the near-neutral stability observed during strong winds and overcast skies; these are highly variable in both time and space.

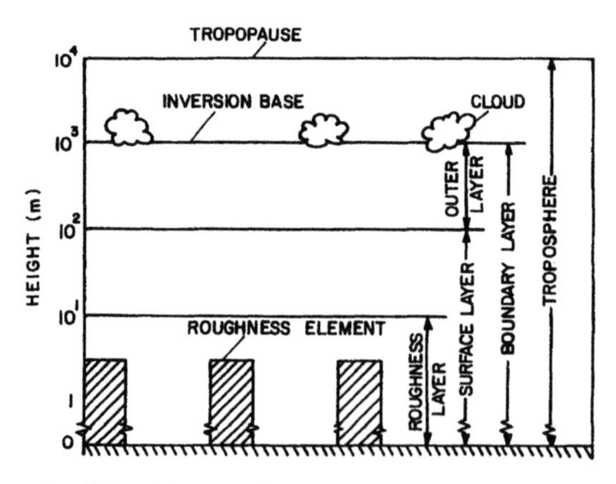

Figure 1.1 Schematic of the planetary boundary layer as the lower part of the troposphere. [From Arya (1982).]

The atmospheric PBL thickness over land surfaces varies over a wide range (several tens of meters to several kilometers) and depends on the rate of heating or cooling of the surface, strength of winds, the roughness and topographical characteristics of the surface, large-scale vertical motion, horizontal advections of heat and moisture, and other factors. In the air pollution literature the PBL height is commonly referred to as the mixing height or depth, since it represents the depth of the layer through which pollutants released from the near-surface sources are eventually mixed. As a result, the PBL is generally dirtier than the free atmosphere above it. The contrast between the two is usually quite sharp over large cities and can be observed from an aircraft as it leaves or enters the PBL.

In response to the strong diurnal cycle of heating and cooling of land surfaces during fair-weather conditions, the PBL thickness and other boundary layer characteristics also display strong diurnal variations. For example, the PBL height over a dry land surface in summer can vary from less than 100 m in the early morning to several kilometers in the late afternoon.

Following sunrise on a clear day, the continuous heating of the land surface by the sun and the resulting thermal mixing in the PBL cause the PBL depth to increase steadily throughout the day and attain a maximum value of the order of 1 km (range $\simeq$0.2–5 km) in the late afternoon. Later in the evening and throughout the night, on the other hand, the radiative cooling of the ground surface results in the suppression or weakening of turbulent mixing and consequently in the shrinking of the PBL depth to a typical value of the order of only 100 m (range $\simeq$20–500 m). Thus the PBL depth waxes and wanes in response to the diurnal heating and cooling cycle. The winds, temperatures, and other properties of the PBL may also be expected to exhibit strong diurnal variations. Diurnal variations of the PBL height and other meteorological variables are found to be much smaller over large lakes, seas and oceans, because of the small diurnal changes of the water surface temperature due to the large heat capacity of the mixed layer in water.

Other temporal variations of the PBL height and structure often occur as a result of the development and the passage of mesoscale and synoptic systems. Generally, the PBL becomes thinner under the influence of large-scale subsidence (downward motion) and the low-level horizontal divergence associated with a high-pressure system (anticyclone). On the other hand, the PBL can grow to be very deep and merge with towering clouds in disturbed weather conditions that are associated with low-pressure systems (cyclones). It is often difficult to distinguish the PBL top from in-cloud circulations under these conditions; the cloud base is generally used as an arbitrary cutoff for the boundary layer top.

Spatial variations of the PBL depth and structure occur as a result of changes in land use and topography of the underlying surface. Spatial variations of meteorological variables influenced by mesoscale and large-scale systems also lead to similar variations in the boundary layer. On a flat and homogeneous surface, however, the PBL is generally considered horizontally homogeneous.

1.1.2 The surface layer

Some investigators limit the scope of micrometeorology to only the so-called atmospheric surface layer, which comprises the lowest one-tenth or so of the PBL and in which the earth's rotational or Coriolis effects can be ignored. Such a restriction may not be desirable, because the surface layer is an integral part of and is much influenced by the PBL as a whole, and the top of this layer is not physically as well defined as the top of the PBL. The latter represents the fairly sharp boundary between the irregular and almost chaotic (turbulent) motions in the PBL and the smooth and streamlined (nonturbulent) flow in the free atmosphere above. The PBL top can easily be detected by ground-based remote-sensing devices, such as acoustic sounder, lidar, etc., and can also be inferred from temperature, humidity, and wind soundings.

However, the surface layer is more readily amenable to observations from the surface, as well as from micrometeorological masts and towers. It is also the layer in which most human beings, animals, and vegetation live and in which most human activities take place. The sharpest variations in meteorological variables with height occur within the surface layer and, consequently, the most significant exchanges of momentum, heat, and mass also occur in this layer. Therefore, it is not surprising that the surface layer has received far greater attention from micrometeorologists and microclimatologists than has the outer part of the PBL.

The lowest part of the surface layer in which the influence of individual roughness elements can readily be discerned is called the roughness layer or the canopy layer (see Figure 1.1). For bare land surfaces, it is quite thin and often ignored. For grasslands and other vegetated surfaces, the height of the roughness layer is proportional to (say, 1.5 times) the average height of vegetation. In built-up (suburban and urban) areas, the height of roughness or canopy layer may depend on the spatial distribution and heights of buildings in the particular area. Over large city centers, the roughness layer may comprise a significant portion of the urban boundary layer, especially at night.

1.1.3 Turbulence

Turbulence refers to the apparently chaotic nature of many flows, which is manifested in the form of irregular, almost random fluctuations in velocity, temperature, and scalar concentrations around their mean values in time and space. Atmospheric turbulence is always manifested in the form of gustiness of winds, so that gustiness can be regarded as a simple measure of turbulence strength or intensity. The motions in the atmospheric boundary layer are almost always turbulent. In the surface layer, turbulence is more or less continuous, while it may be intermittent and patchy in the upper part of the PBL and is sometimes mixed with internal gravity waves. The PBL top is usually defined as the level where turbulence disappears or becomes insignificant.

Near the surface, atmospheric turbulence manifests itself through the flutter of leaves of trees and blades of grass, swaying of branches of trees and plants, irregular movements of smoke and dust particles, generation of ripples and waves on water surfaces, and a variety of other visible phenomena. In the upper part of the PBL, turbulence is manifested by irregular motions of kites and balloons, spreading of smoke and other visible pollutants as they exit tall stacks or chimneys, and fluctuations in the temperature and refractive index encountered in the transmission of sound, light, and radio waves.

1.2 Micrometeorology versus Microclimatology

The difference between micrometeorology and microclimatology is primarily in the time of averaging the variables, such as air velocity, temperature, humidity, etc. While the micrometeorologist is primarily interested in fluctuations, as well as in the short-term (of the order of an hour or less) averages of meteorological variables in the PBL or the surface layer, the microclimatologist mainly deals with the long-term (climatological) averages of the same variables. The latter is also interested in diurnal and seasonal variations, as well as in very long-term trends of meteorological parameters.

In micrometeorology, detailed examination of small-scale temporal and spatial structure of flow and thermodynamic variables is often necessary to gain a better understanding of the phenomena of interest. For example, in order to determine the short-time average concentrations on the ground some distance away from a pollutant source, one needs to know not only the mean winds that are responsible for the mean transport of the pollutant, but also the statistical properties of wind gusts (turbulent fluctuations) that are responsible for spreading or diffusing the pollutant as it moves along the mean wind.

In microclimatology, one deals with long-term averages of meteorological variables in the near-surface layer of the atmosphere. Details of fine structure are not considered so important, because their effects on the mean variables are smoothed out in the averaging process. Still, one cannot ignore the long-term consequences of the small-scale turbulent transfer and exchange processes.

Despite the above-mentioned differences, micrometeorology and microclimatology have much in common, because they both deal with similar atmospheric processes occurring near the surface. Their interrelationship is further emphasized by the fact that long-term averages dealt with in the latter can, in principle, be obtained by the integration in time of the short-time averaged micrometeorological variables. It is not surprising, therefore, to find some of the fundamentals of micrometeorology described in books on microclimatology (e.g., Geiger, 1965; Oke, 1987; Rosenberg *et al.*, 1983; Geiger *et al.*, 1995). Likewise, microclimatological information is found in and serves useful purposes in texts on micrometeorology (e.g., Sutton, 1953; Munn, 1966; Arya, 1988; Stull, 1988; Sorbjan, 1989; Garratt, 1992; Kaimal and Finnigan, 1994).

1.3 Importance and Applications of Micrometeorology

Although the atmospheric boundary layer comprises only a tiny fraction of the atmosphere, the small-scale processes occurring within the PBL are useful to various human activities and are important for the well-being and even survival of life on earth. This is not merely because the air near the ground provides the necessary oxygen to human beings and animals, but also because this air is always in turbulent motion, which causes efficient mixing of pollutants and exchanges of heat, water vapor, etc., with the surface.

1.3.1 Turbulent transfer processes

Turbulence is responsible for the efficient mixing and exchange of mass, heat, and momentum throughout the PBL. In particular, the surface layer turbulence is responsible for exchanging these properties between the atmosphere and the earth's surface. Without turbulence, such exchanges would have been at the molecular scale and minuscule in magnitudes (10^{-3}–10^{-6} times the turbulent transfers that now commonly occur). Nearly all the energy which drives the large-scale weather and general circulation comes through the PBL.

Through the efficient transfer of heat and moisture, the boundary layer turbulence moderates the microclimate near the ground and makes it habitable for animals, organisms, and plants. The atmosphere receives virtually all of its water vapor through turbulent exchanges near the surface. Evaporation from land and water surfaces is not only important in the surface water budget and the hydrological cycle, but the latent heat of evaporation is also an important component of the surface energy budget. This water vapor, when condensed on tiny dust particles and other aerosols (cloud condensation nuclei), leads to the formation of fog, haze, and clouds in the atmosphere.

Besides water vapor, there are other important exchanges of mass within the PBL involving a variety of gases and particulates. Turbulence is important in the exchange of carbon dioxide between plants and animals. Its atmospheric concentration has steadily been rising due to ever increasing use of fossil fuels in energy production and other industry, heating and cooking, and forest clearing by burning. This, in conjunction with similar increasing trends in the concentrations of methane and other radiatively active gases, threatens global climate warming. Many harmful toxic substances are also released into the atmosphere by human activities. Through efficient diffusion of the various pollutants released near the ground and mixing them throughout the PBL and parts of the lower troposphere, atmospheric turbulence prevents the fatal poisoning of life on earth. The quality of air we breathe depends to a large extent on the mixing capability of the PBL turbulence. The boundary layer turbulence also picks up pollen and other seeds of life, spreads them out, and deposits them over wide areas far removed from their origin. It lifts dust, salt particles, and other aerosols from the surface and spreads them throughout the lower atmosphere. Of these, the so-called cloud condensation nuclei are an essential ingredient in the condensation and precipitation processes in the atmosphere.

Through the above-mentioned mass exchange processes between the earth's surface and the atmosphere, the radiation balance and the heat energy budget at or near the earth's surface are also significantly affected. More direct effects of turbulent transfer on the surface heat energy budget are through sensible and latent heat exchanges between the surface and the atmosphere. Over land, the sensible heat exchange is usually more important than the latent heat of evaporation, but the reverse is true over large lakes and the oceans. Heat exchanges through the underlying bodies of water are also turbulent, especially those in the immediate vicinity of the surface.

Turbulent transfer of momentum between the earth and atmosphere is also very important. It is essentially a one-way process in which the earth acts as a sink of atmospheric momentum (relative to earth). In other words, the earth's surface exerts frictional resistance to atmosphere motions and slows them down in the process. The moving air near the surface may be considered to exert an equivalent drag force on the surface. The rougher the surface, the larger would be the drag force per unit area of the surface. Some commonly encountered examples of this are the marked slowdown of surface winds in going from a large lake, bay, or sea to inland areas and from rural to urban areas. Perhaps the most vivid manifestations of the effect of increased surface drag are the rapid weakening and subsequent demise of hurricanes and other tropical storms as they move inland. Another factor responsible for this phenomenon is the marked reduction or cutoff of the available latent heat energy to the storm from the surface.

Over large lakes and oceans, wind drag is responsible for the generation of waves and currents in water, as well as for the movement of sea ice. In coastal areas, wind drag causes storm surges and beach erosion. Over land, drag exerts wind loads on vehicles, buildings, towers, bridges, cables, and other structures. Wind stress over sand surfaces raises dust and creates ripples and sand dunes. When accompanied by strong winds, the surface layer turbulence can be quite discomforting and even harmful to people, animals, and vegetation.

Turbulent exchange processes in the PBL have profound effects on the evolution of local weather. Boundary layer friction is primarily responsible for the low-level convergence and divergence of flow in the regions of lows and highs in surface pressures, respectively. The frictional convergence in a moist boundary layer is also responsible for the low-level convergence of moisture in low-pressure regions. The kinetic energy of the atmosphere is continuously dissipated by small-scale turbulence in the atmosphere. Almost one-half of this loss on an annual basis occurs within the PBL, even though the PBL comprises only a tiny fraction (less than 2%) of the total kinetic energy of the atmosphere.

1.3.2 Applications

In the following text, we have listed some of the possible areas of application of micrometeorology together with the subareas or activities in which micrometeorological information may be especially useful.

1. Air pollution meteorology
    - Atmospheric transport and diffusion of pollutants
    - Atmospheric deposition on land and water surfaces
    - Prediction of local, urban, and regional air quality
    - Selection of sites for power plants and other major industries
    - Selection of sites for monitoring urban and regional air pollution
    - Industrial operations with emissions dependent on meteorological conditions
    - Agricultural operations such as dusting, spraying, and burning
    - Military operations with considerations of obscurity and dispersion of contaminants
2. Mesoscale meteorology
    - Urban boundary layer and heat island
    - Land–sea breezes
    - Drainage and mountain valley winds
    - Dust devils, water spouts, and tornadoes
    - Development of fronts and cyclones
3. Macrometeorology
    - Atmospheric predictability
    - Long-range weather forecasting

- Siting and exposure of meteorological stations
- General circulation and climate modeling

4. Agricultural and forest meteorology
   - Prediction of surface temperatures and frost conditions
   - Soil temperature and moisture
   - Evapotranspiration and water budget
   - Energy balance of a plant cover
   - Carbon dioxide exchanges within the plant canopy
   - Temperature, humidity, and winds in the canopy
   - Protection of crops and shrubs from strong winds and frost
   - Wind erosion of soil and protective measures
   - Effects of acid rain and other pollutants on plants and trees
5. Urban planning and management
   - Prediction and abatement of ground fogs
   - Heating and cooling requirements
   - Wind loading and designing of structures
   - Wind sheltering and protective measures
   - Instituting air pollution control measures
   - Flow and dispersion around buildings
   - Prediction of road surface temperature and possible icing
6. Physical oceanography
   - Prediction of storm surges
   - Prediction of the sea state
   - Dynamics of the oceanic mixed layer
   - Movement of sea ice
   - Modeling large-scale oceanic circulations
   - Navigation
   - Radio transmission

Problems and Exercises

1. Compare and contrast the following terms:
(a) microscale and macroscale atmospheric processes and phenomena;
(b) micrometeorology and microclimatology;
(c) atmospheric boundary layer and the surface layer.

2. On a schematic of the lower atmosphere, including the tropopause, indicate the PBL and the surface layer during typical fair-weather (a) daytime convective and (b) nighttime stable conditions.

3. Discuss the following exchange processes between the earth and the atmosphere and the importance of boundary layer turbulence on them:
(a) sensible heat;
(b) water vapor;
(c) momentum.

Chapter 2 | Energy Budget Near the Surface

2.1 Energy Fluxes at an Ideal Surface

The flux of a property in a given direction is defined as its amount per unit time passing through a unit area normal to that direction. In this chapter, we will be concerned with fluxes of the various forms of heat energy at or near the surface. The SI units of energy flux are $J\ s^{-1}\ m^{-2}$ or $W\ m^{-2}$.

The 'ideal' surface considered here is relatively smooth, horizontal, homogeneous, extensive, and opaque to radiation. The energy budget of such a surface is considerably simplified in that only the vertical fluxes of energy need to be considered.

There are essentially four types of energy fluxes at an ideal surface, namely, the net radiation to or from the surface, the sensible (direct) and latent (indirect) heat fluxes to or from the atmosphere, and the heat flux into or out of the submedium (soil or water). The net radiative flux is a result of the radiation balance at the surface, which will be discussed in more detail in Chapter 3. During the daytime, it is usually dominated by the solar radiation and is almost always directed toward the surface, while at night the net radiation is much weaker and directed away from the surface. As a result, the surface warms up during the daytime, while it cools during the evening and night hours, especially under clear sky and undisturbed weather conditions.

The direct or sensible heat flux at and above the surface arises as a result of the difference in the temperatures of the surface and the air above. Actually, the temperature in the atmospheric surface layer varies continuously with height, with the magnitude of the vertical temperature gradient usually decreasing with height. In the immediate vicinity of the interface and within the so-called molecular sublayer, the primary mode of heat transfer in air is conduction, similar to that in solids. At distances beyond a few millimeters (the thickness of the molecular sublayer) from the interface, however, the primary mode of heat exchange becomes advection or convection involving air motions. The sensible heat flux is usually directed away from the surface during the daytime hours, when the surface is warmer than the air above, and vice versa during the evening and nighttime periods. Thus, the heat flux is down the average temperature gradient.

The latent heat or water vapor flux is a result of evaporation, evapotranspiration, or condensation at the surface and is given by the product of the latent heat of evaporation or condensation and the rate of evaporation or condensation. Evaporation occurs from water surfaces as well as from moist soil and vegetative surfaces, whenever the air above is drier, i.e., it has lower specific humidity than the air in the immediate vicinity of the surface and its transpirating elements (e.g., leaves). This is usually the situation during the daytime. On the other hand, condensation in the form of dew may occur on relatively colder surfaces at nighttime. The water vapor transfer through air does not involve any real heat exchange, except where phase changes between liquid water and vapor actually take place. Nevertheless, evaporation results in some cooling of the surface, which in the surface energy budget is represented by the latent heat flux from the surface to the air above. The ratio of the sensible heat flux to the latent heat flux is called the Bowen ratio.

The heat exchange through the ground medium is primarily due to conduction if the medium is soil, rock, or concrete. Through water, however, heat is transferred in the same way as it is through air, first by conduction in the top few millimeters (molecular sublayer) from the surface and then by advection or convection in the deeper layers (surface layer, mixed layer, etc.) of water in motion. The depth of the submedium, which responds to and is affected by changes in the energy fluxes at the surface on a diurnal basis, is typically less than a meter for land surfaces and several tens of meters for lakes and oceans.

2.2 Energy Balance Equations

2.2.1 The surface energy budget

For deriving a simplified equation for the energy balance at an ideal surface, we assume the surface to be a very thin interface between the two media (air and soil or water), having no mass and heat capacity. The energy fluxes would flow in and out of such a surface without any loss or gain due to the surface. Then, the principle of the conservation of energy at the surface can be expressed as

$$R_N = H + H_L + H_G \tag{2.1}$$

where R_N is the net radiation, H and H_L are the sensible and latent heat fluxes to or from the air, and H_G is the ground heat flux to or from the submedium. Here we use the sign convention that all the radiative fluxes directed toward the surface are positive, while other (nonradiative) energy fluxes directed away from the surface are positive and vice versa.

Equation (2.1) describes how the net radiation at the surface must be balanced by a combination of the sensible and latent heat fluxes to the air and the heat flux to the subsurface medium. During the daytime, the surface receives radiative energy ($R_N > 0$), which is partitioned into sensible and latent heat fluxes to the atmosphere and the heat flux to the submedium. Typically, H, H_L, and H_G are all positive over land surfaces during the day. This situation is schematically shown in Figure 2.1a. Actual magnitudes of the various components of the surface energy budget depend on many factors, such as the type of surface and its characteristics (soil moisture, texture, vegetation, etc.), geographical location, month or season, time of day, and weather. Under special circumstances, e.g., when irrigating a field, H and/or H_G may become negative and the latent heat flux due to evaporative cooling of the surface may exceed the net radiation received at the surface.

At night, the surface loses energy by outgoing radiation, especially during clear or partially overcast conditions. This loss is compensated by gains of heat from air and soil media and, at times, from the latent heat of condensation released during the process of dew formation. Thus, according to our sign

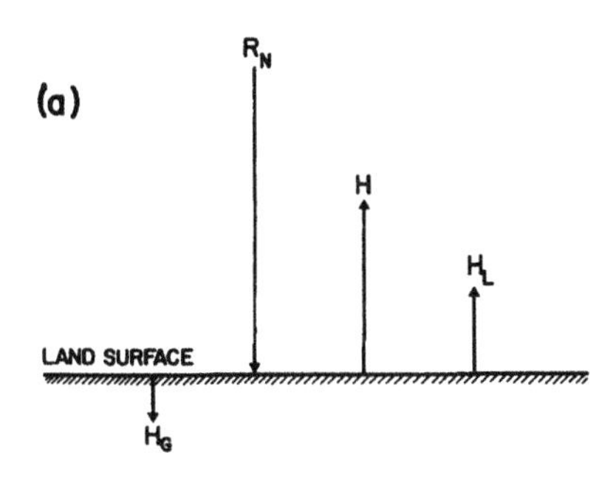

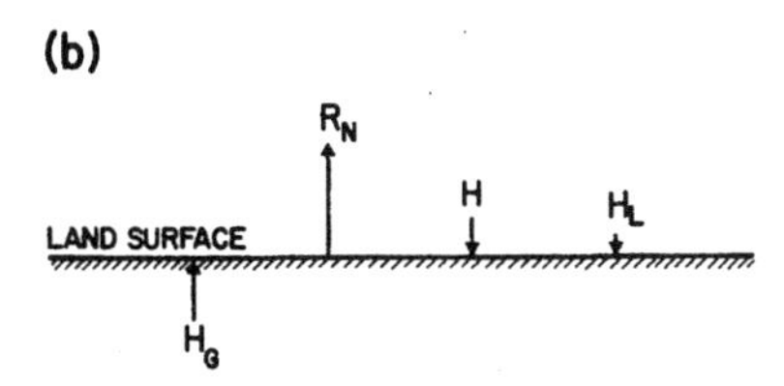

Figure 2.1 Schematic representation of typical surface energy budgets during (a) daytime and (b) nighttime.

convention, all the terms of the surface energy balance [Equation (2.1)] for land surfaces are usually negative during the evening and nighttime periods. Their magnitudes are generally much smaller than the magnitudes of the daytime fluxes, except for H_G. The magnitudes of H_G do not differ widely between day and night, although the direction or sign obviously reverses during the morning and evening transition periods, when other fluxes are also changing their signs (this does not happen simultaneously for all the fluxes, however). Figure 2.1b gives a schematic representation of the surface energy balance during the nighttime.

From the above description of the diurnal cycle of the surface energy budget, it is clear that the fluxes of sensible and latent heat to or from the surface are governed by the diurnal cycle of net radiation. One can interpret Equation (2.1) in terms of the partitioning of net radiation into other fluxes (H, H_L, and H_G). The net radiation may be considered an external forcing, while the sensible, latent and ground heat fluxes are responses to this radiative forcing. Relative measures of this partitioning are the ratios H/R_N, H_L/R_N, and H_G/R_N, which are expected to depend on the various surface, subsurface and meteorological characteristics. Diurnal variations of these ratios are expected to be much smaller than those of the individual fluxes. Of these, the ratio H_G/R_N might be expected to show the least variability, especially for a given land surface, because the above ratio may not be as sensitive to the variations in surface meteorological parameters as H/R_N and H_L/R_N. In simpler parameterizations of the ground heat flux, H_G is assumed to be proportional to R_N or related to R_N through an empirical regression relationship (Doll *et al.*, 1985). The ratio H_G/R_N is found to be larger for nighttime ($R_N < 0$) than for the daytime ($R_N > 0$) conditions. Values exceeding one have been observed during nighttime over urban surfaces, because of positive (upward) sensible heat fluxes even during nighttime.

The simple relationship between the ground heat flux and net radiation allows for the sum of the sensible and latent heat fluxes ($H + H_L$) to be easily determined from the measurement or calculation of net radiation. Further separation of the total heat flux into sensible and latent components can be made if the Bowen ratio, $B = H/H_L$, can be estimated independently (e.g., using the Bowen ratio method to be discussed later). In terms of the Bowen ratio, one can obtain from Equation (2.1)

$$H = \frac{R_N - H_G}{1 + B^{-1}} \tag{2.2}$$

$$H_L = \frac{R_N - H_G}{1 + B} \tag{2.3}$$

The latent heat flux can also be expressed as $H_L = L_e E$, where $L_e \simeq 2.45 \times 10^6$ J kg^{-1} is the latent heat of evaporation/condensation and E is the rate of evaporation/condensation.

Example Problem 1

The average values of the ratio of ground heat flux to net radiation determined from the Wangara experiment data are 0.30 and 0.52 for the daytime and nighttime, respectively. Assuming a Bowen ratio of 5.0, estimate the sensible and latent heat fluxes at that site when the measured net radiation is (a) 250 W m^{-2}, and (b) -55 W m^{-2}.

Solution

(a) $R_N = 250$ W m^{-2}; $H_G = 0.30 \times 250 = 75$ W m^{-2}.
From Equations (2.2) and (2.3), $H = 145.8$ W m^{-2}; $H_L = 29.2$ W m^{-2}.
Check that the above values satisfy Equation (2.1).

(b) $R_N = -55$ W m^{-2}; $H_G = -0.52 \times 55 = -28.6$ W m^{-2}.
From Equations (2.2) and (2.3), $H = -22.0$ W m^{-2}; $H_L = -4.4$ W m^{-2}.
Again, one can check that these values satisfy Equation (2.1).

The energy budgets of extensive water surfaces (large lakes, seas, and oceans) differ from those of land surfaces in several important ways. In the former, the combined value of H_L and H_G balances most of the net radiation, while H plays only a minor role ($H \ll H_L$, or $B \ll 1$). Since the water surface temperature does not respond readily to solar heating due to the large heat capacity and depth of the subsurface mixed layer of a large lake or ocean, the air–water exchanges (H and H_L) do not undergo large diurnal variations.

An important factor to be considered in the energy balance over water surfaces is the penetration of solar radiation to depths of tens of meters. Radiation processes occurring within large bodies of water are not well understood. The radiative fluxes on both sides of the air–water interface must be measured in order to determine the net radiation at the surface. This is not easy to do in the field. Therefore, the surface energy budget as expressed by Equation (2.1) may not be very useful or even appropriate to consider over water surfaces. A better alternative is to consider the energy budget of the whole energy-active water layer.

2.2.2 Energy budget of a layer

An 'ideal,' horizontally homogeneous, plane surface, which is also opaque to radiation, is rarely encountered in practice. More often, the earth's surface has horizontal inhomogeneities at small scale (e.g., plants, trees, houses, and building blocks), mesoscale (e.g., urban–rural differences, coastlines, hills, and valleys), and large scale (e.g., large mountain ranges). It may be partially transparent to radiation (e.g., water, tall grass, and crops). The surface may be sloping or undulating. In many practical situations, it will be more appropriate to consider the energy budget of a finite interfacial layer, which may include the small-scale surface inhomogeneities and/or the upper part of the subsurface medium which might be active in radiative exchanges. This layer must have finite mass and heat capacity which would allow the energy to be stored in or released from the layer over a given time interval. Such changes in the energy storage with time must, then, be considered in the energy budget for the layer.

If the surface is relatively flat and homogeneous, so that the interfacial layer can be considered to be bounded by horizontal planes at the top and the bottom, one can still use a simplified one-dimensional energy budget for the layer:

$$R_N = H + H_L + H_G + \Delta H_S \tag{2.4}$$

where ΔH_S is the change in the energy storage per unit time, per unit horizontal area, over the whole depth of the layer. Thus, the main difference in the energy budget of an interfacial layer from that of an ideal surface is the presence of the rate of energy storage term ΔH_S in the former. Strictly speaking, the various energy fluxes in Equation (2.4), except for ΔH_S, are net vertical fluxes going in or out of the top and bottom faces of the layer. In reality, however, H and H_L are associated only with the top surface (for an interfacial layer involving the water medium, the bottom boundary may be chosen so that there is no significant radiative flux in or out of the bottom surface), as shown schematically in Figure 2.2a.

The rate of heat energy storage in the layer can be expressed as

$$\Delta H_S = \int \frac{\partial}{\partial t}(\rho c T)\, dz \tag{2.5}$$

in which ρ is the mass density, c is the specific heat, T is the absolute temperature of the material at some level z, and the integral is over the whole depth of the layer. When the heat capacity of the medium can be assumed to be constant, independent of z, Equation (2.5) gives a direct relationship between the rate of energy storage and the rate of warming or cooling of the layer.

The energy storage term ΔH_S in Equation (2.4) may also be interpreted as the difference between the energy coming in (H_{in}) and the energy going out (H_{out}) of the layer, where H_{in} and H_{out} represent appropriate combinations of R_N, H, H_L, and H_G, depending on their signs (Oke, 1987, Chapter 2). When the energy input to the layer exceeds the outgoing energy, there is a flux convergence

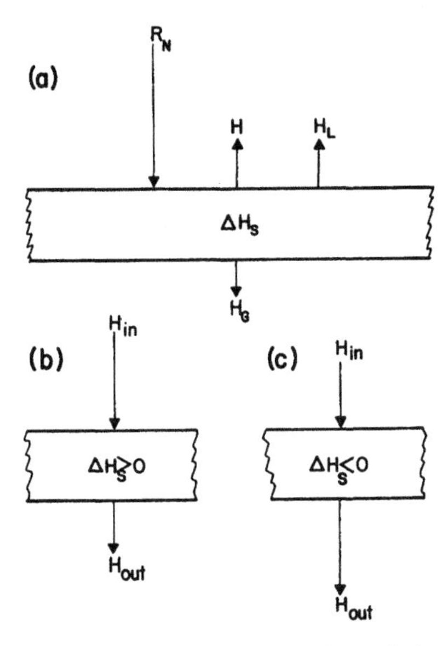

Figure 2.2 Schematic representation of (a) energy budget of a layer, (b) flux convergence, and (c) flux divergence.

($\Delta H_S > 0$) which results in the warming of the layer. On the other hand, when energy going out exceeds that coming in, the layer cools as a result of flux divergence ($\Delta H_S < 0$). In the special circumstances of energy coming in exactly balancing the energy going out, there is no change in the energy storage of the layer ($\Delta H_S = 0$) or its temperature with time. These processes of vertical flux convergence and divergence are schematically shown in Figure 2.2b and c.

Example Problem 2
Over the tropical oceans the Bowen ratio is typically 0.1. Estimate the sensible and latent heat fluxes, as well as the rate of evaporation, in millimeters per day, from the ocean surface, when the net radiation received just above the surface is 400 W m^{-2}, the heat flux to water below 50 m is negligible, the rate of warming of the 50-m-deep oceanic mixed layer is 0.05°C day^{-1}, and the ocean surface temperature is 30°C.

Solution
Here, $R_N = 400$ W m^{-2}; $B = H/H_L = 0.1$.
From Equation (2.5),

$$\Delta H_s = \rho c \int_0^D \frac{\partial T}{\partial t} dz = \rho c D \left(\frac{\partial T}{\partial z}\right)_m$$

where $D = 50$ m is the mixed layer depth, and $(\partial T/\partial z)_m = 0.05°\text{C day}^{-1}$ is the mean (layer-averaged) rate of warming of the mixed layer. Thus,

$$\Delta H_S = 1000 \text{ kg m}^{-3} . 4.18 \times 10^3 \text{ J kg}^{-1} \text{ K}^{-1} . 50 \text{ m} . \frac{0.05}{86\,400} \text{ K s}^{-1} = 121 \text{ W m}^{-2}$$

Substituting in the energy budget Equation (2.4), one obtains

$$H + H_L = 259.1 \text{ W m}^{-2}$$

With the Bowen ratio $H/H_L = 0.1$, one obtains $H = 25.4 \text{ W m}^{-2}$, and $H_L = 253.7 \text{ W m}^{-2}$.
The rate of evaporation,

$$E = H_L/L_e = \frac{253.7 \text{ W m}^{-2}}{2.45 \times 10^6 \text{ J kg}^{-1}}$$
$$= 1.03 \times 10^{-4} \text{ kg m}^{-2} \text{ s}^{-1}$$

The evaporation rate in terms of the height of water column per unit time = $E/\rho_w = 8.9 \text{ mm day}^{-1}$.

2.2.3 Energy budget of a control volume

In the energy budgets of an extensive horizontal surface and interfacial layer, only the vertical fluxes of energy are involved. When the surface under consideration is not flat and horizontal or when there are significant changes (advections) of energy fluxes in the horizontal, then it would be more appropriate to consider the energy budget of a control volume. In principle, this is similar to the energy budget of a layer, but now one must consider the various energy fluxes integrated or averaged over the bounding surface of the control volume. The energy budget equation for a control volume can be expressed as

$$\bar{R}_N = \bar{H} + \bar{H}_L + \bar{H}_G + \Delta H_S \tag{2.6}$$

where the overbar over a flux quantity denotes its average value over the entire area (A) of the bounding surface and the rate of storage is given by

$$\Delta H_S = \frac{1}{A} \int \frac{\partial}{\partial t} (\rho c T)\, dV \tag{2.7}$$

in which the integration is over the control volume V.

In practice, detailed measurements of energy fluxes are rarely available in order to evaluate their net contributions to the energy budget of a large irregular volume. Still, with certain simplifying assumptions, such energy budgets have been used in the context of several large field experiments over the oceans, such as the 1969 Barbados Oceanographic and Meteorological Experiment (BOMEX) and the 1975 Air Mass Transformation Experiment (AMTEX).

The concept of flux convergence or divergence and its relation to warming or cooling of the medium is more general than depicted in Figure 2.2 for a horizontal layer. For a control volume, the direction of flux is unimportant. According to Equation (2.6), the net convergence or divergence of energy fluxes in all directions determines the rate of energy storage and, hence, the rate of warming or cooling of the medium in the control volume (see Oke, 1987, Chapter 2).

2.3 Energy Budgets of Bare Surfaces

A few cases of observed energy budgets will be discussed here for illustrative purposes only. It should be pointed out that relative magnitudes of the various terms in the energy budget may differ considerably for other places, times, and weather conditions.

Measured energy fluxes over a dry lake bed (desert) on a hot summer day are shown in Figure 2.3. This represents the simplest case of energy balance [Equation (2.1)] for a flat, dry, and bare surface in the absence of any evaporation or condensation ($H_L = 0$). It is also an example of thermally extreme climatic environment; the observed maximum difference in the temperatures between the surface and the air at 2 m was about 28°C.

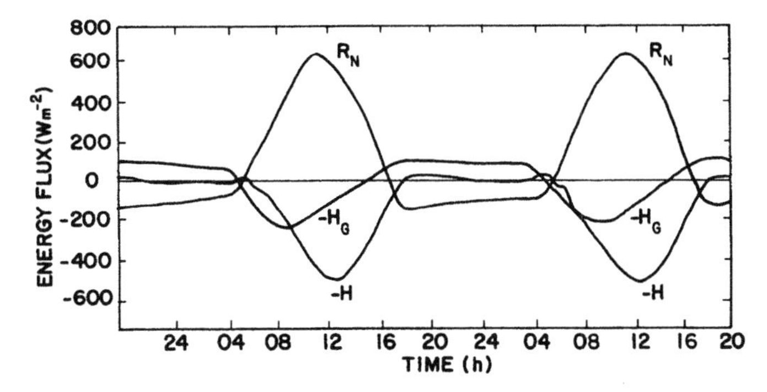

Figure 2.3 Observed diurnal energy budget over a dry lake bed at El Mirage, California, on June 10–11, 1950. [After Vehrencamp (1953).]

Over a bare, dry ground the net radiation is entirely balanced by direct heat exchanges with the surface by both the air and soil media. However, the relative magnitudes of these fluxes may change considerably from day to night, as indicated by measurements in Figure 2.3.

Wetting of the subsurface soil by precipitation or artificial irrigation can dramatically alter the surface energy balance, as well as the microclimate near the surface. The daytime net radiation for the same latitude, season, and weather conditions becomes larger for the wet surface, because of the reduced albedo (reflectivity) and increased absorption of shortwave radiation by the surface. The latent heat flux (H_L) becomes an important or even dominant component of the surface energy balance [Equation (2.1)], while the sensible heat flux to air (H) is considerably reduced. The latter may even become negative during early periods of irrigation, particularly for small areas, due to advective effects. This is called the 'oasis effect' because it is similar to that of warm dry air blowing over a cool moist oasis. There is strong evaporation from the moist surface, resulting in cooling of the surface (due to latent heat transfer). The surface cooling causes a downward sensible heat flux from the warm air to the cool ground. Thus, H_L is positive, while H becomes negative, although much smaller in magnitude. The ground heat flux may also be considerably reduced and even change sign if the surface is cooler than the soil below. Under such conditions, the latent heat flux can be greater than the net radiation. The 'oasis effect' of irrigation disappears as irrigation is applied over large areas and horizontal advections become less important. At later times, the relative importance of H and H_L will depend on the soil moisture content and soil temperature at the surface. As the soil dries up, evaporation rate (E) and $H_L = L_e E$ decrease, while the daytime surface temperature and the sensible heat flux increase with time, for the same environmental conditions. Thus, the daytime Bowen ratio is expected to increase with time after precipitation or irrigation.

A comparison of the surface energy budgets in a dry desert and an irrigated oasis has been given by Budyko (1958). The maximum daytime net radiation in the oasis is 40% larger than that in the dry desert and nearly all of the radiative input to the oasis goes into the latent heat of evaporation. Much smaller positive ground heat flux is nearly balanced by the negative sensible heat flux from air.

2.4 Energy Budgets of Canopies

2.4.1 Vegetation canopies

The growth of vegetation over an otherwise flat surface introduces several complications into the energy balance. First, the ground surface is no longer the most appropriate datum for the surface energy balance, because the radiative, sensible, and latent heat fluxes are all spatially variable within the vegetative canopy. The energy budget of the whole canopy layer [Equation (2.4)] will be more appropriate to consider. For this, measurements of R_N, H, and H_L are needed at the top of the canopy (preferably, well above the tops of plants or trees where horizontal variations of fluxes may be neglected).

Second, the rate of energy storage (ΔH_S) consists of two parts, namely, the rate of physical heat storage and the rate of biochemical heat storage as a result of photosynthesis and carbon dioxide exchange. The latter may not be important on time scales of a few hours to a day, commonly used in micrometeorology. Nevertheless, the rate of heat storage by a vegetative canopy is not easy to measure or calculate.

Third, the latent heat exchange occurs not only due to evaporation or condensation at the surface, but to a large extent due to transpiration from the plant leaves. The combination of evaporation and transpiration is called evapotranspiration; it produces nearly a constant flux of water vapor above the canopy layer.

An example of the observed energy budget over a barley field on a summer day in England is shown in Figure 2.4. Note that the latent heat flux due to evapotranspiration is the dominant energy component, which approximately balances the net radiation, while H and H_G are an order of magnitude smaller. In the late afternoon and evening, H_L even exceeded the net radiation and H became negative (downward heat flux). The rate of heat storage was not measured, but is estimated (from energy balance) to be small in this case.

Forest canopies have similar features to plant canopies, aside from the obvious differences in their sizes and stand architectures. Much larger heights of trees and the associated biomass of the forest canopy suggest that the rate of heat storage may not be insignificant, even over short periods of the order of a day.

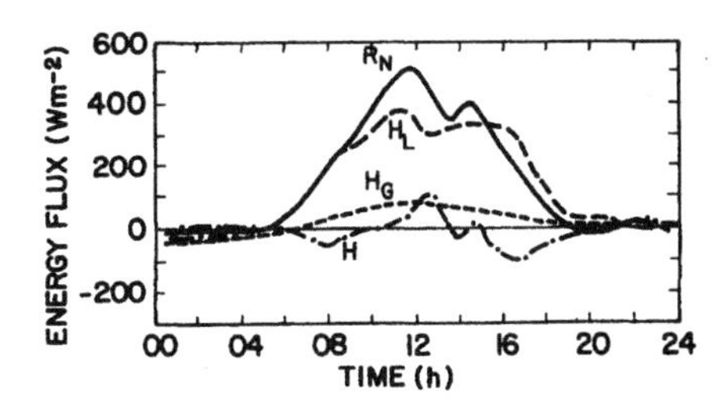

Figure 2.4 Observed diurnal energy budget of a barley field at Rothamsted, England, on July 23, 1963. [From Oke (1987); after Long *et al.* (1964).]

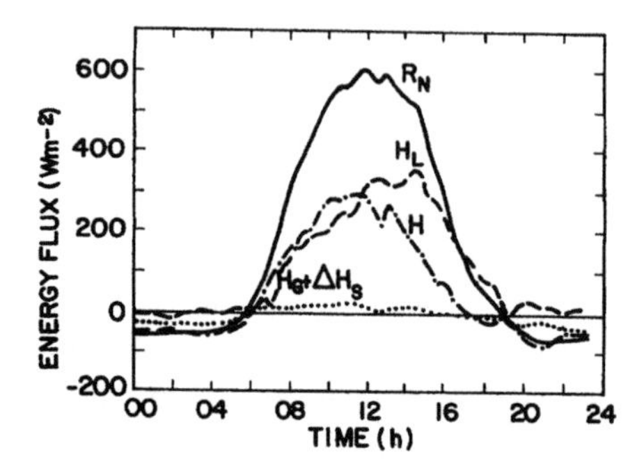

Figure 2.5 Observed energy budget of a Douglas fir canopy at Haney, British Columbia, on July 23, 1970. [From Oke (1987); after McNaughton and Black (1973).]

A typical example of the observed energy budget of a Douglas fir canopy is given in Figure 2.5. In this case, ΔH_S was roughly determined from the estimates of biomass and heat capacity of the trees and measurements of air temperatures within the canopy and, then, added to the measured ground heat flux to get the combined $H_G + \Delta H_S$ term. This term is relatively small during the daytime, but of the same order of magnitude as the net radiation (R_N) at night. Note that, during daytime, R_N is more or less equally partitioned between the sensible and latent heat fluxes to the air. For other examples of measured energy budgets over forests, the reader may refer to Chapter 17 of Munn (1966).

2.4.2 Urban canopies

Urban canopies are much more complex than vegetation canopies, because they represent the large diversity of size, shape, composition, and arrangement of the canopy or 'roughness' elements including buildings, streets, trees, lawns and parks. A useful approach in describing an urban energy budget is to consider a near-surface active layer (urban canopy) or volume (box), whose top is set at or above roof level, and its base at the depth of zero net ground heat flux over the chosen time scale or period (Oke, 1988). Then, one can neglect the extremely complex spatial arrangement of individual canopy elements as energy sources and sinks. The volume or box formulation restricts all energy fluxes, such as R_N, H and H_L, through its top. The internal heat storage change associated with all the canopy elements, the surrounding air and the ground, and internal heat sources due to fuel combustion can be represented as equivalent fluxes through unit area of the top of the box. Thus, an appropriate energy balance for the urban canopy can be expressed as (Oke, 1987, Chapter 8)

$$R_N + H_F = H + H_L + \Delta H_S \tag{2.8}$$

in which H_F is the equivalent heat flux associated with fuel consumption within the urban area or its part (city center, suburb, etc.). Here, we have neglected the net advected heat flow through the sides of the box, assuming that the volume under consideration is surrounded by similar land uses.

In spite of the observed differences in turbidity, temperature, albedo, and emissivity between urban and rural surfaces, the net radiations are not found to be significantly different. But, the addition of the anthropogenic heat flux H_F to R_N makes available substantially more total energy flux for partitioning between the other fluxes on the right-hand side of Equation (2.8). The magnitude of H_F in a city depends on its *per capita* energy use and its population density. Its relative importance in the energy budget of an urban canopy is indicated by the ratio H_F/R_N, which has been estimated for a number of cities in different seasons (Oke, 1987, 1988). The largest values are found in densely inhabited cities in middle and high latitudes and the winter season. Annually averaged values of H_F/R_N for large cities vary between 0.2 (Los Angeles) and 3 (Moscow) with a more typical value of 0.35. Thus, the anthropogenic heat flux is a significant component of the energy budget of any large city. For a particular city, H_F also displays large temporal (both seasonal and diurnal) and spatial variations. In all cases, the central business area or city center appears as the primary heat source, although there may be other localized 'hot spots' in industrial areas. This suggests that H_F can, perhaps, be neglected in the energy budgets of suburban residential areas.

The importance of the sensible heat flux to the energy budget of an urban canopy is well recognized. Built-up urban surfaces (e.g., roofs, streets, and highways) have low surface albedos (0.1–0.2) and readily absorb solar radiation during daytime. With their increased temperatures and enhanced turbulence activity in the urban canopy, a large portion of the available energy is transferred to the atmosphere as sensible heat. The impervious nature of these surfaces reduces the surface water availability for evaporation. Consequently, the latent heat flux is relatively small and the Bowen ratio is large. But, locally, the situation is much different over irrigated lawns or green parks. In urban areas, built-up impervious surfaces and natural pervious surfaces constitute two contrasting surface types with widely different energy budgets on a local basis (Oke, 1987, Chapter 8).

The importance of the heat storage term ΔH_S in the urban canopy budget is also recognized. However, it is almost impossible to determine it from direct measurements, considering the many absorbing elements and surfaces involved in an urban canopy. In most experimental studies of energy budget, ΔH_S is determined as the residual term from the energy balance equation. So far, such studies have been confined to suburban areas where measurements can be made with instruments placed on towers and roof tops.

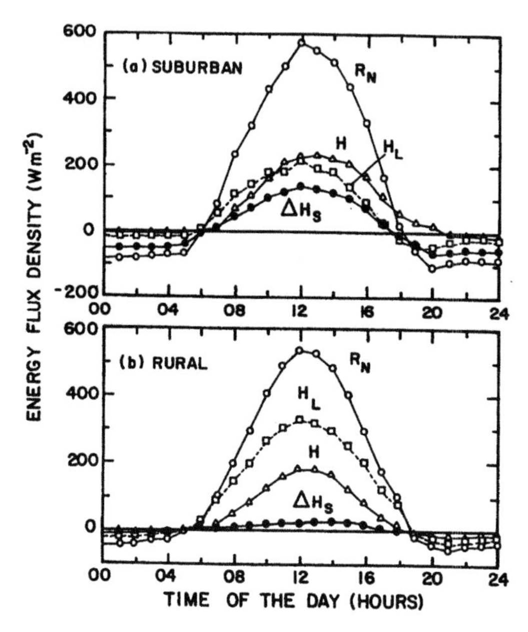

Figure 2.6 Monthly-averaged energy budgets at suburban and rural sites in Greater Vancouver, Canada, during summer. [From Oke (1987).]

An illustration of the suburban energy budget is shown in Figure 2.6 where it is compared with that of a nearby rural surface. The various fluxes shown are monthly averages for suburban and rural sites in Greater Vancouver, Canada, during summer (Oke, 1987, Chapter 8). These are based on direct measurements of R_N and H, calculated ΔH_S, and estimated H_L as the residual in the energy balance equation. The suburban area is an area of fairly uniform one- or two-storey houses (36% built and 64% greenspace), while the 'rural' site is an extensive area of grassland. For the latter, ΔH_S essentially represents H_G. As expected, the heat storage term over the suburban area is a significant part of the energy budget, especially at nighttime. Some day-to-day variations of the energy budget of the same suburban area are shown by Oke (1988). Grimmond and Oke (1995) compare the summertime energy budgets of the suburbs for four North American cities (Chicago, Los Angeles, Sacramento, and Tucson) under clear, cloudy, and all sky conditions. As expected, the magnitudes of fluxes vary between cities; however, the diurnal trends of flux partitioning are similar in terms of the timing of the peaks and changes in sign.

Similar observations of the energy budgets of more heavily built city centers and commercial and industrial districts are lacking. However, we may anticipate both H and ΔH_S playing a much larger role than in the suburbs, and H_L playing a smaller role. Anthropogenic heat flux H_F is expected to be too large to ignore, while the contribution of heat advection to or from surrounding areas may also have to be considered.

2.5 Energy Budgets of Water Surfaces

Water covers more than two-thirds of the earth's surface. Therefore, it is important to understand the energy budget of water surfaces. It is complicated, however, by the fact that water is a fluid with a dynamically active surface and a surface boundary layer or mixed layer in which the motions are generally turbulent. Thus, convective and advective heat transfers within the surface boundary layer in water essentially determine H_G. However, it is not easy to measure these transfers or H_G directly. As discussed earlier, the radiation balance at the water surface is also complicated by the fact that shortwave radiation penetrates to considerable depth in water. Therefore, it will be more appropriate to consider the energy budget of a layer of water extending to a depth where both the convective and radiative heat exchanges become negligible. This will not be feasible, however, for shallow bodies of water with depths less than about 10 m.

Even over large lakes and oceans, simultaneous measurements of all the energy flux terms in Equation (2.4) on a short-term basis are lacking, largely because of the experimental difficulties associated with floating platforms and sea spray. The sensible and latent heat fluxes are the most frequently measured or estimated quantities. Over most ocean areas the latter dominates the former and the Bowen ratio is usually much less than unity (it may approach unity during periods of intense cold air advection over warm waters). The rate of heat storage in the oceanic mixed layer plays an important role in the energy budget. This layer acts as a heat sink ($\Delta H_S > 0$) by day and a heat source ($\Delta H_S < 0$) at night.

The significance of net radiation in the energy budget over water is not so clear. For some measurements of heat balance at sea on a daily basis in the course of a large experiment, the reader may refer to Kondo (1976).

2.6 Applications

The following list includes some of the applications of energy balance at or near the earth's surface.

- Prediction of surface temperature and frost conditions.
- Indirect determination of the surface fluxes of heat (sensible and latent) to or from the atmosphere.

- Estimation of the rate of evaporation from bare ground and water surfaces and evapotranspiration from vegetative surfaces.
- Estimation of the rate of heat storage or loss by an oceanic mixed layer or a vegetative canopy.
- Study of microclimates of the various surfaces.
- Prediction of icing conditions on highways.

In all these and possibly other applications, the various terms in the appropriate energy balance equation, except for the one to be estimated or predicted, have to be measured, calculated, or parameterized in terms of other measured quantities.

Problems and Exercises

1. Describe the typical conditions in which you would expect the Bowen ratio to be as follows:
(a) much less than unity;
(b) much greater than unity;
(c) negative.

2.
(a) Over an ocean surface, the Bowen ratio is estimated to be about 0.2. Estimate the sensible and latent heat fluxes to the atmosphere, as well as the rate of evaporation, in millimeters per day, from the ocean surface, when the net radiation received just above the surface is 600 W m^{-2}, the heat flux to the water below 50 m is negligible, the rate of warming of the 50 m deep oceanic mixed layer is 0.08°C day^{-1}, and the sea surface temperature is 25°C.
(b) What will be the rate of warming or cooling of the 50 m deep oceanic mixed layer at the time of intense cold-air advection when the Bowen ratio is estimated to be 0.5, the net radiation loss from the surface is 50 W m^{-2}, and the rate of evaporation is 20 mm day^{-1}?

3. Giving schematic depictions, briefly discuss the energy balance of an extensive, uniform snowpack for the following conditions:
(a) below-freezing air temperatures;
(b) above-freezing air temperatures.

4.
(a) Give a derivation of Equation (2.5) for the rate of heat storage in a soil layer.
(b) Will the same expression apply to an oceanic mixed layer? Give reasons for your answer.

5. Explain the following terms or concepts used in connection with the energy balance near the surface:
(a) 'ideal' surface;
(b) evaporative cooling;
(c) oasis effect;
(d) flux divergence.

6.
(a) What are the major differences between the energy budgets of a bare soil surface and a vegetative surface?
(b) How would you rewrite the energy balance of a subsurface soil layer and how does it differ from that for the bare soil surface?

7.
(a) Explain the processes and mechanisms involved in the increase of the surface temperature during morning hours after sun rise.
(b) Is the surface energy balance equation (Equation 2.1) valid even when the surface temperature is rising rapidly with time?

8.
(a) Explain the importance of the change in heat storage in the energy budget of an urban canopy.
(b) What are the different elements or components of ΔH_S in Equation (2.8) and why is it so difficult to measure directly?

9.
(a) How would you measure the rate of heat storage in the mixed layer of an ocean?
(b) Discuss the mechanisms or processes involved in the daytime warming of the oceanic mixed layer and qualitatively compare it with that of a subsurface soil layer.

Chapter 3 Radiation Balance Near the Surface

3.1 Radiation Laws and Definitions

The transfer of energy by rapid oscillations of electromagnetic fields is called radiative transfer or simply radiation. These oscillations may be considered as traveling waves characterized by their wavelength or wave frequency c_m/λ, where c_m is the wave speed in a given medium. All electromagnetic waves travel at the speed of light $c \cong 3 \times 10^8$ m s^{-1} in empty space and nearly the same speed in air ($c_m \cong c$). There is an enormous range or spectrum of electromagnetic wavelengths or frequencies. Here we are primarily interested in the approximate range 0.1–100 μm, in which significant contributions to the radiation balance of the atmosphere or the earth's surface occur. This represents only a tiny part of the entire electromagnetic wave spectrum. Of this, the visible light constitutes a very narrow range of wavelengths (0.40–0.76 μm).

The radiant flux density, or simply the radiative flux, is defined as the amount of radiant energy (integrated over all wavelengths) received at or emitted by a unit area of the surface per unit time. The SI unit of radiative flux is W m^{-2}, which is related to the CGS unit cal cm^{-2} min^{-1}, commonly used earlier in meteorology as 1 cal cm^{-2} min^{-1} $\cong$ 698 W m^{-2}.

3.1.1 Blackbody radiation laws

Any body having a temperature above absolute zero emits radiation. If a body at a given temperature emits the maximum possible radiation per unit area of its surface, per unit time, at all wavelengths, it is called a perfect radiator or 'blackbody.' The flux of radiation (R) emitted by such a body is given by the *Stefan–Boltzmann law*

$$R = \sigma T^4 \tag{3.1}$$

where σ is the Stefan–Boltzmann constant = 5.67×10^{-8} W m^{-2} K^{-4}, and T is the surface temperature of the body in absolute (K) units.

Planck's law expresses the radiant energy per unit wavelength emitted by a blackbody as a function of its surface temperature

$$R_\lambda = (2\pi h_p c^2/\lambda^5)[\exp(h_p c/b\lambda T) - 1]^{-1} \tag{3.2}$$

where h_p is Planck's constant = 6.626×10^{-34} J s, and b is the Boltzmann constant = 1.381×10^{-23} J K^{-1}. Note that the total radiative flux is given by

$$R = \int_0^\infty R_\lambda \, d\lambda \tag{3.3}$$

Equation (3.2) can be used to calculate and compare the spectra of blackbody radiation at various body surface temperatures. The wavelength at which R_λ is maximum turns out to be inversely proportional to the absolute temperature and is given by *Wien's law*

$$\lambda_{max} = 2897/T \tag{3.4}$$

when λ_{max} is expressed in micrometers.

From the above laws, it is clear that the radiative flux emitted by a blackbody varies in proportion to the fourth power of its surface temperature, while the wavelengths making the most contribution to the flux, especially λ_{max}, change inversely proportional to T.

3.1.2 Shortwave and longwave radiations

The spectrum of solar radiation received at the top of the atmosphere is well approximated by the spectrum of a blackbody having a surface temperature of about 6000 K. Thus, the sun may be considered as a blackbody with an equivalent surface temperature of about 6000 K and $\lambda_{max} \cong 0.48$ μm.

The observed spectra of terrestrial radiation near the surface, particularly in the absence of absorbing substances such as water vapor and carbon dioxide, can also be approximated by the equivalent blackbody radiation spectra given by Equation (3.2). However, the equivalent blackbody temperature (T_{eb}) is expected to be smaller than the actual surface temperature (T); the difference, $T - T_{eb}$, depends on the radiative characteristics of the surface, which will be discussed later.

Idealized or equivalent blackbody spectra of solar ($T_{eb} = 6000$ K) and terrestrial ($T_{eb} = 287$ K) radiations, both normalized with respect to their peak flux per unit wavelength, are compared in Figure 3.1. Note that almost all the solar energy flux is confined to the wavelength range 0.15–4.0 μm, while the

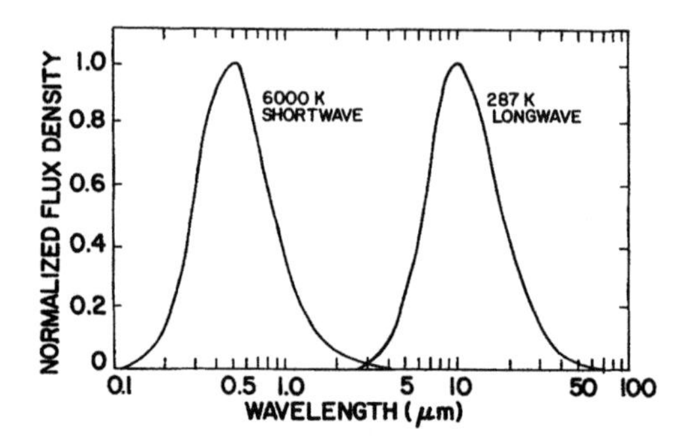

Figure 3.1 Calculated blackbody spectra of flux density of solar and terrestrial radiation, both normalized by their peak flux density. [From Rosenberg *et al.* (1983).]

terrestrial radiation is mostly confined to the range 3–100 μm. Thus, there is very little overlap between the two spectra, with their peak wavelengths (λ_{max}) separated by a factor of almost 20. In meteorology, the above two ranges of wavelengths are characterized as shortwave and longwave radiations, respectively.

3.1.3 Radiative properties of natural surfaces

Natural surfaces are not perfect radiators or blackbodies, but are, in general, gray bodies. They are generally characterized by several different radiative properties, which are defined as follows.

Emissivity is defined as the ratio of the energy flux emitted by the surface at a given wavelength and temperature to that emitted by a blackbody at the same wavelength and temperature. In general, the emissivity may depend on the wavelength and will be denoted by ε_λ. For a blackbody, $\varepsilon_\lambda = 1$ for all wavelengths.

Absorptivity is defined as the ratio of the amount of radiant energy absorbed by the surface material to the total amount of energy incident on the surface. In general, absorptivity is also dependent on wavelength and will be denoted by α_λ. A perfect radiator is also a perfect absorber of radiation, so that $\alpha_\lambda = 1$ for a blackbody.

Reflectivity is defined as the ratio of the amount of radiation reflected to the total amount incident upon the surface and will be denoted by r_λ.

Transmissivity is defined as the ratio of the radiation transmitted to the subsurface medium to the total amount incident upon the surface and will be denoted by t_λ.

It is clear from the above definitions that

$$\alpha_\lambda + r_\lambda + t_\lambda = 1 \tag{3.5}$$

so that absorptivity, reflectivity, and transmissivity must have values between zero and unity.

Kirchoff's law states that for a given wavelength, absorptivity of a material is equal to its emissivity, i.e.,

$$\alpha_\lambda = \varepsilon_\lambda \tag{3.6}$$

Natural surfaces are in general radiatively 'gray,' but in certain wavelength bands their emissivity or absorptivity may be close to unity. For example, in the so-called atmospheric window (8 to 14 μm wavelength band), water, wet soil, and vegetation have emissivities of 0.97–0.99.

In studying the radiation balance for the energy budget near the earth's surface, we are more interested in the radiative fluxes integrated over all the wavelengths of interest, rather than in their wavelength-dependent spectral decomposition. For this, an overall or integrated emissivity and an integrated reflectivity, pertaining to radiation in a certain range of wavelengths, are used to characterize the surface. In particular, the term 'albedo' is used to represent the integrated reflectivity of the surface for shortwave (0.15–4 μm) radiation, while the overall emissivity (ε) of the surface refers primarily to the longwave (3–100 μm) radiation. For natural surfaces, the emission of shortwave radiation is usually neglected, and the emission of longwave radiative flux is given by the modified Stefan–Boltzmann law

$$R_{\mathrm{L}} = -\varepsilon\sigma T^4 \tag{3.7}$$

where the negative sign is introduced in accordance with our sign convention for radiative fluxes. Typical values of surface albedo (a) and emissivity (ε) for different types of surfaces are given in Table 3.1.

The albedo of water is quite sensitive to the sun's zenith angle and may approach unity when the sun is near the horizon (rising or setting sun). It is also dependent on the wave height or sea state. To a lesser extent, the solar zenith angle is also found to affect albedos of soil, snow, and ice surfaces. Although snow has a high albedo, deposition of dust and aerosols, including man-made pollutants such as soot, can significantly lower the albedo of snow. The albedo of a soil surface is strongly dependent on the wetness of the soil.

Infrared emissivities of natural surfaces do not vary over a wide range; most surfaces have emissivities greater than 0.9, while only in a few cases values

Table 3.1 Radiative properties of natural surfaces.

Surface type	Other specifications	Albedo (a)	Emissivity (ε)
Water	Small zenith angle	0.03–0.10	0.92–0.97
	Large zenith angle	0.10–1.00	0.92–0.97
Snow	Old	0.40–0.70	0.82–0.89
	Fresh	0.45–0.95	0.90–0.99
Ice	Sea	0.30–0.45	0.92–0.97
	Glacier	0.20–0.40	
Bare sand	Dry	0.35–0.45	0.84–0.90
	Wet	0.20–0.30	0.91–0.95
Bare soil	Dry clay	0.20–0.40	0.95
	Moist clay	0.10–0.20	0.97
	Wet fallow field	0.05–0.07	
Paved	Concrete	0.17–0.27	0.71–0.88
	Black gravel road	0.05–0.10	0.88–0.95
Grass	Long (1 m)	0.16	0.90
	Short (0.02 m)	0.26	0.95
Agricultural	Wheat, rice, etc.	0.18–0.25	0.90–0.99
	Orchards	0.15–0.20	0.90–0.95
Forests	Deciduous	0.10–0.20	0.97–0.98
	Coniferous	0.05–0.15	0.97–0.99

Compiled from Sellers (1965), Kondratyev (1969), and Oke (1987).

become less than 0.8. Green, lush vegetation and forests are characterized by the highest emissivities, approaching close to unity.

3.2 Shortwave Radiation

The ultimate source of all shortwave radiation received at or near the earth's surface is the sun. A large part of it comes directly from the sun, while other parts come in the forms of reflected radiation from the surface and clouds, and scattered radiation from atmospheric particulates or aerosols.

3.2.1 Solar radiation

As a measure of the intensity of solar radiation, the solar constant (S_0) is defined as the flux of solar radiation falling on a surface normal to the solar beam at the outer edge of the atmosphere, when the earth is at its mean distance from the sun. An accurate determination of its value has been sought by many investigators. The best estimate of $S_0 = 1368 \text{ W m}^{-2}$ is based on a series of measurements from high-altitude platforms such as the Solar Maximum Satellite; other values proposed in the literature fall in the range 1350–1400 W m^{-2}. There are speculations that the solar constant may vary with changes in the sun spot activity, which has a predominant cycle of about 11 years. There are also suggestions of very long-term variability of the solar constant which may have been partially responsible for long-term climatic fluctuations in the past.

The actual amount of solar radiation received at a horizontal surface per unit area over a specified time is called insolation. It depends strongly on the solar zenith angle γ and also on the ratio (d/d_m) of the actual distance to the mean distance of the earth from the sun. The combination of the so-called inverse-square law and Lambert's cosine law gives the flux density of solar radiation at the top of the atmosphere as

$$R_0 = S_0 \, (d_m/d)^2 \cos \gamma \tag{3.8}$$

Then, insolation for a specified period of time between t_1 and t_2 is given by

$$I_0 = \int_{t_1}^{t_2} R_0(t) \, dt \tag{3.9}$$

Thus, one can determine the daily insolation from Equation (3.9) by integrating the solar flux density with time over the daylight hours.

For a given calendar day/time and latitude, the solar zenith angle (γ) and the ratio d_m/d can be determined from standard astronomical formulas or tables, and the solar flux density and insolation at the top of the atmosphere can be evaluated from Equations (3.8) and (3.9). These are also given in Smithsonian Meteorological Tables.

The solar flux density (R_s) and insolation (I) received at the surface of the earth may be considerably smaller than their values at the top of the atmosphere because of the depletion of solar radiation in passing through the atmosphere. The largest effect is that of clouds, especially low stratus clouds. In the presence of scattered moving clouds, R_s becomes highly variable.

The second important factor responsible for the depletion of solar radiation is atmospheric turbidity, which refers to any condition of the atmosphere, excluding clouds, which reduces its transparency to shortwave radiation. The reduced transparency is primarily due to the presence of particulates, such as pollen, dust, smoke, and haze. Turbidity of the atmosphere in a given area may result from a combination of natural sources, such as wind erosion, forest fires, volcanic eruptions, sea spray, etc., and various man-made sources of aerosols. Particles in the path of a solar beam reflect part of the radiation and scatter the other part.

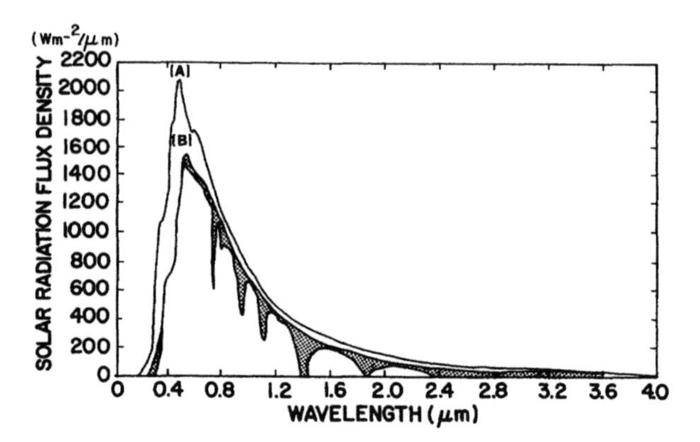

Figure 3.2 Observed flux density of solar radiation at the top of the atmosphere (curve A) and at sea level (curve B). The shaded areas represent absorption due to various gases in a clear atmosphere. [From Liou (1983.]

Large, solid particles reflect more than they scatter light, affecting all visible wavelengths equally. Therefore, the sky appears white in the presence of these particles. Even in an apparently clear atmosphere, air molecules and very small (submicrometer size) particles scatter the sun's rays. According to Rayleigh's scattering law, scattering varies inversely as the fourth power of the wavelength. Consequently, the sky appears blue because there is preferred scattering of blue light (lowest wavelengths of the visible spectrum) over other colors.

Even in a cloud-free nonturbid atmosphere, atmospheric gases, such as oxygen, ozone, carbon dioxide, water vapor, and nitrous oxides, absorb solar radiation in selected wavelength bands. For example, much of the ultraviolet radiation is absorbed by stratospheric ozone and oxygen, and water vapor and carbon dioxide have a number of absorption bands at wavelengths larger than 0.8 μm.

The combined effects of scattering and absorption of solar radiation in a clear atmosphere can be seen in Figure 3.2. Here, curve A represents the measured spectrum of solar radiation at the top of the atmosphere, curve B represents the measured spectrum at sea level, and the shaded areas represent the absorption of energy by various gases, primarily O_3, O_2, CO_2, and H_2O. The unshaded area between curve A and the outer envelope of the shaded area represents the depletion of solar radiation due to scattering.

3.2.2 Reflected radiation

A significant fraction (depending on the surface albedo) of the incoming shortwave radiation is reflected back by the surface. Shortwave reflectivities or albedos for the various natural and man-made surfaces are given in Table 3.1. Knowledge of surface albedos and of their possible changes (seasonal, as well as long term) due to man's activities, such as deforestation, agriculture, and urbanization, has been of considerable interest to climatologists. Satellites now permit systematic and frequent observations of surface albedos over large regions of the world. Any inadvertent or intentional changes in the local, regional, or global albedo may cause significant changes in the surface energy balance and hence in the micro- or macroclimate. For example, deliberate modification of albedo by large-scale surface covering has been proposed as a means of increasing precipitation in certain arid regions. It has also been suggested that large, frequent oil spills in the ice- and snow-covered arctic region could lead to significant changes in the regional energy balance and climate.

Snow is the most effective reflector (high albedo) of shortwave radiation. On the other hand, water is probably the poorest reflector (low albedo), while ice falls between snow and water. Since large areas of earth are covered by water, sea ice, and snow, significant changes in the snow and ice cover are likely to cause perceptible changes in the regional and, possibly, global albedo.

Most bare rock, sand, and soil surfaces reflect 10–45% of the incident shortwave radiation, the highest value being for desert sands. Albedos of most vegetative surfaces fall in the range 10–25%. Wetting of the surface by irrigation or precipitation considerably lowers the albedo, while snowfall increases it. Albedos of vegetative surfaces, being sensitive to the solar elevation angle, also show some diurnal variations, with their minimum values around noon and maximum values near sunrise and sunset (see Rosenberg *et al.*, 1983, Chapter 1).

Shortwave reflectivities of clouds vary over a wide range, depending on the cloud type, height, and size, as well as on the angle of incidence of radiation (Welch *et al.*, 1980). Thick stratus, stratocumulus, and nimbostratus clouds are good reflectors ($a = 0.6$–0.8, at normal incidence), large cumulus clouds are moderate reflectors ($a = 0.2$–0.5), and small cumuli are poor reflectors ($a < 0.2$). Only a small part of solar radiation may reach the ground in cloudy conditions. A significant part of the reflected radiation is likely to undergo multiple reflections between the surfaces and bases of clouds, thereby increasing the effective albedo of the surface.

3.2.3 Diffuse radiation

Diffuse or sky radiation is that portion of the solar radiation that reaches the earth's surface after having been scattered by molecules and suspended particulates in the atmosphere. In cloudy conditions it also includes the portion of the shortwave radiation which is reflected by the clouds. It is the incoming shortwave radiation in shade. Before sunrise and after sunset, all shortwave radiation is in diffuse form. The ratio of diffuse to the total incoming shortwave radiation varies diurnally, seasonally, and with latitude. In high latitudes, diffuse radiation is very important; even in midlatitudes it constitutes 30–40% of the total incoming solar radiation. Cloudiness considerably increases the ratio of diffuse to total solar radiation.

3.3 Longwave Radiation

The longwave radiation R_L received at or near the surface has two components: (1) outgoing radiation $R_{L\uparrow}$ from the surface, and (2) incoming radiation $R_{L\downarrow}$ from the atmosphere, including clouds. These are discussed separately in the following sections.

3.3.1 Terrestrial radiation

All natural surfaces radiate energy, depending on their emissivities and surface temperatures, whose flux density is given by Equation (3.7). As discussed earlier in Section 3.1, emissivities for most natural surfaces range from 0.9 to 1.0. Thus knowing the surface temperature and a crude estimate of emissivity, one can determine terrestrial radiation from Equation (3.7) to better than 10% accuracy. Difficulties arise, however, in measuring or even defining the surface temperature, especially for vegetative surfaces. In such cases, it may be more appropriate to determine the apparent surface temperature, or equivalent blackbody temperature T_{eb} from measurements of terrestrial radiation near the surface using Equation (3.1).

In passing through the atmosphere, a large part of the terrestrial radiation is absorbed by atmospheric gases, such as water vapor, carbon dioxide, nitrogen oxides, methane, and ozone. In particular, water vapor and CO_2 are primarily responsible for absorbing the terrestrial radiation and reducing its escape to the space (the so-called greenhouse effect).

3.3.2 Atmospheric radiation

From our discussion earlier, it is clear that the atmosphere absorbs much of the longwave terrestrial radiation and a significant part of the solar radiation. The atmospheric gases and aerosols which absorb energy also radiate energy, depending on the vertical distributions of their concentrations or mixing ratios and air temperature as functions of height. An important aspect of the atmospheric radiation is that absorption and emission of radiation by various gases occur in a series of discrete wavelengths or bands of wavelengths, rather than continuously across the spectrum, as shown in Figure 3.3. All layers of the atmosphere participate, to varying degrees, in absorption and emission of radiation, but the atmospheric boundary layer is most important in this, because the largest concentrations of water vapor, CO_2, and other gases occur in this layer.

Clouds, when present, are the major contributors to the incoming longwave radiation to the surface. They radiate like blackbodies ($\varepsilon \cong 1$) at their respective cloud base temperatures. However, some of the radiation is absorbed by water vapor, CO_2, and other absorbing gases before reaching the earth's surface.

Computation of incoming longwave radiation from the atmosphere is tedious and complicated, even when the distributions of water vapor, CO_2, cloudiness, and temperature are measured. It is preferable to measure $R_{L\downarrow}$ directly with an appropriate radiometer.

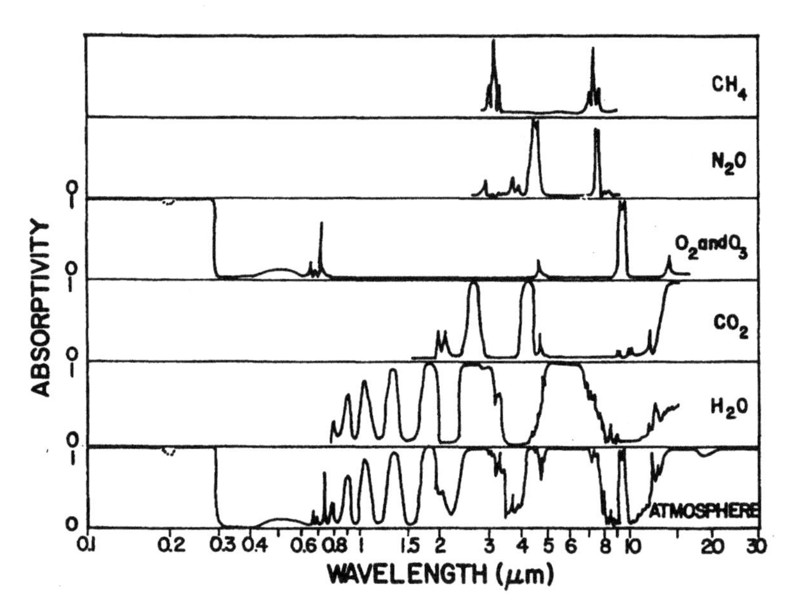

Figure 3.3 Absorption spectra of water vapor, carbon dioxide, oxygen and ozone, nitrogen oxide, methane and the atmosphere. [From Fleagle and Businger (1980).]

3.4 Radiation Balance Near the Surface

The net radiation flux R_N in Equations (2.1) and (2.2) is a result of radiation balance between shortwave (R_S) and longwave (R_L) radiations at or near the surface, which can be written as

$$R_N = R_S + R_L \tag{3.10}$$

Further, expressing shortwave and longwave radiation balance terms as

$$R_S = R_{S\downarrow} + R_{S\uparrow} \tag{3.11}$$

$$R_L = R_{L\downarrow} + R_{L\uparrow} \tag{3.12}$$

the overall radiation balance can also be written as

$$R_N = R_{S\downarrow} + R_{S\uparrow} + R_{L\downarrow} + R_{L\uparrow} \tag{3.13}$$

where the downward and upward arrows denote incoming and outgoing radiation components, respectively.

The incoming shortwave radiation ($R_{S\downarrow}$) consists of both the direct-beam solar radiation and the diffuse radiation. It is also called insolation at the ground and can be easily measured by a solarimeter. It has strong diurnal variation (almost sinusoidal) in the absence of fog and clouds.

The outgoing shortwave radiation ($R_{S\uparrow}$) is actually the fraction of $R_{S\uparrow}$ that is reflected by the surface, i.e.,

$$R_{S\uparrow} = -aR_{S\downarrow} \tag{3.14}$$

where a is the surface albedo. Thus, for a given surface, the net shortwave radiation $R_S = (1 - a)R_{S\downarrow}$ is essentially determined by insolation at the ground.

The incoming longwave radiation ($R_{L\downarrow}$) from the atmosphere, in the absence of clouds, depends primarily on the distributions of temperature, water vapor, and carbon dioxide. It does not show a significant diurnal variation. The outgoing terrestrial radiation ($R_{L\uparrow}$), being proportional to the fourth power of the surface temperature in absolute units, shows stronger diurnal variation, with its maximum value in the early afternoon and minimum value at dawn. The two components are usually of the same order of magnitude, so that the net longwave radiation (R_L) is generally a small quantity.

Under clear skies, $|R_L| \ll R_S$ during the bright daylight hours and an approximate radiation balance is given by

$$R_N \cong R_S = (1 - a)R_{S\downarrow} \tag{3.15}$$

which can be used to determine net radiation from simpler measurements or calculation of solar radiation at the surface. At nighttime, however, $R_{S\downarrow} = 0$, and the radiation balance becomes

$$R_N = R_L = R_{L\downarrow} + R_{L\uparrow} \tag{3.16}$$

Frequently at night $R_{L\downarrow} < -R_{L\uparrow}$, so that R_N or R_L is usually negative, implying radiative cooling of the surface.

Near the sunrise and sunset times, all the components of the radiation balance are of the same order of magnitude and Equation (3.10) or (3.13) will be more appropriate than the simplified Equation (3.15) or (3.16).

Example Problem 1

The following radiation measurements were made over a dry, bare field during a very calm and clear spring night:

Outgoing longwave radiation from the surface = 400 W m^{-2}.

Incoming longwave radiation from the atmosphere = 350 W m^{-2}.

(a) Calculate the equivalent blackbody surface temperature, as well as the actual surface temperature if the surface emissivity is 0.95.

(b) Estimate the ground heat flux, making appropriate assumptions about other fluxes.

Solution

(a) $R_{L\uparrow} = \sigma T_{eb}^4 = -400$ W m^{-2}; $\sigma = 5.67 \times 10^{-8}$ W m^{-2} K^{-4}, so that $T_{eb} = (400/\sigma)^{1/4} \cong 289.8$ K, is the equivalent blackbody temperature.
Also, $R_{L\uparrow} = -\varepsilon\sigma T_S^4$, so that $T_S = (400/\varepsilon\sigma)^{1/4} \cong 293.6$ K, is the actual surface temperature.

(b) Using the energy budget equation (Equation 2.1), $H_G = R_N - (H + H_L)$.
Very calm night implies $H + H_L \cong 0$, so that
$H_G = R_N = R_{L\uparrow} + R_{L\downarrow}$, using Equation (3.16).
Thus, $H_G = -400 + 350 = -50$ W m^{-2}, is the ground heat flux.

3.5 Observations of Radiation Balance

For the purpose of illustration, some examples of measured radiation balance over different types of surfaces are provided. Figure 3.4 represents the various

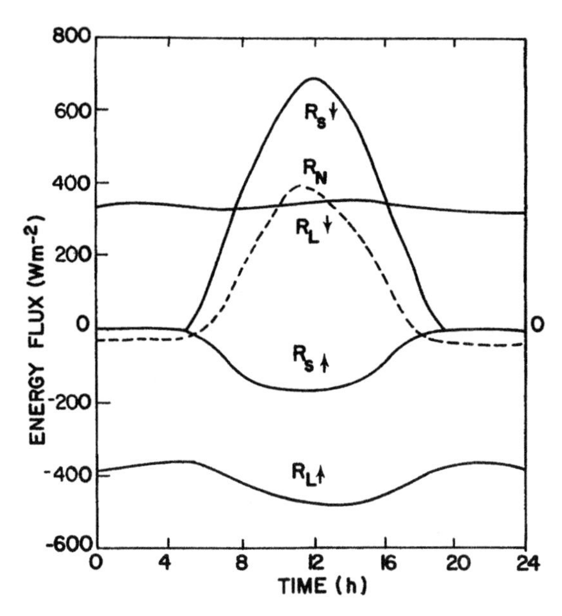

Figure 3.4 Observed radiation budget over a 0.2 m stand of native grass at Matador, Saskatchewan, on 30 July 1971. [From Oke (1987); after Ripley and Redmann (1976).]

components of Equation (3.10) for a clear August day in England over a thick stand of grass, when the diurnal variation of grass temperature was about 20°C. Note that in this case about one-fourth of the incident solar radiation was reflected back by the surface, while about three-fourths was absorbed. The net radiation is slightly less than R_S, even during the midday hours, because of the net loss due to longwave radiation. The diurnal variation of $R_{L\downarrow}$ is much less than that of $R_{L\uparrow}$ (28% variation in $R_{L\uparrow}$ is consistent with the observed diurnal range of 20°C in surface temperature).

Figure 3.5 shows the diurnal variation of radiation budget components over Lake Ontario on a clear day in August. Note that the measured shortwave radiation ($R_{S\downarrow}$) near the lake surface is only about two-thirds of the computed solar radiation (R_0) at the top of the atmosphere. Of this, 25–30% is in the form of diffuse-beam radiation (R_D) at midday and the rest as direct-beam solar radiation (not plotted). The outgoing shortwave radiation ($R_{S\uparrow}$) is relatively small, due to the low albedo ($a \cong 0.07$) of water. Both the incoming and outgoing longwave radiation components are relatively constant with time, due to small diurnal variations in the temperatures of the lake surface and the air above it. The net longwave radiation is a constant energy loss throughout the period of observations. The net radiation (R_N) is dominated by $R_{S\downarrow}$ during the day and is equal to R_L at night.

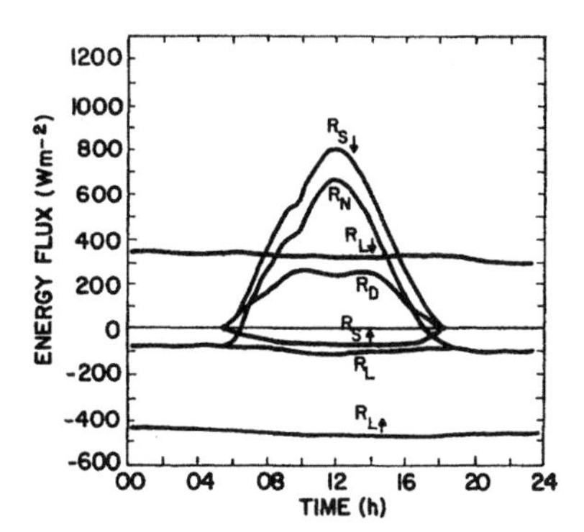

Figure 3.5 Observed radiation budget over Lake Ontario under clear skies on 28 August 1969. [From Oke (1987); after Davies *et al.* (1970).]

3.6 Radiative Flux Divergence

The concept of energy flux convergence or divergence and its relation to cooling or warming of a layer of the atmosphere of submedium has been explained in Chapter 1. Here, we discuss the significance of net radiative flux convergence or divergence to warming or cooling in the lowest layer of the atmosphere, namely, the PBL.

The rate of warming or cooling of a layer of air due to change of net radiation with height can be calculated from the principle of conservation of energy. Considering a thin layer between the levels z and $z + \Delta z$, where the net radiative fluxes are $R_N(z)$ and $R_N(z + \Delta z)$, as shown in Figure 3.6, we get

$$\rho c_p \Delta z (\partial T/\partial t)_R = R_N(z + \Delta z) - R_N(z) = (\partial R_N/\partial z)\Delta z$$

or

$$(\partial T/\partial t)_R = (1/\rho c_p)(\partial R_N/\partial z) \tag{3.17}$$

where $(\partial T/\partial t)_R$ is the rate of change of temperature due to radiation and $\partial R_N/\partial z$ represents the convergence or divergence of net radiation. Radiative flux convergence occurs when R_N increases with height ($\partial R_N/\partial z > 0$) and divergence occurs when $\partial R_N/\partial z < 0$. The former leads to warming and the latter to cooling of the air. Warming or cooling of air due to radiative flux convergence or divergence is only a part of the total warming or cooling. Contributions of warm or cold air advection and turbulent exchange of sensible heat will be discussed later in Chapter 5.

In the daytime, during clear skies, net radiation is dominated by the net shortwave radiation (R_S), which usually does not vary with height in the PBL.

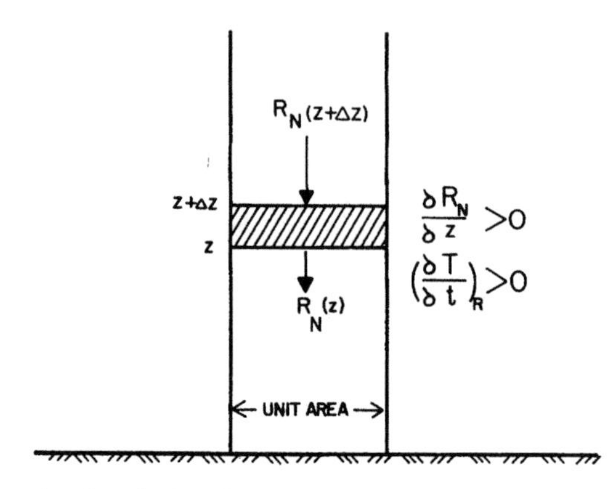

Figure 3.6 Schematic of radiative flux convergence or divergence in the lower atmosphere.

Therefore, changes in air temperature with time due to radiation are insignificant or negligible; the observed daytime warming of the PBL is largely due to the convergence of sensible heat flux. This may not be true, however, in the presence of fog and clouds in the PBL, when radiative flux convergence or divergence may also become important.

At night, the net radiation is entirely due to the net longwave radiation (R_L). Both the terrestrial and atmospheric radiations and, hence, R_L generally vary with height, because the concentrations of water vapor, CO_2, and other gases that absorb and emit longwave radiation vary strongly with height in the PBL. Frequently, there is significant radiative flux divergence (which implies that radiative heat loss increases with height) within the PBL, especially during clear and calm nights. A significant part of cooling in the PBL may be due to radiation, while the remainder is due to the divergence of sensible heat flux. In many theoretical studies of the nocturnal stable boundary layer (SBL), radiative flux divergence has erroneously been ignored; its importance in the determination of thermodynamic structure of the SBL has recently been pointed out (Garratt and Brost, 1981).

In the presence of strong radiative flux divergence or convergence, radiation measurements at some height above the surface may not be representative of their surface values. Corrections for the flux divergence need to be applied to such measurements.

Example Problem 2

The following measurements were made at night from a meteorological tower in the middle of a large farm:

Net radiation at the 5 m level $= -75$ W m^{-2}.

Net radiation at the 100 m level = -150 W m^{-2}.
Sensible heat flux at the surface = -45 W m^{-2}.
Planetary boundary layer height = 90 m.
Calculate the average rate of cooling in the PBL due to the divergence of (a) net radiation and (b) sensible heat flux.

Solution

(a) The rate of change of air temperature due to radiative flux divergence

$$\left(\frac{\partial T}{\partial t}\right)_R = \frac{1}{\rho c_p}\frac{\partial R_N}{\partial z} = \frac{1}{1200}\left(\frac{-150+75}{100-5}\right)$$

$$= -6.58 \times 10^{-4}\ \mathrm{K\ s^{-1}}$$

$$= -2.37\ \mathrm{K\ h^{-1}}$$

(b) From the definition of the PBL, the sensible heat flux at the top of the PBL (90 m) must be zero. The rate of change of air temperature due to sensible heat flux divergence is given by

$$\left(\frac{\partial T}{\partial t}\right)_H = -\frac{1}{\rho c_p}\frac{\partial H}{\partial z} = -\frac{1}{1200}\left(\frac{0+45}{90}\right)$$

$$= -4.17 \times 10^{-4}\ \mathrm{K\ s^{-1}}$$

$$= -1.50\ \mathrm{K\ h^{-1}}$$

Thus, the total rate of cooling in the PBL was 3.87 K h^{-1}.

3.7 Applications

The radiation balance at or near the earth's surface has the following applications:

- Determining climate near the ground.
- Determining radiative properties, such as albedo and emissivity of the surface.
- Determining net radiation, which is an important component of the surface energy budget.
- Defining and determining the apparent (equivalent blackbody) surface temperature for a complex surface.

- Parameterizing the surface heat fluxes to soil and air in terms of net radiation.
- Determining radiative cooling or warming of the PBL.

These and other applications are particularly important in determining the microclimates of various natural and urban areas.

Problems and Exercises

1. The following measurements were made over a short grass surface on a winter night when no evaporation or condensation occurred:
Outgoing longwave radiation from the surface = 365 W m^{-2}.
Incoming longwave radiation from the atmosphere = 295 W m^{-2}.
Ground heat flux from the soil = 45 W m^{-2}.
(a) Calculate the apparent (equivalent blackbody) temperature of the surface.
(b) Calculate the actual surface temperature if surface emissivity is 0.92.
(c) Estimate the sensible heat flux to or from air.

2.
(a) Estimate the combined sensible and latent heat fluxes from the surface to the atmosphere, given the following observations:
Incoming shortwave radiation = 800 W m^{-2}.
Heat flux to the submedium = 150 W m^{-2}.
Albedo of the surface = 0.35.
(b) What would be the result if the surface albedo were to drop to 0.07 after irrigation?

3. The following measurements or estimates were made of the radiative fluxes over a short grass surface during a clear sunny day:
Incoming shortwave radiation = 675 W m^{-2}.
Incoming longwave radiation = 390 W m^{-2}.
Ground surface temperature = 35°C.
Albedo of the surface = 0.20.
Emissivity of the surface = 0.92.
(a) From the radiation balance equation, calculate the net radiation at the surface.
(b) What would be the net radiation after the surface is thoroughly watered so that its albedo drops to 0.10 and its effective surface temperature reduces to 25°C?
(c) Qualitatively discuss the effect of watering on other energy fluxes to or from the surface.

4. Show that the variation of about 28% in terrestrial radiation in Figure 3.4 is consistent with the observed range of 10–30°C in surface temperatures.

5. Explain the nature and causes of depletion of the solar radiation in passing through the atmosphere.

6. Discuss the consequences of the absorption of longwave radiation by atmospheric gases and the so-called greenhouse effect.

7. Discuss the merits of the proposition that net radiation R_N can be deduced from measurements of solar radiation $R_{S\downarrow}$ during the daylight hours, using the empirical expression

$$R_N = AR_{S\downarrow} + B$$

where A and B are constants. On what factors are A and B expected to depend?

8.
(a) Discuss the importance and consequences of the radiative flux divergence at night above a grass surface.
(b) If the net longwave radiation fluxes at 1 and 10 m above the surface are -135 and -150 W m^{-2}, respectively, calculate the rate of cooling or warming in °C h^{-1} due to radiation alone.

9. The following measurements were made at night from a meteorological tower:

Net radiation at the 2 m level = -125 W m^{-2}.
Net radiation at the 100 m level = -165 W m^{-2}.
Sensible heat flux at the surface = -75 W m^{-2}.
Planetary boundary layer height = 80 m.

Calculate the average rate of cooling in the PBL due to the following:
(a) the radiative flux divergence;
(b) the sensible heat flux divergence.

Chapter 4 Soil Temperatures and Heat Transfer

4.1 Surface Temperature

An 'ideal' surface, defined in Chapter 2, may be considered to have a uniform surface skin temperature (T_s) which varies with time only in response to the time-dependent energy fluxes at the surface. An uneven or nonhomogeneous surface is likely to have spatially varying surface temperature. The surface temperature at a given location is essentially given by the surface energy balance, which in turn depends on the radiation balance, atmospheric exchange processes in the immediate vicinity of the surface, presence of vegetation or plant cover, and thermal properties of the subsurface medium. Here, we distinguish between the surface skin temperature and the near-surface air temperature. The latter is measured at standard meteorological stations at the screen height of 1–2 m. The temperature at the air–soil interface will be referred to as the surface skin temperature or simply as the surface temperature (T_s).

The direct *in situ* measurement of surface temperature is made very difficult by the extremely large temperature gradients that commonly occur near the surface in both the air (temperature gradients of 10–20 K mm^{-1} in air are not uncommon very close to a bare heated surface) and the soil media, by the finite dimensions of the temperature sensor, and by the difficulties of ventilating and shielding the sensor when it is placed at the surface. Therefore, the surface skin temperature is often determined by extrapolation of measured temperature profiles in soil and air, with the knowledge of their expected theoretical behaviors. Another, perhaps better, method of determining the surface temperature, when emissivity of the surface is known, is through remote sensors, such as a downward-looking radiometer which measures the flux of outgoing longwave radiation from the surface and, hence, T_s, using the modified Stefan–Boltzmann Equation (3.7). If the surface emissivity is not known with sufficient accuracy, this method will give the apparent (equivalent blackbody) surface temperature. In this way, surface or apparent surface temperatures are being routinely monitored by weather satellites. Measurements are made in narrow regions of the spectrum in which water vapor and CO_2 are transparent. The technique has proved to be fairly reliable for measuring sea-surface temperatures, but not so reliable for land-surface temperatures.

The times of minimum and maximum in surface temperatures as well as the diurnal range are of considerable interest to micrometeorologists. On clear days, the maximum surface temperature is attained typically an hour or two after the time of maximum insolation, while the minimum temperature is reached in early morning hours. The maximum diurnal range is achieved for a relatively dry and bare surface, under relatively calm air and clear skies. For example, on bare soil in summer, midday surface temperatures of 50–60°C are common in arid regions, while early morning temperatures may be only 10–20°C. The bare-surface temperatures also depend on the texture of the soil. Fine-textured soils (e.g., clay) have greater heat capacities and smaller diurnal temperature range, as compared to coarse soils (e.g., sand).

The presence of moisture at the surface and in the subsurface soil greatly moderates the diurnal range of surface temperatures. This is due to the increased evaporation from the surface, and also due to increased heat capacity and thermal conductivity of the soil. Over a free water surface, on the average, about 80% of the net radiation is utilized for evaporation. Over a wet, bare soil a substantial part of net radiation goes into evaporation in the beginning, but this fraction reduces as the soil surface dries up. Increased heat capacity of the soil further slows down the warming of the upper layer of the soil in response to radiative heating of the surface. The ground heat flux is also reduced by evaporation.

The presence of vegetation on the surface also reduces the diurnal range of surface temperatures. Part of the incoming solar radiation is intercepted by plant surfaces, reducing the amount reaching the surface. Therefore surface temperatures during the day are uniformly lower under vegetation than over a bare soil surface. At night the outgoing longwave radiation is also partly intercepted by vegetation, but the latter radiates energy back to the surface. This slightly slows down the radiative cooling of the surface. Vegetation also enhances the latent heat exchange due to evapotranspiration. It increases turbulence near the surface which provides more effective exchanges of sensible and latent heat between the surface and the overlying air. The combined effect of all these processes is to significantly reduce the diurnal range of temperatures of vegetative surfaces.

4.2 Subsurface Temperatures

Subsurface soil temperatures are easier to measure than the skin surface temperature. It has been observed that the range or amplitude of diurnal variation of soil temperatures decreases exponentially with depth and becomes

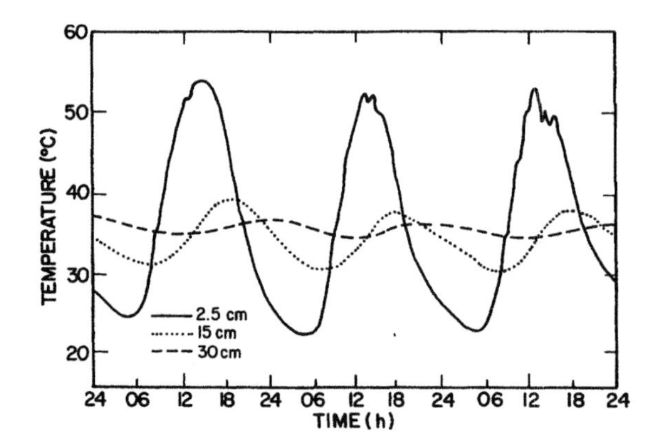

Figure 4.1 Observed diurnal course of subsurface soil temperatures at various depths in a sandy loam soil with bare surface. [From Deacon (1969); after West (1952).]

insignificant at a depth of the order of 1 m or less. An illustration of this is given in Figure 4.1, which is based on measurements in a dry, sandy loam soil.

Soil temperatures depend on a number of factors, which also determine the surface temperature. The most important are the location (latitude) and the time of the year (month or season), net radiation at the surface, soil texture and moisture content, ground cover, and surface weather conditions. Temperature may increase, decrease, or vary nonmonotonically with depth, depending on the season and the time of day.

In addition to the 'diurnal waves' present in soil temperatures in the top layer, daily or weekly averaged temperatures show a nearly sinusoidal 'annual wave' which penetrates to much greater depths (~ 10 m) in the soil. This is illustrated in Figure 4.2.

A simple theory for explaining the observed diurnal and annual temperature waves in soils and variations of their amplitudes and phase with depth will be presented later. First, we define and discuss the thermal properties of soils which influence soil temperatures and heat transfer.

4.3 Thermal Properties of Soils

Thermal properties relevant to the transfer of heat through a medium and its effect on the average temperature or distribution of temperatures in the medium are the mass density, specific heat, heat capacity, thermal conductivity, and thermal diffusivity.

The specific heat (c) of a material is defined as the amount of heat absorbed or released in raising or lowering the temperature of a unit mass of the material by

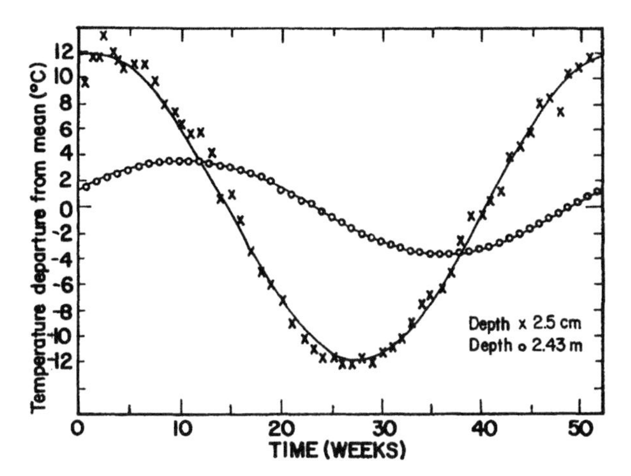

Figure 4.2 Annual temperature waves in the weekly averaged subsurface soil temperatures at two depths in a sandy loam soil. Fitted solid curves are sine waves. [From Deacon (1969); after West (1952).]

1°. The product of mass density (ρ) and specific heat is called the heat capacity per unit volume, or simply heat capacity (C).

Through solid media and still fluids, heat is transferred primarily through conduction, which involves molecular exchanges. The rate of heat transfer or heat flux in a given direction is found to be proportional to the temperature gradient in that direction, i.e., the heat flux in the z direction

$$H = -k(\partial T/\partial z) \tag{4.1a}$$

in which the proportionality factor k is known as the thermal conductivity of the medium. The ratio of thermal conductivity to heat capacity is called the thermal diffusivity α_h. Then, Equation (4.1) can also be written as

$$H/\rho c = -\alpha_h(\partial T/\partial z) \tag{4.1b}$$

where $\alpha_h = k/\rho c = k/C$.

Heat is transferred through soil, rock, and other subsurface materials (except for water in motion) by conduction, so that the above-mentioned molecular thermal properties also characterize the submedium. Table 4.1 gives typical values of these for certain soils, as well as for the reference air and water media, for comparison purposes.

Note that air has the lowest heat capacity, as well as the lowest thermal conductivity of all the natural materials, while water has the highest heat

Table 4.1 Molecular thermal properties of natural materials[a].

Material	Condition	Mass density ρ ($kg\ m^{-3} \times 10^3$)	Specific heat c ($J\ kg^{-1}\ K^{-1} \times 10^3$)	Heat capacity C ($J\ m^{-3}\ K^{-1} \times 10^6$)	Thermal conductivity k ($W\ m^{-1}\ K^{-1}$)	Thermal diffusivity α_h ($m^2\ s^{-1} \times 10^{-6}$)
Air	20°C, Still	0.0012	1.01	0.0012	0.025	20.5
Water	20°C, Still	1.00	4.18	4.18	0.57	0.14
Ice	0°C, Pure	0.92	2.10	1.93	2.24	1.16
Snow	Fresh	0.10	2.09	0.21	0.08	0.38
Snow	Old	0.48	2.09	0.84	0.42	0.05
Sandy soil	Fresh	1.60	0.80	1.28	0.30	0.24
(40% pore space)	Saturated	2.00	1.48	2.96	2.20	0.74
Clay soil	Dry	1.60	0.89	1.42	0.25	0.18
(40% pore space)	Saturated	2.00	1.55	3.10	1.58	0.51
Peat soil	Dry	0.30	1.92	0.58	0.06	0.10
(80% pore space)	Saturated	1.10	3.65	4.02	0.50	0.12
Rock	Solid	2.70	0.75	2.02	2.90	1.43

[a]After Oke (1987) and Garratt (1992).

capacity. Thermal diffusivity of air is very large because of its low density. Thermal properties of both air and water depend on temperature. Most soils consist of particles of different sizes and materials with a significant fraction of pore space, which may be filled with air and/or water. Thermal properties of soils, therefore, depend on the properties of the solid particles and their size distribution, porosity of the soil, and the soil moisture content. Of these, the soil moisture content is a short-term variable, changes of which may cause significant changes in heat capacity, conductivity, and diffusivity of soils (Oke, 1987, Chapter 2; Rosenberg *et al.*, 1983, Chapter 2).

Addition of water to an initially dry soil increases its heat capacity and conductivity markedly, because it replaces air (a poor heat conductor) in the pore space. Both the heat capacity and thermal conductivity are monotonic increasing functions of soil moisture content. The former increases almost linearly with the soil moisture content, while the latter approaches a constant value with increasing soil moisture. Their ratio, thermal diffusivity, for most soils increases with moisture content initially, attains a maximum value, and then falls off with further increase in soil moisture content. It will be seen later that thermal diffusivity is a more appropriate measure of how rapidly surface temperature changes are transmitted to deeper layers of submedium. Soils with high thermal diffusivities allow rapid penetration of surface temperature changes to large depths. Thus, for the same diurnal heat input or output, their temperature ranges are less than for soils with low thermal diffusivities.

4.4 Theory of Soil Heat Transfer

Here we consider a uniform conducting medium (soil), with heat flowing only in the vertical direction. Let us consider the energy budget of an elemental volume consisting of a cylinder of horizontal cross-section area ΔA and depth Δz, bounded between the levels z and $z + \Delta z$, as shown in Figure 4.3.

Heat flow in the volume at depth $z = H\Delta A$.

Heat flow out of the volume at $z + \Delta z = [H + (\partial H/\partial z)\,\Delta z]\Delta A$.

The net rate of heat flow in the control volume $= -(\partial H/\partial z)\Delta z\,\Delta A$.

The rate of change of internal energy within the control volume $= (\partial/\partial t)$ $(\Delta A \Delta z C_s T)$, where C_s is the volumetric heat capacity of soil.

According to the law of conservation of energy, if there are no sources or sinks of energy within the elemental volume, the net rate of heat flowing in the volume should equal the rate of change of internal energy in the volume, so that

$$(\partial/\partial t)(C_s T) = -\partial H/\partial z \qquad (4.2)$$

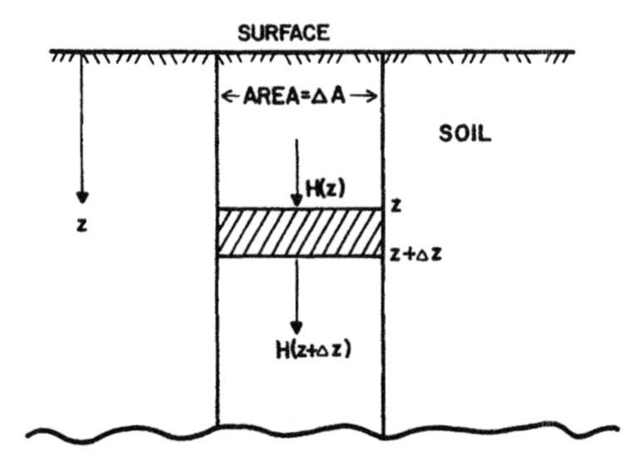

Figure 4.3 Schematic of heat transfer in a vertical column of soil below a flat, horizontal surface.

Further, assuming that the heat capacity of the medium does not vary with time, and substituting from Equation (4.1) into (4.2), we obtain Fourier's equation of heat conduction:

$$\partial T/\partial t = (\partial/\partial z)\,[(k/C_s)(\partial T/\partial z)] = (\partial/\partial z)[\alpha_h(\partial T/\partial z)] \tag{4.3}$$

The one-dimensional heat conduction equation derived here can easily be generalized to three dimensions by considering the net rate of heat flow in an elementary control volume $\Delta x \Delta y \Delta z$ from all the directions. In our applications involving heat transfer through soils, however, we will be primarily concerned with the one-dimensional Equation (4.2) or (4.3).

Equation (4.2) can be used to determine the ground heat flux H_G in the energy balance equation from measurements of soil temperatures as functions of time, at various depths below the surface. The method is based on the integration of Equation (4.2) from $z = 0$ to D

$$H_G = H_D + \int_0^D \frac{\partial}{\partial t}(C_s T)\, dz \tag{4.4}$$

where D is some reference depth where the soil heat flux H_D is either zero (e.g., if at $z = D$, $\partial T/\partial z = 0$) or can be easily estimated [e.g., using Equation (4.1)]. The former is preferable whenever feasible, because it does not require a knowledge of thermal conductivity, which is more difficult to measure than heat capacity.

4.5 Thermal Wave Propagation in Soils

The solution of Equation (4.3), with given initial and boundary conditions, is used to study theoretically the propagation of thermal waves in soils and other substrata. For any arbitrary prescription of surface temperature as a function of time and soil layers, the solution to Equation (4.3) can be obtained numerically. Much about the physics of thermal wave propagation can be learned, however, from a simple analytic solution which is obtained when the surface temperature is specified as a sinusoidal function of time and the subsurface medium is assumed to be homogeneous throughout the depth of wave propagation

$$T_s = T_m + A_s \sin[(2\pi/P)(t - t_m)] \tag{4.5}$$

Here, T_m is the mean temperature of the surface or submedium, A_s and P are the amplitude and period of the surface temperature wave, and t_m is the time when $T_s = T_m$, as the surface temperature is rising.

The solution of Equation (4.3) satisfying the boundary conditions that at $z = 0$, $T = T_s(t)$, and as $z \to \infty$, $T \to T_m$, is given by

$$T = T_m + A_s \exp(-z/d) \sin[(2\pi/P)(t - t_m) - z/d] \tag{4.6}$$

which the reader may verify by substituting in Equation (4.3). Here, d is the damping depth of the thermal wave, defined as

$$d = (P\alpha_h/\pi)^{1/2} \tag{4.7}$$

Note that the period of thermal wave in the soil remains unchanged, while its amplitude decreases exponentially with depth ($A = A_s \exp(-z/d)$); at $z = d$ the wave amplitude is reduced to about 37% of its value at the surface and at $z = 3d$ the amplitude decreases to about 5% of the surface value. The phase lag relative to the surface wave increases in proportion to depth (phase lag $= z/d$), so that there is a complete reversal of the wave phase at $z = \pi d$. The corresponding lag in the time of maximum or minimum in temperature is also proportional to depth (time lag $= zP/2\pi d$).

The results of the above simple theory would be applicable to the propagation of both the diurnal and annual temperature waves through a homogeneous submedium, provided that the thermal diffusivity of the medium remains constant over the whole period, and the surface temperature wave is nearly sinusoidal. For the diurnal period, the latter condition is usually not satisfied especially during the nighttime period when the wave is observed to be more asymmetric around its minimum value (Rosenberg *et al.*, 1983, Chapter 2). Also, disturbed weather conditions with clouds and precipitation are likely to alter the surface temperature wave, as well as thermal diffusivity due to changes in the soil moisture content. For the annual period, the assumption of constant thermal diffusivity may be even more questionable, except for bare soils in arid regions. Note that the damping depth, which is a measure of the extent of thermal wave propagation, for the annual wave is expected to be $\sqrt{365} = 19.1$ times the damping depth for the diurnal wave.

Example Problem 1

Calculate the damping depth and the depth of thermally active soil layer where there is a complete reversal of the phase of the diurnal thermal wave from that at the surface for the following types of soils:
(a) dry sandy soil; (b) saturated sandy soil (40% pore space);
(c) dry clay soil; (d) dry peat soil.
What will be the amplitude of the thermal wave at the depth of the phase reversal relative to that at the surface?

Solution

Using the thermal diffusivities given in Table 4.1, Equation (4.7) for damping depth d, and the depth of phase reversal as πd, one can obtain the following results:

(a) For dry sandy soil, $\alpha_h = 0.24 \times 10^{-6}\ \mathrm{m^2\ s^{-1}}$.
Damping depth $d = (24 \times 3600 \times 0.24 \times 10^{-6}/\pi)^{1/2} = 0.081$ m.
Depth of phase reversal $= \pi d = 0.25$ m.
(b) For saturated sandy soil, $\alpha_h = 0.74 \times 10^{-6}\ \mathrm{m^2\ s^{-1}}$.
Damping depth $d = 0.143$ m.
Depth of phase reversal $= \pi \mathrm{d} = 0.45$ m.
(c) For dry clay soil, $\alpha_h = 0.18 \times 10^{-6}\ \mathrm{m^2\ s^{-1}}$.
Damping depth $d = 0.070$ m.
Depth of phase reversal $= \pi d = 0.22$ m.
(d) For dry peat soil, $\alpha_h = 0.10 \times 10^{-6}\ \mathrm{m^2\ s^{-1}}$.
Damping depth $d = 0.052$ m.
Depth of phase reversal $= \pi d = 0.16$ m.

In all the cases the amplitude of the thermal wave at $z = \pi d$, relative to that at the surface, is

$$A/A_s = \exp(-\pi\, d/d) = \exp(-\pi) \cong 0.042$$

One can conclude that dry peat soil offers the maximum resistance (minimum d) and saturated sandy soil the minimum resistance (maximum d) to the propagation of the diurnal thermal wave.

The observed temperature waves (Figures 4.1 and 4.2) in bare, dry soils are found to conform well to the pattern predicted by the theory. In particular, the observed annual waves show nearly perfect sinusoidal forms. The plots of wave amplitude (on log scale) and phase or time lag as functions of depth (both on linear scale) are well represented by straight lines (Figure 4.4) whose slopes determine the damping depth and, hence, thermal diffusivity of the soil.

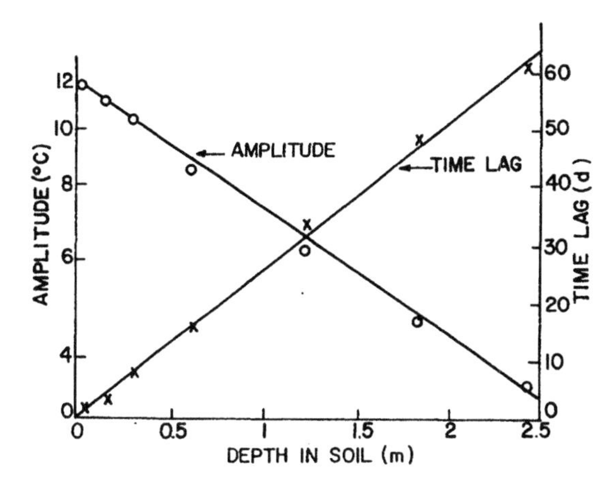

Figure 4.4 Variations of amplitude and time lag of the annual soil temperature waves with depth in the soil. [From Deacon (1969).]

Example Problem 2

Using the weekly averaged data on soil temperatures at different depths obtained by West (1952), the amplitude and time lag of thermal waves are plotted as functions of depth in soil in Figure 4.4. Estimate the damping depth and thermal diffusivity of the soil from the best-fitted lines through the data points.

Solution

Note that according to Equation (4.6), the amplitude of thermal wave decreases exponentially with depth, i.e.,

$$A = A_s \exp(-z/d)$$

or

$$\ln A = \ln A_s - \frac{z}{d}$$

Therefore, a plot of ln A (A on a log scale) against z should result in a straight line with a slope of $-1/d$. The slope of the best-fitted straight line through the amplitude data in Figure 4.4 can be estimated, from which $d \cong 2.05$ m.

Thermal diffusivity $\alpha_h = \pi d^2/P \cong 0.42 \times 10^{-6}\ \mathrm{m^2\ s^{-1}}$.

Also, according to the thermal wave equation (Equation 4.6),

$$\text{Time lag} = Pz/2\pi d$$

Therefore, a plot of time lag versus z should also result in a straight line with a slope of $P/2\pi d$. The slope of the best-fitted line throughout the time lag data points in Figure 4.4 can be estimated as 25.6 days m^{-1}, from which

$$d = P/(2\pi \times \text{slope}) \cong 2.27 \text{ m}$$

$$\alpha_h = \pi d^2/P \cong 0.51 \times 10^{-6} \text{ m}^2 \text{ s}^{-1}$$

The two estimates of d, based on amplitude and time lag data, are in fairly good agreement. The agreement between the estimated damping depths based on the top 0.30 m of the soil temperature data for the diurnal period is found to be much better (Deacon, 1969).

4.6 Measurement and Parameterization of Ground Heat Flux

4.6.1 Measurement by heat flux plate

There is no direct method of measuring the ground heat flux H_G (the heat flux through soil at the surface). Theoretically, the heat conduction equation [Equation (4.1a)] is valid right up to the surface. In practice, however, the difficulty of accurately measuring the near-surface soil temperature gradient and soil thermal conductivity renders this approach impractical. Even at a small depth below the surface, where the soil temperature gradient can be accurately measured, thermal conductivity of the soil is highly variable and not easy to measure. However, heat flux plates are often used to measure the soil heat flux directly. The plate consists of material of known thermal conductivity, and temperature difference across its lower and upper faces is measured by a differential thermopile. The faces are often covered by thin metal plates in order to protect them and to ensure good thermal contact with the soil. The heat flux plate should be small, so that it does not disrupt the soil and has a low heat capacity for quick response to any changes with time. In a field experiment, several heat flux plates may be buried horizontally in the soil at a depth of 10–50 mm. Thus, the average soil heat flux H_1 at a small depth z_1 can be measured.

In order to estimate H_G from the measurement of H_1, one can use Equation (4.4) with $D = z_1$ so that

$$H_G = H_1 + \int_0^{z_1} \frac{\partial}{\partial t}(C_s T) dz = H_1 + \Delta H_s \tag{4.8}$$

where ΔH_s is the rate of heat storage in the top soil layer of depth z_1. For the thin soil layer, the above heat storage term will be significant only when there is rapid warming or cooling of the top soil layer (e.g., around the times of sunrise and sunset). The average warming or cooling rate can be determined from the continuous measurement of soil temperature as a function of time in the middle of the layer.

An indirect method of estimating H_G from the measurement of soil temperatures at various depths, at regular time intervals, has already been discussed in Section 4.4. It is based on the application of Equation (4.4) in which the last term represents the rate of heat storage in a deeper layer of depth D. The soil heat flux at the bottom of this layer is also estimated, rather than measured with a heat flux plate.

4.6.2 Simple parameterizations

In some climate and general circulation models, the ground heat flux is simply neglected. The rationale for this is provided by the observation that, averaged over the entire diurnal cycle, the net ground heat flux is often near zero. This is because the heating of the thermally active soil layer during the day is approximately balanced by the cooling at night. For shorter periods used in micrometeorology, however, H_G is a significant component of the energy balance near the surface, especially at night. A simple parameterization of the same, discussed in Chapter 2, is to specify H_G as a constant fraction of the net radiation R_N during the day and a different (larger) fraction at night. The actual values of the ratio H_G/R_N for the day and night periods depend on the surface and soil characteristics and are determined empirically.

Alternatively, the ground heat flux can also be assumed as a constant fraction of the sensible heat flux into air. Both of these simple parameterizations imply that the sign of the ground heat flux is always the same as that of net radiation or sensible heat flux in air. This is not particularly true during the morning and evening transition periods when different energy fluxes change signs at different times. These simple parameterizations are also not applicable over water surfaces.

4.6.3 Soil heat transfer models

An expression for the ground heat flux can be obtained from Equation (4.6) as

$$H_G = -k\left(\frac{\partial T}{\partial z}\right)_{z=0} = (2\pi C_s k/P)^{1/2} A_s \sin\left[\frac{2\pi}{P}(t - t_m) + \frac{\pi}{4}\right] \tag{4.9}$$

which predicts that the maximum in surface temperature should lag behind the maximum in ground heat flux by $P/8$, or 3 h for the diurnal period and 1.5 months for the annual period. The above equation also indicates that the amplitude of ground heat flux wave is proportional to the square root of the product of heat capacity and thermal conductivity and inversely proportional to the square root of the period; it is also proportional to the amplitude of the surface temperature wave.

The above predicted sinusoidal variation of the ground heat flux is not usually consistent with direct measurements or indirect estimates of H_G from the energy budget equation. In particular, the nighttime variation of H_G with time is much flatter or more nearly uniform (see Munn, 1966, Chapter 5). There are several reasons for this disagreement. First, the assumed sinusoidal variation of the surface temperature in Equation (4.5), which constitutes the upper boundary condition for the solution of Equation (4.3), may not be realistic. Secondly, the simplifying assumption of the homogeneity of the submedium is often not realized. Thermal properties may vary with depth because of the layer structure of the submedium, variations in the soil moisture content, and presence of roots and surface vegetation. Thermal properties may also vary with time in response to irrigation, precipitation, and evaporation. In such cases, numerical models of soil heat and moisture transfer with varying degrees of complication are used, in conjunction with the surface energy balance equation.

4.6.4 Force–restore method

A more widely used parameterization of the energy exchange with the subsurface medium in atmospheric models is based on the two-layer approximation of the submedium, in which a shallow thermally active layer of soil overlies a thick constant-temperature slab. The depth d_g of the near-ground layer is carefully chosen, based on knowledge of the damping depth. The lower layer is assumed to be of sufficient thickness that, over the time scales of interest, the soil heat flux at the bottom is zero. The soil heat flux at $z = d_g$ can be parameterized as

$$H_g = \mu_g(T_g - T_m) \tag{4.10}$$

where T_g and T_m are the layer-averaged temperatures of the upper and lower layers, respectively, and μ_g is a heat transfer coefficient.

The energy balance equation near the surface, including the near-ground soil layer of depth d_g, can be written as

$$R_N = H + H_L + H_g + \Delta H_s \tag{4.11}$$

in which the rate of energy storage can be expressed as

$$\Delta H_s = C_s d_g \frac{\partial T_g}{\partial t} = C_g \frac{\partial T_g}{\partial t} \tag{4.12}$$

where $C_g = C_s d_g$ is the heat capacity per unit area of the near-ground soil layer.

Substituting from Equations (4.10) and (4.12) into (4.11), one obtains a prognostic equation for the temperature of the near-ground layer as

$$C_g \frac{\partial T_g}{\partial t} = (R_N - H - H_L) - \mu_g (T_g - T_m) \tag{4.13}$$

This prognostic equation for the ground temperature is generally referred to as the force–restore method, because the net radiative and atmospheric forcing $(R_N - H - H_L)$ at the surface is modified by the deep soil flux which tends to restore the near-ground temperature to that of deep soil. Note that, if the surface forcing term is removed, the restoring term in Equation (4.13) will cause T_g to move exponentially towards T_m.

Equation (4.13) has simply been extended to predict the diurnal variation of the ground surface temperature as

$$C_g \frac{\partial T_s}{\partial t} = (R_N - H - H_L) = \mu_s (T_s - T_m) \tag{4.14}$$

in which the soil properties are appropriately chosen or specified as (Garratt, 1992, Chapter 8)

$$C_g = C_s(\alpha_h/2\Omega)^{1/2} \tag{4.15}$$

$$\mu_s = \Omega C_g \tag{4.16}$$

where Ω is the angular rotational speed of the earth. Note that Equation (4.15) with $C_g = C_s d_g$ implies that $d_g = d/2$. Thus, the most appropriate depth of the near-ground layer in the above two-layer model is half of the damping depth for the diurnal period $P = 2\pi/\Omega$.

The above expressions for C_g and μ_s are strictly based on the assumption of a sinusoidal surface temperature and, hence, ground heat flux forcing, as expressed in Equations (4.5) and (4.9), respectively. Since the surface forcing is not really sinusoidal in nature, slightly modified expressions

$$C_g = 0.95 C_s(\alpha_h/2\Omega)^{1/2} \tag{4.17}$$

$$\mu_s = 1.18 \Omega C_g \tag{4.18}$$

are more commonly used in numerical prediction models of the ground surface temperature (Garratt, 1992).

4.7 Applications

Some of the applications of the knowledge of surface and soil temperatures and soil heat transfer are as follows:

- Study of surface energy budget and radiation balance.
- Prediction of surface temperature and frost conditions.
- Determination of the rate of heat storage or release by the submedium.
- Study of microenvironment of plant cover, including the root zone.
- Environmental design of underground structures.
- Determination of the depth of permafrost zone in high latitudes.

Problems and Exercises

1. You want to measure the temperature of a bare soil surface using fine temperature sensors on a summer day when the sensible heat flux to air might be typically 600 W m^{-2} and the ground heat flux 100 W m^{-2}. Will you do this by extrapolation of measurements near the surface in air or in soil? Give reasons for your choice. Also estimate temperature gradients near the surface in both the air ($k = 0.03$ W m^{-1} K^{-1}) and the soil ($k = 0.30$ W m^{-1} K^{-1}).

2. Discuss the effect of vegetation or plant cover on the diurnal range of surface temperatures, as compared to that over a bare soil surface, and explain why green lawns are much cooler than roads and driveways on sunny afternoons.

3. What is the physical significance of the damping depth in the propagation of thermal waves in a submedium and on what factors does it depend?

4. What are the basic assumptions underlying the simple theory of soil heat transfer presented in this chapter? Discuss the situations in which these assumptions may not be valid.

5. By substituting from Equation (4.6) into (4.3), verify that the former is a solution of Fourier's equation of heat conduction through a homogeneous soil medium, and then, derive the expression [Equation (4.9)] for the ground heat flux.

6. Compare and contrast the alternative methods of determining the ground heat flux from measurements of soil temperatures and thermal properties.

7. The following soil temperatures (°C) were measured during the Great Plains Field Program at O'Neil, Nebraska:

		Depth (m)				
Time (h)	Day	0.025	0.05	0.10	0.20	0.40
0435	August 31	25.54	26.00	26.42	26.32	24.50
0635		24.84	25.30	25.84	25.97	24.48
0835		25.77	25.35	25.42	25.56	24.36
1035		29.42	27.36	25.98	25.39	24.34
1235		33.25	30.32	27.62	25.57	24.27
1435		35.25	32.63	29.52	26.11	24.24
1635		34.84	33.20	30.62	26.88	24.26
1835		32.63	32.05	30.62	27.41	24.32
2035		30.07	30.20	29.91	27.68	24.47
2235		28.42	28.74	28.84	27.57	24.64
0035	September 1	27.09	27.50	27.84	27.22	24.73
0235		26.09	26.60	27.06	26.87	24.78
0435		25.30	25.83	26.40	26.53	24.84

(a) Plot on a graph temperature waves as functions of time and depth.
(b) Plot on a graph the vertical soil temperature profiles at 0435, 0835, 1235, 1635, 2035, and 0035 h.
(c) Determine the damping depth and thermal diffusivity of the soil from the observed amplitudes, as well as from the times of temperature maxima (taken from the smoothed temperature waves) as functions of depth.
(d) Estimate the amplitude of the surface temperature wave and the time of maximum surface temperature from extrapolation of the soil temperature data.

8. From the soil temperature data given in the above problem, calculate the ground heat flux at 0435 and 1635 h, using Equation (4.4) with the measured soil heat capacity of 1.33×10^6 J m^{-3} K^{-1} and the average value of thermal diffusivity determined in Problem 7(c) above.

Chapter 5 | Air Temperature and Humidity in the PBL

5.1 Factors Influencing Air Temperature and Humidity

The following factors and processes influence the vertical distribution (profile) of air temperature in the atmospheric boundary layer.

- Type of air mass and its temperature just above the PBL, which depend on the synoptic situation and the large-scale circulation pattern.
- Thermal characteristics of the surface and submedium, which influence the diurnal range of surface temperatures.
- Net radiation at the surface and its variation with height, which determine the radiative warming or cooling of the surface and the PBL.
- Sensible heat flux at the surface and its variation with height, which determine the rate of warming or cooling of the air due to convergence or divergence of sensible heat flux.
- Latent heat exchanges during evaporation and condensation processes at the surface and in air, which influence the surface and air temperatures, respectively.
- Warm or cold air advection as a function of height in the PBL.
- Height or the PBL to which turbulent exchanges of heat are confined.

Similarly, the factors influencing the specific humidity or mixing ratio of water vapor in the PBL are as follows:

- Specific humidity of air mass just above the PBL.
- Type of surface, its temperature, and availability of moisture for evaporation and/or transpiration.
- The rate of evapotranspiration or condensation at the surface and the variation of water vapor flux with height in the PBL.
- Horizontal advection of water vapor as a function of height.
- Mean vertical motion in the PBL and possible cloud formation and precipitation processes.
- The PBL depth through which water vapor is mixed.

The vertical exchanges of heat and water vapor are primarily through turbulent motions in the PBL; the molecular exchanges are important only in a very thin (of the order of 1 mm or less) layer adjacent to the surface. In the atmospheric PBL, turbulence is generated mechanically (due to surface friction and wind shear) and/or convectively (due to surface heating and buoyancy). These two types of turbulent motions are called forced convection and free convection, respectively. The combination of the two is sometimes referred to as mixed convection, although more often the designation 'forced' or 'free' is used, depending on whether a mechanical or convective exchange process dominates.

The convergence or divergence of sensible heat flux leads to warming or cooling of air, similar to that due to net radiative flux divergence or convergence. The rate of warming is equal to the vertical gradient of $-H/\rho c_p$, i.e., $\partial T/\partial t = -(\partial H/\partial z)/\rho c_p$. Thus a gradient of 1 W m^{-3} in sensible heat flux will produce a change in air temperature at the rate of about 3°C h^{-1}. Similarly, divergence or convergence of water vapor flux leads to decrease or increase of specific humidity with time.

Horizontal advections of heat and moisture become important only when there are sharp changes in the surface characteristics (e.g., during the passage of a front) in the horizontal. Mean vertical motions forced by changes in topography often lead to rapid changes in temperature and humidity of air, as well as to local cloud formation and precipitation processes, as moist air rises over the mountain slopes. Equally sharp changes occur on the lee sides of mountains. Even over a relatively flat terrain, the subsidence motion often leads to considerable warming and drying of the air in the PBL.

5.2 Basic Thermodynamic Relations and the Energy Equation

The pressure in the lower atmosphere decreases with height in accordance with the hydrostatic equation

$$\partial P/\partial z = -\rho g \tag{5.1}$$

Equation (5.1) is strictly valid only for the atmosphere at rest. It is also a good approximation for large-scale atmospheric flows. For small mesoscale and microscale motions, however, the deviations from the hydrostatic equation due to vertical motions and accelerations may also become significant. But, in thermodynamical relations discussed here, the atmosphere is assumed to be in a hydrostatic equilibrium.

Another fundamental relationship between the commonly used thermodynamic variables is the equation of state or the ideal gas law

$$P = R\rho T = (R_*/m)\rho T \tag{5.2}$$

in which R is the specific gas constant, $R_* = 8.314 \text{ J K}^{-1} \text{ mole}^{-1}$ is the absolute gas constant, and m is the mean molecular mass of air.

Some of the thermodynamic relations follow from the first law of thermodynamics which is essentially a statement about the conservation of energy. When applied to an air parcel, the law states that an increase in the internal energy, dU, of the parcel can occur only through an addition of heat, dH, to the parcel and/or by performing work, dW, on the parcel. This can be expressed mathematically as

$$dU = dH + dW \tag{5.3a}$$

where each term may be considered to represent unit volume of the parcel.

Experimental evidence suggests that the internal energy per unit volume of an ideal gas is proportional to its absolute temperature, so that $dU = \rho c_p \, dT$ for a constant pressure process, and $dU = \rho c_v \, dT$ for a constant volume process. Here, c_p and c_v are the specific heat capacities at constant pressure and volume, respectively. Their difference is equal to the specific gas constant, that is $c_p - c_v = R$ (for dry air, $c_p = 1005 \text{ J K}^{-1} \text{ kg}^{-1}$ and $R = 287.04 \text{ J K}^{-1} \text{ kg}^{-1}$).

Expressing the work done on the air parcel in terms of pressure and volume, one can derive several different expressions of the first law of thermodynamics (Hess, 1959; Fleagle and Businger, 1980). Here, we give only the one involving changes in temperature and pressure

$$\rho c_p dT = dH + dP \tag{5.3b}$$

which implies that the parcel temperature may change due to an addition of heat and/or a change in pressure.

If a parcel of air is lifted upward in the atmosphere, its pressure will decrease in response to decreasing pressure of its surroundings. This will lead to a decrease in the temperature and, possibly, to a decrease in the density of the parcel (due to expansion of the parcel) in accordance with Equation (5.2). If there is no exchange of heat between the parcel and the surrounding environment, the above process is called adiabatic. Vertical turbulent motions in the PBL, which carry air parcels up and down, are rapid enough to justify the assumption that changes in thermodynamic properties of such parcels are adiabatic ($dH = 0$). The rate of change of temperature of such a parcel with height can be calculated from Equations (5.1) and (5.3b) and is characterized by the adiabatic lapse rate (Γ)

$$\Gamma = -(\partial T/\partial z)_{ad} = g/c_p \tag{5.4}$$

This is also the temperature lapse rate (rate of decrease of temperature with height) in an adiabatic atmosphere, which under dry or unsaturated conditions amounts to 0.0098 K m^{-1} or nearly 10 K km^{-1}.

The relationship between the changes of temperature and pressure in a parcel moving adiabatically, as well as those in an adiabatic atmosphere, are also given by Equations (5.2) and (5.3b), with $dH = 0$:

$$dT/T = (R/c_p)(dP/P) \tag{5.5}$$

integration of which gives the Poisson equation

$$T = T_0(P/P_0)^k \tag{5.6}$$

where T_0 is the temperature corresponding to the reference pressure P_0 and the exponent $k = R/c_p \cong 0.286$.

Equation (5.6) is used to relate the potential temperature Θ, defined as the temperature which an air parcel would have if it were brought down to a pressure of 1000 mbar adiabatically from its initial state, to the actual temperature T as

$$\Theta = T(1000/P)^k \tag{5.7}$$

where P is in millibars. The potential temperature has the convenient property of being conserved during vertical movements of an air parcel, provided heat is not added or removed during such excursions. Then, the parcel may be identified or labeled by its potential temperature. In an adiabatic atmosphere, potential temperature remains constant with height. For a nonadiabatic or diabatic atmosphere, it is easy to show from Equation (5.7) that, to a good approximation,

$$\frac{\partial \Theta}{\partial z} = \frac{\Theta}{T}\left(\frac{\partial T}{\partial z} + \Gamma\right) \cong \frac{\partial T}{\partial z} + \Gamma \tag{5.8}$$

This approximation is particularly useful in the PBL where potential and actual temperatures, in absolute units, do not usually differ by more than 10%. The above relationship is often used to express the difference in the potential temperatures between any two height levels as

$$\Delta\Theta = \Delta T + \Gamma \Delta z \tag{5.9}$$

An integral version of the same relationship,

$$\Theta - \Theta_0 = T - T_0 + \Gamma z \tag{5.10}$$

can be used to convert a temperature sounding (profile) into a potential temperature profile. The near-surface value Θ_0 can be calculated from Equation (5.7) if the surface pressure is known. A distinct advantage of using Equation (5.10) for calculating potential temperatures in the PBL is that one does not need to measure or estimate pressure at each height level.

The above relations are applicable to both dry and moist atmospheres, so long as water vapor does not condense and one uses the appropriate values of thermodynamic properties for the moist air. In practice, it is found to be more convenient to use constant dry-air properties such as R and c_p, but to modify some of the relations or variables for the moist air.

5.2.1 Moist unsaturated air

There are more than a dozen variables used by meteorologists that directly or indirectly express the moisture content of the air (Hess, 1959, Chapter 4). Of these the most often used in micrometeorology is the specific humidity (Q), defined as the ratio, $M_w/(M_w + M_d)$, of the mass of water vapor to the mass of moist air containing the water vapor. In magnitude, it does not differ significantly from the mixing ratio, defined as the ratio, M_w/M_d, of the mass of water vapor to the mass of dry air containing the vapor. Both are directly related to the water vapor pressure (e), which is the partial pressure exerted by water vapor and is a tiny fraction of the total pressure (P) anywhere in the PBL. To a good approximation,

$$Q \cong m_w e/m_d P = 0.622e/P \tag{5.11}$$

where m_w and m_d are the mean molecular masses of water vapor and dry air, respectively.

The water vapor pressure is always less than the saturation vapor pressure (e_s), which is related to the temperature through the Clausius–Clapeyron equation (Hess, 1959, Chapter 4)

$$\frac{de_s}{e_s} = \frac{m_w L_e}{R_*}\frac{dT}{T^2} \tag{5.12}$$

where L_e is the latent heat of vaporization or condensation. Integration of Equation (5.12) and using the empirical fact (condition) that at $T = 273.2$ K,

$e_s = 6.11$ mbar, yields

$$\ln \frac{e_s}{6.11} = \frac{m_w L_e}{R_*}\left(\frac{1}{273.2} - \frac{1}{T}\right) \tag{5.13}$$

in which e_s is in millibars. A plot of e_s versus T is called the saturation water vapor pressure curve and is given in many meteorological textbooks and monographs.

Using Dalton's law of partial pressures ($P = P_d + e$), the equation of state for the moist air can be written as

$$P = \frac{R_*}{m}\rho T = \frac{R_*}{m_d}\frac{M_d}{V}T + \frac{R_*}{m_w}\frac{M_w}{V}T$$

or, after some algebra,

$$P = \frac{R_*}{m_d}\rho T\left[1 + \left(\frac{m_d}{m_w} - 1\right)Q\right] \tag{5.14}$$

After comparing Equation (5.14) to Equation (5.2) it becomes obvious that the factor in the brackets on the right-hand side of Equation (5.14) is the correction to be applied to the specific gas constant for dry air, R, in order to obtain the specific gas constant for the moist air. Instead of taking R as a humidity-dependent variable, in practice, the above correction is applied to the temperature in defining the virtual temperature

$$T_v = T\left[1 + \left(\frac{m_d}{m_w} - 1\right)Q\right] \cong T(1 + 0.61Q) \tag{5.15}$$

so that the equation of state for the moist air is given by

$$P = (R_*/m_d)\rho T_v = R\rho T_v \tag{5.16}$$

The virtual temperature is obviously the temperature which dry air would have if its pressure and density were equal to those of the moist air. Note that the virtual temperature is always greater than the actual temperature and the difference between the two, $T_v - T \cong 0.61QT$, may become as large as 7 K, which may occur over warm tropical oceans, and is usually less than 2 K over mid-latitude land surfaces.

For moist air one can define the virtual potential temperature, Θ_v, similar to that for dry air and use Equations (5.7) and (5.8) with Θ and T replaced by Θ_v and T_v, respectively.

The mixing ratio of water vapor in air being always very small (<0.04), the specific heat capacity is not significantly different from that of dry air. Therefore, changes in state variables of moving air parcels in an unsaturated atmosphere are not significantly different from those in a dry atmosphere, so long as the processes remain adiabatic. In particular, the adiabatic lapse rate given by Equation (5.4) is also applicable to most unsaturated air parcels moving in a moist environment, provided T in Equation (5.4) is replaced by T_v.

5.2.2 *Saturated air*

In saturated air, a part of the water vapor may condense and, consequently, the latent heat of condensation may be released. If condensation products (e.g., water droplets and ice crystals) remain suspended in an upward-moving parcel of saturated air, the saturated lapse rate, or the rate at which the saturated air cools as it expands with the increase in height, must be lower than the dry adiabatic lapse rate Γ, because of the addition of the latent heat of condensation in the former. If there is no exchange of heat between the parcel and the surrounding environment, such a process is called moist adiabatic.

From the first law of thermodynamics for the moist adiabatic process and the equation of state, the moist adiabatic or saturated lapse rate is given by (Hess, 1959):

$$\Gamma_s = -\left(\frac{\partial T}{\partial z}\right)_s = g\left(c_p + L_e\frac{dQ_s}{dT}\right)^{-1} \tag{5.17}$$

where dQ_s/dT is the slope of the saturation-specific humidity versus temperature curve, as given by Equations (5.12) and (5.13). Unlike the dry adiabatic lapse rate, however, the saturated adiabatic lapse rate is quite sensitive to temperature. For example, at $T = 273.2$ K, $\Gamma_s = 0.0069$ K m^{-1} and at $T = 303$ K, $\Gamma_s = 0.0036$ K m^{-1}, both at the reference sea-level pressure of 1000 mbar.

In the real atmospheric situation, some of the condensation products are likely to fall out of an upward-moving parcel, while others may remain suspended as cloud particles in the parcel. As a result of these changes in mass and composition of the parcel, some exchange of heat with the surrounding air must occur. Therefore the processes within the parcel are not truly adiabatic, but are called pseudoadiabatic. The mass of condensation products being only a tiny fraction of the total mass of parcel, however, the pseudoadiabatic lapse rate is not much different from the moist adiabatic lapse rate.

The difference between the dry adiabatic and moist adiabatic lapse rates accounts for some interesting orographic phenomena, such as the formation of clouds and possible precipitation on the upwind slopes of mountains and warm and dry downslope winds (e.g., Chinook and Foen winds) near the lee side base (Rosenberg *et al.*, 1983, Chapter 3).

Example Problem 1

A moist air parcel with a temperature of 20°C and specific humidity of 10 g kg^{-1} is lifted adiabatically from the upwind base of a mountain, where the pressure is 1000 mb, to its top, at 3000 m above the base, and is then brought down to the base on the other side of the mountain. Using the appropriate thermodynamic relations or diagrams, estimate the following parameters:

(a) The height z_s above the base where the air parcel would become saturated and the saturated adiabatic lapse rate at this level.
(b) The temperature of the parcel at the mountain top.
(c) The temperature of the parcel at the downwind base.

Solution

(a) First, we determine the temperature $T(z_s)$ at which the parcel would become saturated from Equation (5.13)

$$\ln\left(\frac{e_s}{6.11}\right) = 5404\left[\frac{1}{273.2} - \frac{1}{T(z_s)}\right]$$

where, from Equation (5.11),

$$e_s = Q_s P/0.622 \cong 0.0161 P(z_s)$$

The pressure and temperature of the parcel at the height z_s are related through the Poisson equation [Equation (5.6)] as

$$T(z_s) = T_0[P(z_s)/P_0]^{0.286}$$

The above expressions form a closed loop which can be solved, using an iterative procedure, to determine the variables at z_s as follows:

$$T(z_s) = 285.9 \text{ K};\ P(z_s) = 917 \text{ mb};\ e_s = 14.73 \text{ mb}$$

Then the height z_s is given by

$$z_s = (293.2 - 285.9)/0.0098 \simeq 745 \text{ m}$$

Alternatively, z_s, $T(z_s)$, and $P(z_s)$ can also be determined from an appropriate thermodynamic diagram.

The saturated adiabatic lapse rate can be determined from Equation (5.17) in which

$$\frac{dQ_s}{dT} = 0.622\frac{m_w L_e}{R_*}\frac{e_s}{PT^2} \cong 6.48 \times 10^{-4}\ \mathrm{K}^{-1}$$

Then,

$$\Gamma_s = 9.816/(1010 + 2.45 \times 648) \cong 0.0038\ \mathrm{K\ m}^{-1}$$

(b) Assuming a constant saturated adiabatic lapse rate up to the top of the mountain at $z_t = 3000$ m, the temperature of the parcel there is given by

$$T(z_t) = T(z_s) - \Gamma_s(z_t - z_s) \cong 277.4\ \mathrm{K}$$

However, this is only an approximation, because in reality dQ_s/dT is expected to vary with height. A more accurate estimate of the temperature at the mountain top can be obtained by following the moist adiabatic for $T = 286$ from the 745 m height level to 3000 m on a suitable thermodynamic diagram or chart.

(c) Assuming that the parcel remained unsaturated during its descent on the other side of the mountain, the temperature at the downwind base is given by

$$T_b = 277.4 + 0.0098 \times 3000 = 306.8\ \mathrm{K}$$

which is 13.6 K higher than the air temperature at the upwind base.

This exercise shows that, in going up and down a high mountain, air temperature can increase substantially, provided that the lifting condensation level is well below the mountain top.

5.2.3 Thermodynamic energy equation

Using the principle of the conservation of energy within an elementary volume of air, one can derive the following differential equation for the variation of temperature (Kundu, 1990, Chapter 4):

$$\frac{DT}{Dt} = \alpha_h \nabla^2 T + S_H \tag{5.18}$$

where α_h is thermal diffusivity and S_H is the source term representing the rate of change of temperature within the elementary volume. Here, the total derivative DT/Dt represents the sum of the local change of temperature with time and advective changes in the same due to flow, while the ∇^2 (Laplacian) operator has the usual meaning. In deriving the simplified Equation (5.18), the so-called Boussinesq's approximations have been used and any change in temperature due to radiative flux divergence has been neglected. A similar equation can also be written for potential temperature.

5.3 Local and Nonlocal Concepts of Static Stability

The variations of temperature and humidity with height in the PBL lead to density stratification with the consequence that an upward- or downward-moving parcel of air will find itself in an environment whose density will, in general, differ from that of the parcel, after accounting for the adiabatic cooling or warming of the parcel. In the presence of gravity this density difference leads to the application of a buoyancy force on the parcel which would accelerate or decelerate its vertical movement. If the vertical motion of the parcel is enhanced, i.e., the parcel is accelerated farther away from its equilibrium position by the buoyancy force, the environment is called statically unstable. On the other hand, if the parcel is decelerated and is moved back to its equilibrium position, the atmosphere is called stable or stably stratified. When the atmosphere exerts no buoyancy force on the parcel at all, it is considered neutral. In general the buoyancy force or acceleration exerted on the parcel may vary with height and so does the static stability.

One can derive an expression for the buoyant acceleration, a_b, on a parcel of air by using the Archimedes principle and determining the net upward force due to buoyancy

$$a_b = g\left(\frac{\rho - \rho_p}{\rho_p}\right) \tag{5.19}$$

in which the subscript p refers to the parcel. Further, using the equation of state for the moist air, Equation (5.19) can also be written as

$$a_b = -g\left(\frac{T_v - T_{vp}}{T_v}\right) \tag{5.20}$$

5.3.1 *Local static stability*

Expressions (5.19) and (5.20) for the buoyant acceleration on a parcel in a stratified environment are valid irrespective of the origin or initial position of the parcel. An alternative, but approximate, expression for a_b in terms of the local gradient of virtual temperature or potential temperature is often given in the literature as

$$a_b \cong -\frac{g}{T_v}\left(\frac{\partial T_v}{\partial z} + \Gamma\right)\Delta z = -\frac{g}{T_v}\frac{\partial \Theta_v}{\partial z}\Delta z \qquad (5.21)$$

This expression can be derived from Equation (5.20) by assuming that any parcel displacement, Δz, from its equilibrium position is small and by expanding T_v and T_{vp} in the Taylor series around their equilibrium value T_{ve}.

Equation (5.21) gives a criterion, as well as a quantitative measure, of static stability of the atmosphere in terms of $\partial T_v/\partial z$ or $\partial \Theta_v/\partial z$. In particular, the static stability parameter is defined as

$$s = (g/T_v)(\partial \Theta_v/\partial z) \qquad (5.22)$$

Thus, qualitatively, the static stability is often divided into three broad categories:

1. Unstable, when $s < 0$, $\partial \Theta_v/\partial z < 0$, or $\partial T_v/\partial z < -\Gamma$.
2. Neutral, when $s = 0$, $\partial \Theta_v/\partial z = 0$, or $\partial T_v/\partial z = -\Gamma$.
3. Stable, when $s > 0$, $\partial \Theta_v/\partial z > 0$, or $\partial T_v/\partial z > -\Gamma$.

The magnitude of s provides a quantitative measure of static stability. It is quite obvious from Equation (5.21) that vertical movements of parcels up or down are generally enhanced in an unstable atmosphere, while they are suppressed in a stably stratified environment. Note that, for an upward-moving parcel, $a_b > 0$ in a statically unstable environment and $a_b < 0$ in a stable environment.

On the basis of virtual temperature gradient or lapse rate (LR), relative to the adiabatic lapse rate Γ, atmospheric layers are variously characterized as follows:

1. Superadiabatic, when $LR > \Gamma$.
2. Adiabatic, when $LR = \Gamma$.
3. Subadiabatic, when $0 < LR < \Gamma$.
4. Isothermal, when $\partial T_v/\partial z = 0$.
5. Inversion, when $\partial T_v/\partial z > 0$.

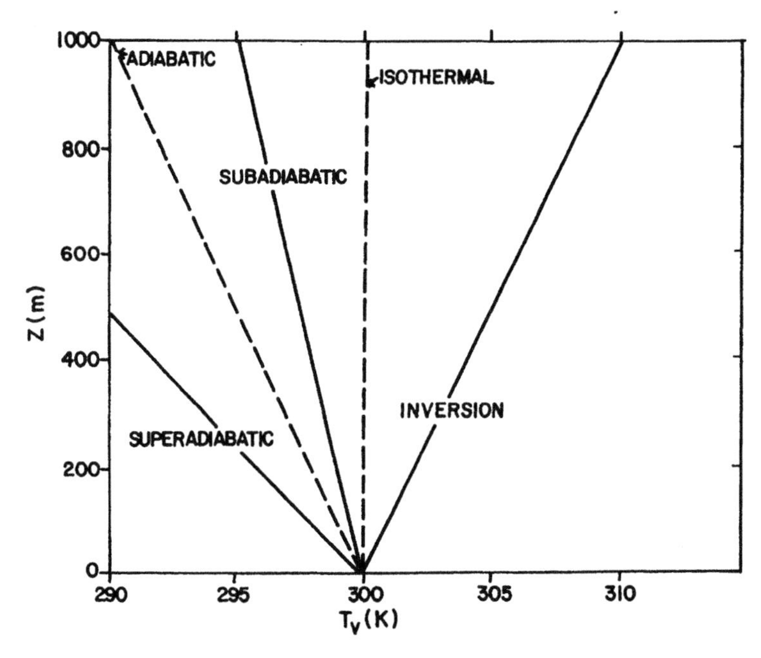

Figure 5.1 Schematic of the various stability categories on the basis of virtual temperature gradient.

Figure 5.1 gives a schematic of these static stability categories in the lower part of the PBL. Here, the surface temperature is assumed to be the same and virtual temperature profiles are assumed to be linear for convenience only; actual profiles are usually curvilinear, as will be shown later.

This traditional view of static stability, based on the local gradient of virtual temperature or potential temperature has some serious limitations and is not very satisfactory, especially when static stability is used as a measure of turbulent mixing and diffusion. For example, it is well known that in the bulk of the convective boundary layer (CBL), $\partial\Theta_v/\partial z \cong 0$, or is slightly positive. Its characterization as a neutral or slightly stable layer, according to the traditional definition of static stability based on the local gradient, would be very misleading because the convective mixed layer (the middle part of the CBL where $\partial\Theta_v/\partial z \cong 0$) has all the attributes of an unstable boundary layer, such as upward heat flux, vigorous mixing, and large thickness. It would be correctly recognized as an unstable layer if one considers the positive buoyancy forces or accelerations that would act in all the warm air parcels originating in the superadiabatic layer near the surface. For these parcels, experiencing large displacements, before they arrive in the mixed layer, Equation (5.21) is not valid and the local static stability parameter becomes irrelevant. Thus, s is not a good measure of buoyant accelerations or forces on air parcels undergoing large displacements from their equilibrium positions. The differences in densities or virtual temperatures between parcels and the environment are the simplest and most direct measures of buoyancy forces and should be considered in any stability characterization of the environment. The original criterion for determining static stability in terms of buoyancy forces on parcels coming from all possible initial positions appears to be basically sound. It is only when one considers the alternative criterion in terms of the local lapse rate, $\partial\Theta_v/\partial z$, or s that ambiguities arise in certain situations.

5.3.2 Nonlocal static stability

In order to remove the ambiguities associated with the concept of local static stability, Stull (1988, Chapter 5; 1991) has proposed a nonlocal characterization of stability. For this, an environmental sounding of thermodynamic variables should be taken over a deep layer between the surface and the level where vertical parcel movements are likely to become insignificant (e.g., strong inversion or the tropopause). To determine the stability of the various layers, air parcels should be displaced up and down from all possible starting points within the whole domain. In practice, one needs to consider parcels starting from only the relative minima and maxima of the virtual potential temperature sounding or profile. Air parcel movement up or down from the initial position should be based on the parcel buoyancy and not on the local lapse rate. The parcel buoyancy at any level is determined by the difference in virtual temperatures of the parcel and the environment at that level. Buoyancy forces warm air parcels to rise and cold parcels to sink. Displaced air parcels that would continue to move farther away from their starting level should be tracked all the way to the level where they would become neutrally buoyant ($T_{vp} = T_v$, or $\Theta_{vp} = \Theta_v$). After parcel movements from all the salient levels (minima and maxima in the Θ_v profile) have been tracked, nonlocal static stability of different portions (layers) of the sounding domain should be determined in the following order (Stull, 1991):

1. *Unstable*: Those regions in which parcels can enter and transit under their own buoyancy. Individual air parcels need not traverse the whole unstable region. Many parcels would traverse only a part of the unstable region or layer.
2. *Stable*: Those regions of subadiabatic lapse rate that are not unstable.
3. *Neutral*: Those regions of adiabatic lapse rate that are not unstable.
4. *Unknown*: Those top or bottom portions of the sounding that are apparently stable or neutral, but that do not end at a material surface, such as the

ground surface, strong inversion, or the tropopause. Above or below the known sounding region might be a cold or warm layer, respectively, that could provide a source of buoyant air parcels.

The above method or procedure for determining the nonlocal stability is illustrated in Figure 5.2 for the various hypothetical, but practically feasible, soundings (Θ_v profiles). Note that in some of these cases, especially those with unstable regions, the traditional characterizations of stability based on the local lapse rate or $\partial\Theta_v/\partial z$ would be quite different from the broader, nonlocal characterization used here. The latter are based on the buoyant movements (indicated by vertical dashed lines and arrows in Figure 5.2) of air parcels, some of which are undergoing large displacements.

When compared with traditional local static stability measures, the nonlocal characterization of stability is more consistent with the empirical evidence that unstable and convective boundary layers have strong and efficient mixing, while the stable boundary layer and other inversion layers have relatively weak mixing associated with them. The former is recommended as the preferred choice for

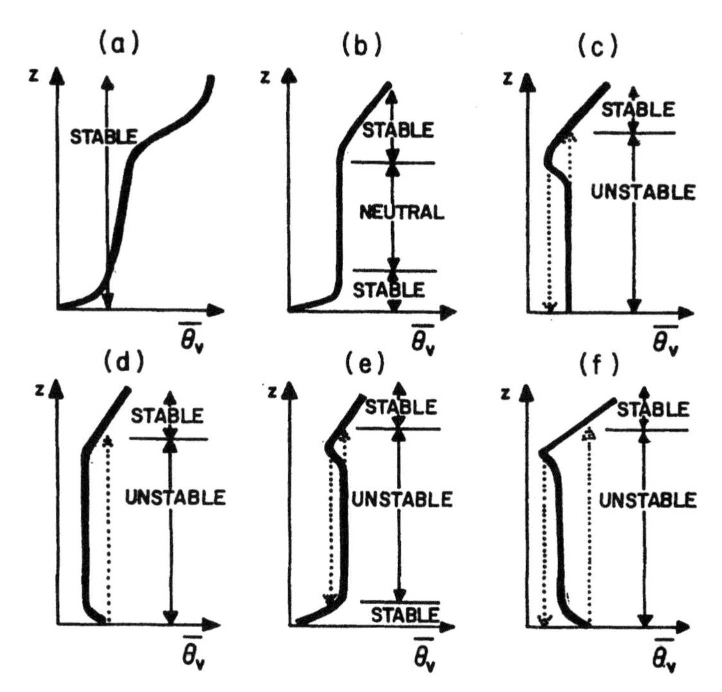

Figure 5.2 Nonlocal stability characterization for the various hypothetical virtual potential temperature profiles. [After Stull (1988).]

applications in micrometeorology and air pollution meteorology (Arya, 1999, Chapter 2).

Example Problem 2

The following measurements were made from the research vessel FLIP during the 1969 Barbados Oceanographic and Meteorological Experiment (BOMEX):

Height above sea-surface (m):	2.1	3.6	6.4	11.4
Air temperature (°C):	27.74	27.72	27.69	27.62
Specific humidity (g kg^{-1}):	18.02	17.83	17.54	17.29

(a) Calculate the virtual temperature at each level and its difference from the actual temperature.
(b) Calculate and compare the gradients of potential temperature and virtual potential temperature between adjacent measurement levels.
(c) Calculate the local static stability parameter, s, and the percentage contribution of specific humidity gradient to the same at different height levels.

Solution

(a) Using $T_v = T(1 + 0.61Q)$ and $T_v - T = 0.61QT$, where T and T_v are in absolute units (K) and Q is dimensionless, we obtain the following:

z (m):	2.1	3.6	6.4	11.4
T_v (K):	304.25	304.19	304.11	303.99
$(T_v - T)$ (K):	3.31	3.21	3.22	3.17

(b) Estimating the gradients between adjacent levels as

$$\frac{\partial T}{\partial z} \cong \frac{\Delta T}{\Delta z}; \quad \frac{\partial T_v}{\partial z} \cong \frac{\Delta T_v}{\Delta z}$$

and using the approximate relations

$$\frac{\partial \Theta}{\partial z} \cong \frac{\partial T}{\partial z} + \Gamma; \quad \frac{\partial \Theta_v}{\partial z} \cong \frac{\partial T_v}{\partial z} + \Gamma$$

We can estimate and compare the above gradients as follows:

Mean height (m):	2.85	5.00	8.90
$\partial\Theta/\partial z$ (K m^{-1}):	−0.0035	−0.0009	−0.0042
$\partial\Theta_v/\partial z$ (K m^{-1}):	−0.0302	−0.0188	−0.0142

Note that magnitudes of $\partial\Theta_v/\partial z$ over the tropical ocean are much larger than those of $\partial\Theta/\partial z$.

(c) Using the definition

$$s = \frac{g}{T_v}\frac{\partial\Theta_v}{\partial z},$$

it can easily be calculated. For determining the relative contribution of the specific humidity gradient to s, we can write

$$s = \frac{g}{T_v}\left(\frac{\partial T_v}{\partial z} + \Gamma\right) = \frac{g}{T_v}\left(\frac{\partial T}{\partial z} + 0.61T\frac{\partial Q}{\partial z} + \Gamma\right)$$

$$= \frac{g}{T_v}\left(\frac{\partial\Theta}{\partial z} + 0.61T\frac{\partial Q}{\partial z}\right)$$

The second term represents the contribution of the specific humidity gradient to s. The relative contribution of the same is given by

$$0.61T\frac{\partial Q}{\partial z}\bigg/\frac{\partial\Theta_v}{\partial z} = 1 - \frac{\partial\Theta/\partial z}{\partial\Theta_v/\partial z}$$

Mean height (m):	2.85	5.00	8.90
Static stability (s^{-2}):	-9.74×10^{-4}	-6.02×10^{-4}	-4.54×10^{-4}
% contribution of $\partial Q/\partial z$:	88.4	95.2	70.4

This example illustrates that it is extremely important to include the specific humidity gradients in any estimates of static stability over water surfaces. The stability based on temperature alone would be characterized as near-neutral, while it is significantly unstable, because of the stronger negative humidity gradients.

5.4 Mixed Layers, Mixing Layers and Inversions

In moderate to extremely unstable conditions, typically encountered during midday and afternoon periods over land, mixing within the bulk of the boundary layer is intense enough to cause conservable scalars, such as Θ and Θ_v, to be distributed uniformly, independent of height. The layer over which this occurs is called the mixed layer, which is an adiabatic layer overlying a superadiabatic surface layer. To a lesser extent, the specific humidity, momentum, and concentrations of pollutants from distant sources also have uniform distributions in the mixed layer. Mixed layers are found to occur most persistently over the tropical and subtropical oceans.

Sometimes, the term 'mixing layer' is used to designate a layer of the atmosphere in which there is significant mixing, even though the conservable properties may not be uniformly mixed. In this sense, the turbulent PBL is also a mixing layer, a large part of which may become well mixed (mixed layer) during unstable conditions. Mixing layers are also found to occur sporadically or intermittently in the free atmosphere above the PBL, whenever there are suitable conditions for the breaking of internal gravity waves and the generation of turbulence in an otherwise smooth, streamlined flow.

The term 'inversion' in common usage applies to an atmospheric condition when air temperature increases with height. Micrometeorologists sometimes refer to inversion as any stable atmospheric layer in which potential temperature (more appropriately, Θ_v) increases with height. The strength of inversion is expressed in terms of the layer-averaged $\partial\Theta_v/\partial z$ or the difference $\Delta\Theta_v$ across the given depth of the inversion layer.

Inversion layers are variously classified by meteorologists according to their location (e.g., surface and elevated inversions), the time (e.g., nocturnal inversion), and the mechanism of formation (e.g., radiation, evaporation, advection, subsidence, frontal and sea-breeze inversions). Because vertical motion and mixing are considerably inhibited in inversions, a knowledge of the frequency of occurrence of inversions and their physical characteristics (e.g., location, depth, and strength of inversion) is very important to air pollution meteorologists. Pollutants released within an inversion layer often travel long distances without much mixing and spreading. These are then fumigated to ground level as soon as the inversion breaks up to the level of the ribbonlike thin plume, resulting in large ground-level concentrations. Low-level and ground inversions act as lids, preventing the downward diffusion or spread of pollutants from elevated sources. Similarly, elevated inversions put an effective cap on the upward spreading of pollutants from low-level sources. An inversion layer usually caps the daytime unstable or convective boundary layer, confining the pollutants largely to the PBL. Consequently, the PBL depth is commonly referred to as the mixing depth in the air quality literature.

The most persistent and strongest surface inversions are found to occur over the Antarctica and Arctic regions; the tradewind inversions are probably the most persistent elevated inversions capping the PBL.

5.5 Vertical Temperature and Humidity Profiles

A typical sequence of observed potential temperature profiles at 3 h intervals during the course of a day is shown in Figure 5.3. These were obtained from

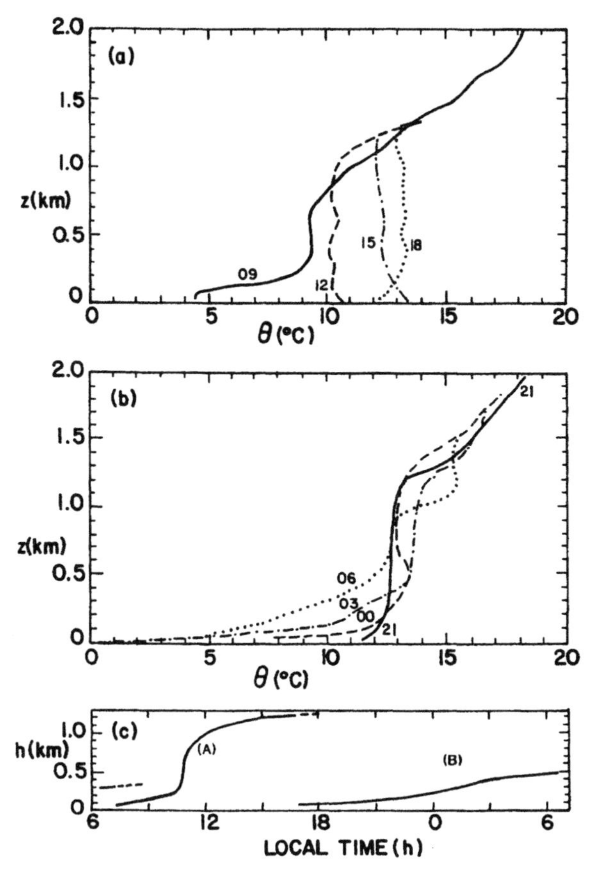

Figure 5.3 Diurnal variations of potential temperature profiles and inversion heights during (a) day 33 and (b) days 33–34 of the Wangara Experiment. (c) Curve A, inversion base height in the daytime CBL. Curve B, surface inversion height in the nighttime SBL. [After Deardorff (1978).]

radiosonde measurements during the 1967 Wangara Experiment near Hay, New South Wales, Australia, on a day when there were clear skies, very little horizontal advections of heat and moisture, and no frontal activity within 1000 km.

Note that, before sunrise and at the time of the minimum in the near-surface temperature, the Θ profile is characterized by nocturnal inversion, which is produced as a result of radiative cooling of the surface. The nocturnal boundary layer is stably stratified, so that vertical turbulent exchanges are greatly suppressed. It is generally referred to as the stable boundary layer (SBL). Shortly after sunrise, surface heating leads to an upward exchange of sensible heat and subsequent warming of the lowest layer due to heat flux convergence. This process progressively erodes the nocturnal inversion from below and replaces it with an unstable or convective boundary layer (CBL) whose depth h grows with time (see Figure 5.3c). The rate of growth of h usually attains maximum a few hours after sunrise when all the surface-based inversion has been eroded, and slows down considerably in the late afternoon.

The daytime unstable or convective boundary layer has a three-layered structure. The potential temperature decreases with height within the shallow surface layer. In this layer, turbulence is generated by both wind shear and buoyancy. The surface layer is followed by the mixed layer in which potential temperature remains more or less uniform. The mixed layer comprises the bulk of the CBL. This layer is dominated by buoyancy-generated turbulence, except during morning and evening transition hours when shear effects also become important. Above the mixed layer is a shallow transition layer in which potential temperature increases with height. In this stable layer, turbulence decreases rapidly with height. From the temperature sounding alone, the top of the PBL, including the transition layer, is not as distinct as the top of the mixed layer which is also the base of the inversion. Therefore, the height of inversion (z_i) is often used as an approximation for the PBL height (h), although the former can be substantially (10–30%) smaller than the latter.

Just before sunset, there is net radiative loss of energy from the surface and, consequently, an inversion forms at the surface. The nocturnal inversion deepens in the early evening hours and sometimes throughout the night, as a result of a radiative and sensible heat flux divergence. The surface inversion height (h_i) may not correspond to the height, h, of the stable boundary layer (SBL) based on significant turbulence activity. The inversion height is more easily detected from temperature soundings and is depicted in Figure 5.3c. The PBL height can be directly measured from remote-sensing instruments such as sodar and lidar or indirectly estimated from both temperature and wind profiles. Such measurements indicate that, unlike the surface inversion height, the PBL height (h) generally decreases with time during early evening hours and attains a constant value which is usually less than h_i. Sometimes, the PBL height oscillates around some average value throughout the night.

The SBL is often capped by the remnant of the previous afternoon's mixed layer. Mixing processes in this so-called residual layer are likely to be very weak. Strong wind shears may generate some intermittent and sporadic turbulence, which is not connected to the SBL.

Observed vertical profiles of specific humidity for the same Wangara day 33 are shown in Figure 5.4. The surface during this period of drought was rather dry, with little or no growth of vegetation (predominantly dry grass, legume, and cottonbush). In the absence of any significant evapotranspiration from the surface, the specific humidity profiles are nearly uniform in the daytime PBL,

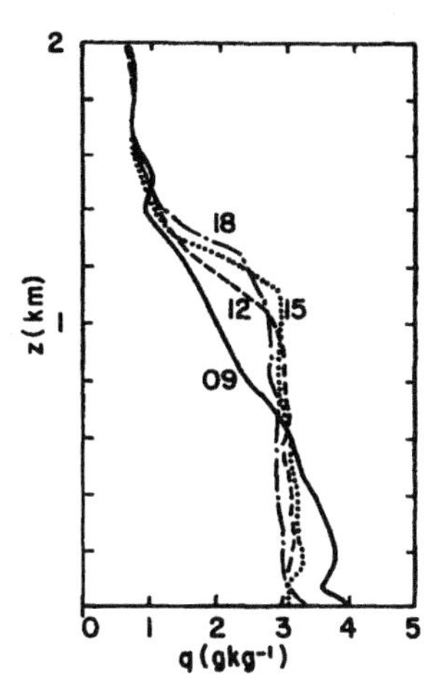

Figure 5.4 Diurnal variation of specific humidity profiles during day 33 of the Wangara Experiment. [From Andre *et al.* (1978).]

with the value of Q changing largely in response to the evolution of the mixed layer.

In other situations where there is dry air aloft, but plenty of moisture available for evaporation at the surface, the evolution of Q profiles might be quite different from those shown in Figure 5.4. In response to the increase in the rate of evaporation or evapotranspiration during the morning and early afternoon hours, Q may actually increase and attain its maximum value sometime in the afternoon and, then, decrease thereafter, as the evaporation rate decreases.

The evaporation from the surface, although strongest by day, may continue at a reduced rate throughout the night. Under certain conditions, when winds are light and the surface temperature falls below the dew point temperature of the moist air near the surface, the water vapor is likely to condense and deposit on the surface in the form of dew. The resulting inversion in the Q profile leads to a downward flux of water vapor and further deposition.

Temperature and humidity profiles over open oceans do not show any significant diurnal variation when large-scale weather conditions remain unchanged, because the sea-surface temperature changes very little (typically, less than 0.5°C), if at all, between day and night. Figure 5.5 shows the observed profiles of T and Q from two research vessels (with a separation distance of about 750 km) during the 1969 Atlantic Tradewind Experiment (ATEX). Here, only the averaged profiles for the two periods designated as 'undisturbed' (February 7–12) and 'disturbed' (February 13–17) weather are given. During the undisturbed period, the temperature and humidity structure of the PBL is characterized by a mixed layer (depth $\cong$ 600 m), followed by a shallow

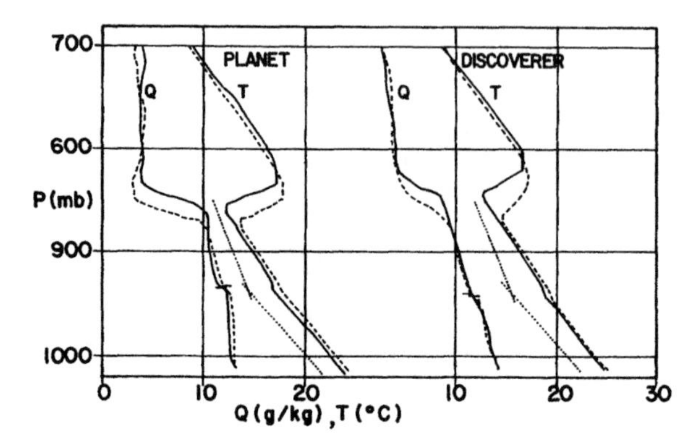

Figure 5.5 Observed mean vertical profiles of temperature and specific humidity at two research vessels during ATEX. First ('undisturbed') period: solid lines; second ('disturbed') period: dashed lines. The dotted lines indicate the dry adiabatic and saturated adiabatic lapse rates. [After Augstein *et al.* Copyright © (1974) by D. Reidel Publishing Company. Reprinted by permission.]

isothermal transition layer (depth ≅ 50 m) and a deep cloud layer extending to the base of the trade wind inversion. The same features can also be seen during the disturbed period, but with larger spatial differences.

At the other extreme, Figure 5.6 shows the measured mean temperature profiles in the surface inversion layer at the Plateau Station, Antarctica (79°15′S, 40°30′E), in three different seasons (periods) and for different stability classes (identified here as (1) through (8) in the order of increasing stability and defined from both the temperature and wind data). These include some of the most stable conditions ever observed on the earth's surface.

5.6 Diurnal Variations

As a result of the diurnal variations of net radiation and other energy fluxes at the surface, there are large diurnal variations in air temperature and specific humidity in the PBL over land surfaces. The diurnal variations of the same over large lakes and oceans are much smaller and often negligible.

The largest diurnal changes in air temperature are experienced over desert areas where strong large-scale subsidence limits the growth of the PBL and the air is rapidly heated during the daytime and is rapidly cooled at night. The mean diurnal range of screen-height temperatures is typically 20°C, with mean maximum temperatures of about 45°C during the summer months (Deacon, 1969).

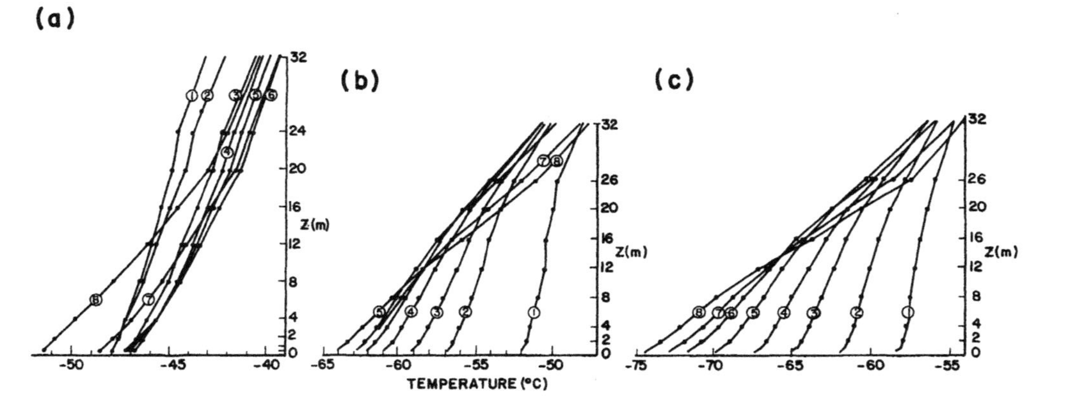

Figure 5.6 Observed mean temperature profiles in the surface inversion layer at Plateau Station, Antarctica, for three different periods grouped under different stability classes. (a) Sunlight periods, 1967; (b) transitional periods, 1967; (c) dark season, 1967. [After Lettau *et al.* (1977).]

The diurnal range of temperatures is reduced by the presence of vegetation and the availability of moisture for evaporation. Even more dramatic reductions occur in the presence of cloud cover, smoke, haze, and strong winds. The diurnal range of temperatures decreases rapidly with height in the PBL and virtually disappears at the maximum height of the PBL or inversion base, which is reached in the late afternoon. Some examples of observed diurnal variations of air temperature at different heights and under different cloud covers and wind conditions are given in Figure 5.7.

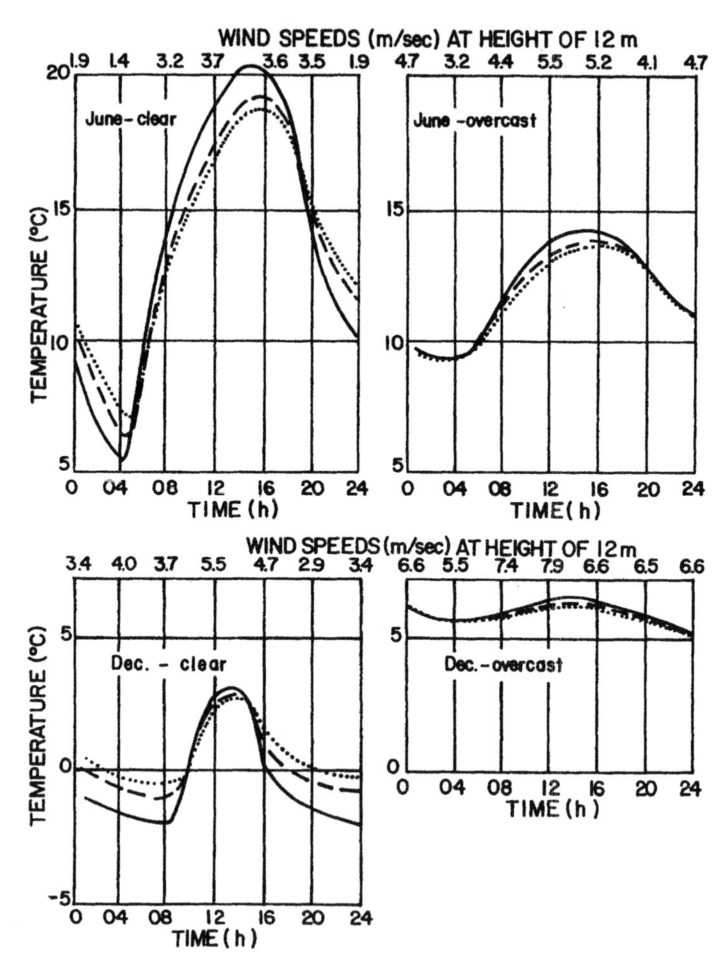

Figure 5.7 Diurnal variation of air temperature at three heights (solid lines: 1.2 m; dashed lines: 7 m; dotted lines: 17 m) over downland in Southern England for various combinations of clear and overcast days in June and December. Wind speeds at 12 m height are also indicated. [From Deacon (1969); after Johnson (1929).]

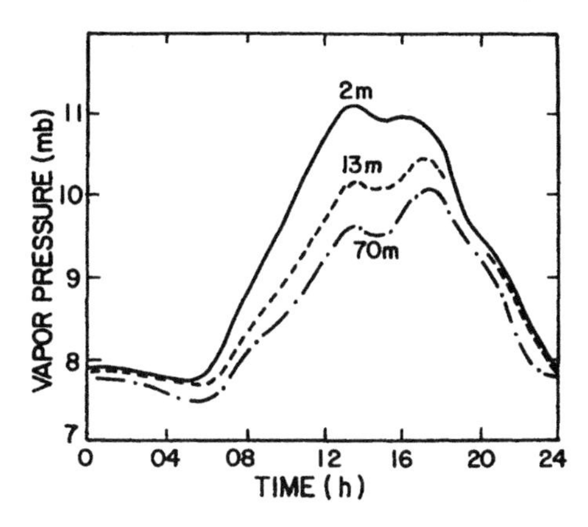

Figure 5.8 Diurnal variation of water vapor pressure at three heights at Quickborn, Germany, on clear May days. [From Deacon (1969).]

The diurnal variation of specific humidity depends on the diurnal course of evapotranspiration and condensation, surface temperatures, mean winds, turbulence, and the PBL height. Large diurnal changes in surface temperature generally lead to large variations of specific humidity, due to the intimate relationship between the saturation vapor pressure and temperature. Figure 5.8 illustrates the diurnal course of water vapor pressure at three different heights. Here the secondary minimum of vapor pressure in the afternoon is believed to be caused by further deepening of the mixed layer after evapotranspiration has reached its maximum value. This phenomenon may be even more marked at other places and times.

5.7 Applications

The knowledge of air temperature and humidity in the PBL is useful and may have applications in the following areas:

- Determining the static stability of the lower atmosphere and identifying mixed layers and inversions.
- Determining the PBL height, as well as the depth and strength of surface or elevated inversion.
- Predicting air temperature and humidity near the ground and possible fog and frost conditions.
- Determining the sensible heat flux and the rate of evaporation from the surface.
- Calculating atmospheric radiation.

- Estimating diffusion of pollutants from surface and elevated sources and possible fumigation.
- Designing heating and air-conditioning systems.

Problems and Exercises

1. How do the mechanisms of heat transfer in the atmosphere differ from those in a subsurface medium, such as soil and rock?

2.

(a) Derive the following equation for the variation of air temperature with time, in the absence of any horizontal and vertical advections of heat:

$$\rho c_p(\partial T/\partial t) = (\partial R_N/\partial z) - (\partial H/\partial z)$$

(b) Indicate how the above equation may be used to estimate the sensible heat flux at the surface, H_0, during the daytime when the rate of warming is nearly constant through the depth of the PBL.

(c) Temperature measurements at the top of a 20 m tower indicated a rate of warming of 1.8 °C h^{-1} during the mid-day period when the PBL height was estimated to be 850 m from a minisonde sounding. Estimate the surface heat flux at the time of observations.

3.

(a) Derive an expression for the adiabatic lapse rate in a moist, unsaturated atmosphere and discuss the effect of moisture on the same.

(b) Explain why the saturated adiabatic lapse rate must be smaller than the dry diabatic lapse rate.

4.

(a) Derive the equation of state for the moist air and hence define the virtual temperature T_v.

(b) Show that, to a good approximation,

$$\partial T_v/\partial z = (\partial T/\partial z) + 0.61T(\partial Q/\partial z)$$

5.

(a) Show that the buoyant acceleration on a parcel of air in a thermally stratified environment is given by

$$a_b = -(g/T_v)(T_v - T_{vp})$$

(b) Considering a small vertical displacement, Δz, from the equilibrium position of the parcel, show that

$$a_{\mathrm{b}} \cong -(g/T_{\mathrm{v}})(\partial\Theta_{\mathrm{v}}/\partial z)\Delta z$$

6. A moist air parcel at a temperature of 30°C and specific humidity of 25 g kg^{-1} is lifted adiabatically from the upwind base of a tropical mountain island, where the pressure is 1000 mb, to the top, at 2000 m above the base, and is then brought down to the base on the other side of the mountain. Using appropriate equations or tables, calculate the following parameters:

(a) The height above the base where the air parcel would become saturated and the saturated adiabatic lapse rate at this level.
(b) The temperature of the parcel at the top of the mountain.
(c) The temperature of the parcel at the downwind base, assuming that the parcel remained unsaturated during its entire descent.

7. The following pressure, temperature and specific humidity measurements were made from radiosonde soundings on day 11 of the Wangara PBL experiment:

Height (m)	Pressure (mb)	Temperature (°C)	Specific humidity (g kg^{-1})
Surface (1.5)	1013	11.8	6.5
50	1007	10.9	6.1
100	1001	10.1	5.8
150	995	9.6	5.6
200	989	9.2	5.5
300	977	8.4	5.3
400	965	7.4	5.1
500	954	6.4	4.9
550	948	6.0	4.9
600	942	5.6	4.8
650	936	5.1	4.7
700	931	4.9	4.6
750	925	4.9	4.5
800	919	4.9	4.5
850	914	5.0	4.5
900	908	5.0	4.5
1000	897	4.8	4.4

(a) Calculate and plot virtual temperature as a function of height.
(b) Calculate and plot virtual potential temperature as a function of height.

(c) Characterize the various layers on the basis of local and nonlocal stability and discuss the differences between the two.
(d) Estimate the height of the inversion base (z_i), as well as the PBL height (h). Discuss the sensitivity of the latter to the near-surface temperature.

8. Discuss the importance of specific humidity gradient in the determination of the static stability parameter s over different types of surfaces. Under what conditions can one ignore specific humidity in the determination of static stability?

Chapter 6 Wind Distribution in the PBL

6.1 Factors Influencing Wind Distribution

The magnitude and direction of near-surface winds and their variations with height in the PBL are of considerable interest to micrometeorologists. The following factors influence the wind distribution in the PBL:

- Large-scale horizontal pressure and temperature gradients in the lower atmosphere, which drive the PBL flow.
- The surface roughness characteristics, which determine the surface drag and momentum exchange in the lower part of the PBL.
- The earth's rotation, which makes wind turn with height.
- The diurnal cycle of heating and cooling of the surface, which determines the thermal stratification of the PBL.
- The PBL depth, which determines wind shears in the PBL.
- Entrainment of the free atmospheric air into the PBL, which determines the momentum, heat, and moisture exchanges at the top of the PBL, as well as the PBL height.
- Horizontal advections of momentum and heat, which affect both the wind and temperature distributions in the PBL.
- Large-scale horizontal convergence or divergence and the resulting mean vertical motion at the top of the PBL.
- Presence of clouds and precipitation in the PBL, which influence its thermal stratification.
- Surface topographical features, which give rise to local or mesoscale circulations.

Some of these factors will be discussed in more detail in the following text, while others are considered outside the scope of this introductory text.

6.2 Geostrophic and Thermal Winds

The PBL is essentially driven by large-scale atmospheric motions which are set up in response to spatial variations of air pressure and temperature. The geostrophic winds are related to or defined in terms of the pressure gradients as

$$U_g = -\frac{1}{\rho f}\frac{\partial P}{\partial y}; \quad V_g = \frac{1}{\rho f}\frac{\partial P}{\partial x} \tag{6.1}$$

in which U_g and V_g are the components of geostrophic wind vector **G** in the x and y directions, respectively, and f is the Coriolis parameter, which is related to the rotational speed of the earth (Ω) and latitude (ϕ) as

$$f = 2\Omega \sin \phi \tag{6.2}$$

The geostrophic winds are the winds which would occur as a result of the simple geostrophic balance between the pressure gradient and Coriolis (rotational) forces in a frictionless atmosphere with no advective and local accelerations. Equation (6.1) implies that **G** must be parallel to the isobars with low pressure to the left in the northern hemisphere ($f > 0$) (see Figure 6.1) and low pressure to the right in the southern hemisphere ($f < 0$). The absence of advective accelerations further implies that, locally, isobars are equispaced, parallel, straight lines. The geostrophic balance is often assumed to occur, i.e., $U = U_g$ and $V = V_g$, at the top of and outside the PBL. Within the PBL, however, actual

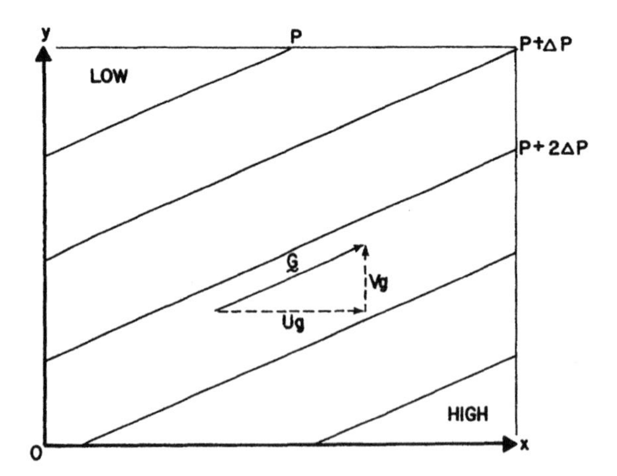

Figure 6.1 Schematic of relationship between geostrophic winds and isobars for the northern hemisphere.

winds differ from the geostrophic winds because of the surface friction and the vertical exchange of momentum.

The horizontal pressure gradients or geostrophic winds may vary with height in response to horizontal temperature gradients. The vertical gradients of geostrophic winds, i.e., geostrophic wind shears, are given by the thermal wind equations

$$\begin{aligned} \frac{\partial U_g}{\partial z} &= -\frac{g}{fT}\frac{\partial T}{\partial y} + \frac{U_g}{T}\frac{\partial T}{\partial z} \cong -\frac{g}{fT}\frac{\partial T}{\partial y} \\ \frac{\partial V_g}{\partial z} &= \frac{g}{fT}\frac{\partial T}{\partial x} + \frac{V_g}{T}\frac{\partial T}{\partial z} \cong \frac{g}{fT}\frac{\partial T}{\partial x} \end{aligned} \tag{6.3}$$

which can be derived from Equation (6.1) in conjunction with the hydrostatic equation and the equation of state (Hess, 1959, Chapter 12). The approximations on the right-hand side of Equation (6.3) ignore the stability-dependent terms, which may account for no more than 10% variation in geostrophic winds per kilometer of height. The approximate thermal wind equations imply that the geostrophic wind shear vector $\partial \mathbf{G}/\partial z$ must be parallel to the isotherms with colder air to the left in the northern hemisphere (Figure 6.2). A part of the geostrophic wind shear is due to the normal (climatological) decrease of temperature in going toward the poles, while a substantial part may be due to local temperature gradients created by surface topographical features (e.g., land–sea, urban–rural, and mountain–valley contrasts) and/or the synoptic weather situation. Note that a horizontal temperature gradient of only 1°C per 100 km will cause a geostrophic wind shear of about 3.3 m s^{-1} km^{-1} or 0.0033 s^{-1} in middle latitudes.

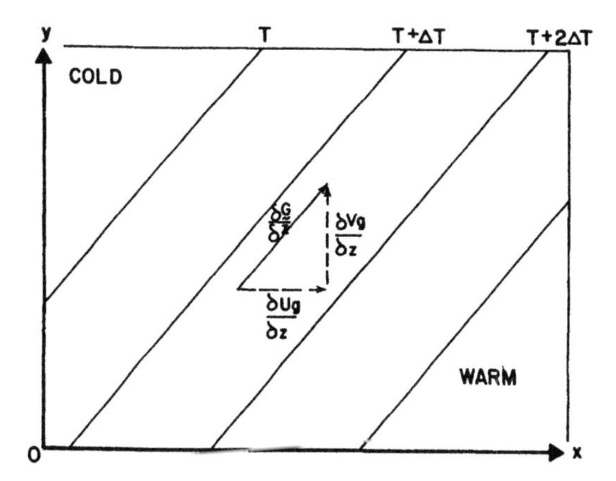

Figure 6.2 Schematic of relationship between thermal winds or geostrophic shears and isotherms for the northern hemisphere.

The changes in geostrophic wind components over a given layer depth, Δz, can be calculated from the finite-difference form of Equation (6.3):

$$\Delta U_g = -\frac{\Delta z g}{fT}\frac{\partial T}{\partial y}; \quad \Delta V_g = \frac{\Delta z g}{fT}\frac{\partial T}{\partial x} \qquad (6.4)$$

Horizontal temperature gradients or geostrophic shears are generally assumed to be constant (independent of height) in the PBL. It follows from Equation (6.3) or (6.4) that geostrophic winds must vary with height in the presence of finite temperature gradients. Figure 6.3 illustrates for the northern hemisphere the relationship between the geostrophic winds at the surface and the top of the PBL (taking $\Delta z = h$) for different orientations of the surface geostrophic wind

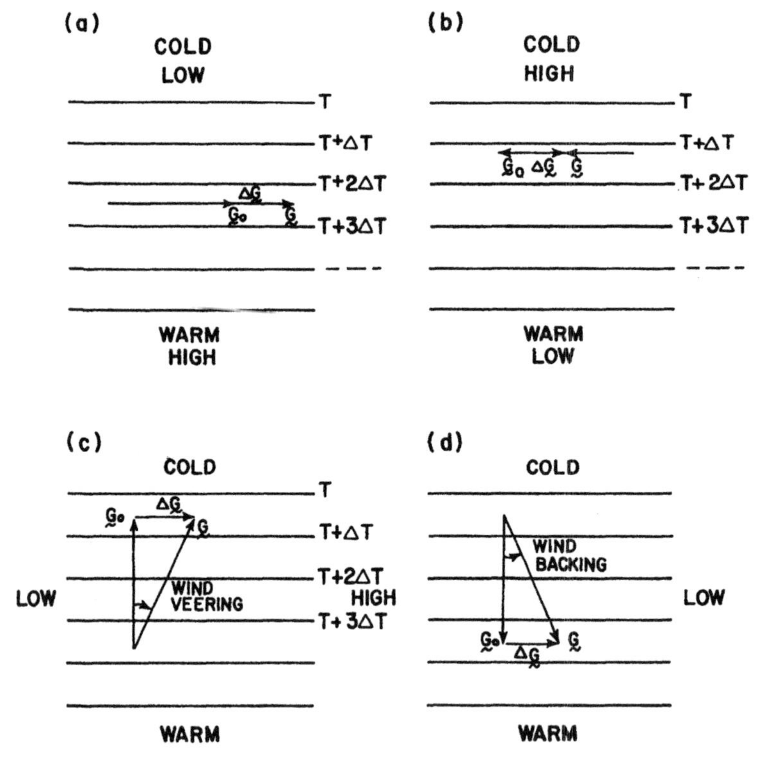

Figure 6.3 Schematic of the relationship between the geostrophic winds at the surface and the top of the PBL for different orientations of surface isobars or $\mathbf{G}_0$ and surface isotherms: (a) parallel geostrophic flow around a 'cold low' or a 'warm high'; (b) parallel flow around a 'cold high', or a 'warm low'; (c) veering geostrophic flow in warm advection; (d) backing flow in cold advection.

$\mathbf{G}_0$ with respect to the isotherms. Again, isobars and isotherms are taken as parallel straight lines for the ideal case of no advective acceleration.

Note that when $\mathbf{G}_0$ is parallel to the isotherms (no temperature advection), the effect of the horizontal temperature gradient is to increase or decrease the magnitude of geostrophic winds, without causing any change in the direction. This will be the rare situation when isobars are parallel to the isotherms throughout the layer. In general, however, the surface isobars are not parallel to the isotherms and the surface geostrophic wind has a component from cold toward warm temperatures or vice versa. Such a flow tends to change the vertical temperature distribution due to horizontal advection of cold or warm air. Figure 6.3 shows that both the magnitude and the direction of the geostrophic wind may change with height in the presence of heat advection. The geostrophic wind turns cyclonically (backs) with height in the case of cold air advection, and it turns anticyclonically (veers) with height in the case of warm advection. Actual winds within the PBL may also be expected to back or veer with height in response to the backing and veering of geostrophic winds.

Equations (6.1) and (6.4) express the geostrophic and thermal winds in terms of horizontal pressure and temperature gradients on a horizontal (constant-level) surface. These can be used with the information provided on a surface chart depicting isobars and isotherms. The geostrophic and thermal winds in meteorology are often expressed in terms of the heights (z_p) of constant-pressure surfaces which are represented on upper-air (e.g., 850 mb, 700 mb, etc.) charts. On any constant-pressure surface, the geostrophic and thermal winds can be expressed as

$$U_g = -\frac{g}{f}\frac{\partial z_p}{\partial y}; \quad V_g = \frac{g}{f}\frac{\partial z_p}{\partial x} \tag{6.5}$$

$$\Delta U_g = -\frac{g}{f}\frac{\partial(\Delta z_p)}{\partial y}; \quad \Delta V_g = \frac{g}{f}\frac{\partial(\Delta z_p)}{\partial x} \tag{6.6}$$

where Δz_p is the thickness of the layer between the p_1 and p_2 constant-pressure surfaces and ΔU_g and ΔV_g are the corresponding thermal wind components. The hypsometric equation can be used to express Δz_p in terms of p_1 and p_2, and the average temperature of the layer. Equation (6.5) implies that the geostrophic wind is proportional to the slope of the constant-pressure surface.

The atmosphere is called barotropic, when there are no geostrophic shears or horizontal temperature gradients, and baroclinic, when there are significant geostrophic shears. In the same way, atmospheric boundary layers may be classified as barotropic and baroclinic PBLs. The term baroclinicity (or baroclinity) is often used as a synonym for geostrophic wind shear. It is not difficult to see that the presence of geostrophic shears or baroclinity in the PBL is more a rule than the exception.

In strongly curved flows about low- and high-pressure systems, the centripetal force alters the normal geostrophic balance between pressure gradient and Coriolis forces. Consequently, the actual wind can be significantly more or less than the geostrophic wind in strongly cyclonic or anticyclonic flows. The so-called 'gradient wind' provides a much better approximation to the actual winds in strongly curved flow systems (Hess, 1959, Chapter 12; Holton, 1992, Chapter 3).

Example Problem 1

Surface isobars at a 30°N latitude site during a mid-day period run east–west, indicating a horizontal pressure gradient of 3 mb per 100 km toward the south. At this time, surface isotherms are nearly perpendicular to isobars indicating a horizontal temperature gradient of 1.5°C per 100 km toward the west. A temperature sounding indicated the PBL height of 2000 m and a temperature drop of 25°C across the PBL from the near-surface temperature of 35°C. The observed surface pressure is 1020 mb. Calculate the geostrophic winds (both magnitude and direction) at (a) the surface and (b) top of the PBL.

Solution

The Coriolis parameter for the site is given by

$$f = 2\Omega \sin\phi = 2 \times \frac{2\pi}{24 \times 3600} \times \sin 30^\circ \cong 0.73 \times 10^{-4}\ \mathrm{s}^{-1}$$

Air density near the surface can be determined from the equation of state

$$\rho = \frac{P}{RT} = \frac{1020 \times 10^2}{287.04 \times 308.2} \cong 1.15\ \mathrm{kg\ m}^{-3}$$

(a) In order to calculate the surface geostrophic wind from the observed pressure gradient, we use a geographical coordinate system with the x axis pointed toward the east and the y axis toward the north. In this coordinate system,

$$\frac{\partial P}{\partial x} = 0; \quad \frac{\partial P}{\partial y} = -3 \times 10^{-5}\ \mathrm{mb\ m}^{-1} = -3 \times 10^{-3}\ \mathrm{Pa\ m}^{-1}$$

Therefore

$$V_g = 0; \quad U_g = -\frac{1}{\rho f}\frac{\partial P}{\partial y} = \frac{10^4}{1.15 \times 0.73} \times 3 \times 10^{-3}$$
$$\cong 35.74 \text{ ms}^{-1}$$

Thus, the magnitude of the surface geostrophic wind, $G_0 = 35.74 \text{ m s}^{-1}$, and its direction is from the west (270°).

(b) Before calculating the geostrophic wind at the top of the PBL (h = 2000 m), we need to calculate the thermal wind components from Equation (6.4) with $\Delta z = h$ = 2000 m. From the observations, we have

$$\frac{\partial T}{\partial x} = -1.5 \times 10^{-5} \text{ K m}^{-1}; \quad \frac{\partial T}{\partial y} = 0$$

The average temperature in the PBL, T = 295.7 K.
Thus,

$$\Delta V_g = \frac{hg}{fT}\frac{\partial T}{\partial x} \cong 13.63 \text{ m s}^{-1}; \quad \Delta U_g = 0$$

Then, the geostrophic wind components at the top of the PBL can be determined as

$$U_{gh} = U_{g0} + \Delta U_g = 35.74 \text{ m s}^{-1}$$
$$V_{gh} = V_{g0} + \Delta V_g = -13.63 \text{ m s}^{-1}$$

Therefore, the magnitude of the geostrophic wind,

$$G_h = [(35.74)^2 + (13.63)^2]^{1/2} \cong 38.25 \text{ m s}^{-1}$$

and its direction is 291° (NW).

6.3 Friction Effects on the Balance of Forces

Because air is a viscous fluid, its velocity relative to the earth's surface must vanish right at the surface. This does not happen abruptly, but the retarding influence of the surface on atmospheric winds pervades the whole depth of the PBL. Consequently, the wind speed decreases gradually and the mean momentum is transferred downward in going toward the surface. Turbulence provides an efficient mechanism for the transfer of momentum from one level to another in the vertical. Also, vertical exchange of horizontal momentum results in the so-called frictional force on any fluid element. Because of this, the simple geostrophic balance between the pressure gradient and Coriolis forces, which may exist outside the PBL, cannot be expected to occur within the PBL.

Figure 6.4 is a schematic of the balance of pressure gradient, Coriolis, and friction forces on fluid elements at different levels in a barotropic PBL. The corresponding geostrophic and actual wind vectors are also shown with the stipulation that the Coriolis force must always be normal to **V** and proportional to wind speed, while the friction force must be perpendicular to the ageostrophic wind vector (**V** − **G**). These conditions follow from the equation of mean motion in the absence of local and advective accelerations.

$$2\rho\boldsymbol{\Omega} \times (\mathbf{V} - \mathbf{G}) = \partial\boldsymbol{\tau}/\partial z \tag{6.7}$$

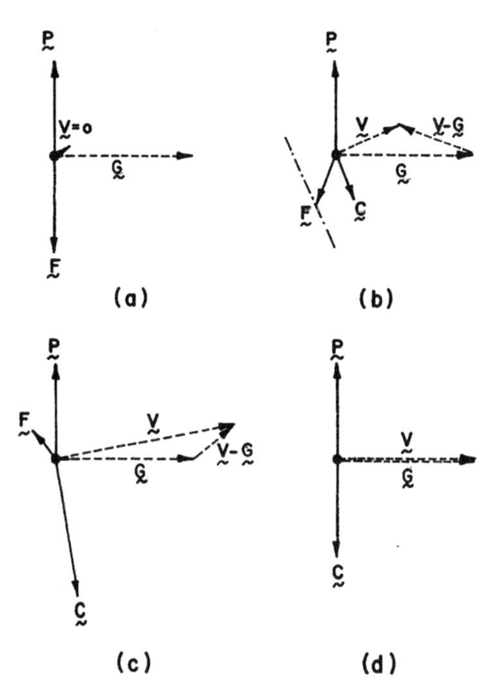

Figure 6.4 Schematic of the balance of forces on a fluid parcel at different heights in a barotropic PBL: (a) at the surface; (b) in the surface layer; (c) in the middle of the PBL; (d) at the top of the PBL. Actual and geostrophic velocities are also indicated. [After Arya (1986).]

Note that in a rotating frame of reference or in the presence of directional shear, the friction force on a fluid element need not be parallel and opposite to the velocity vector, as commonly depicted in many textbook schematics of the force balance in the friction layer. Figure 6.4a shows that, right at the surface where the Coriolis force disappears, the friction force must exactly balance the pressure gradient force and, hence, make a large angle with the direction of near-surface wind or surface shear. As wind speed increases with height, with little change in the wind direction through the surface layer, the balance of forces requires that the friction force decreases and rotates anticyclonically with increasing height (Figure 6.4b). The same tendency continues through the outer part of the PBL also, as wind veers and increases in magnitude (it often becomes super-geostrophic in the middle of the PBL), as shown in Figure 6.4c. Near the top of the PBL, the friction force becomes small in magnitude and highly variable in direction, in response to the changes in the direction of $\mathbf{V} - \mathbf{G}$, as velocity oscillates about the geostrophic equilibrium. Ignoring any such oscillations, as well as any inertial oscillations that may be present in the real atmosphere, the force balance at the top of the PBL is depicted in Figure 6.4d.

The above considerations of the balance of forces suggest that winds must veer with height in a barotropic PBL. The difference between the wind direction at some height and that near the surface is called the veering angle, which normally increases with height in a barotropic PBL. In the presence of thermal winds (geostrophic wind shear), however, actual winds may veer or back with increasing height, in response to the veering or backing of geostrophic winds and the frictional veering. In practice, frictional veering may be estimated as the difference between the actual wind veering and the geostrophic veering.

The angle between the directions of geostrophic and actual winds is called the cross-isobar angle (α) of the flow, which normally decreases with height and vanishes at the top of the PBL (assuming that geostrophic balance occurs there). The surface cross-isobar angle (α_0) which near-surface winds make with surface isobars is of considerable interest to meteorologists. It is found to depend on the surface roughness, latitude, and the surface geostrophic wind, as well as on the stability, baroclinity, and height of the PBL. For the neutral, barotropic PBL in middle and high latitudes, α_0 is found to range from about 10° for relatively smooth (e.g., water, ice, and snow) surfaces to about 35° for very rough (e.g., forests and urban areas) surfaces. It increases with increasing stability and decreasing latitude, so that a wide range of $\alpha_0 \cong 0$–45° occur in the stratified barotropic PBL. In the presence of thermal winds (baroclinic PBL), the range of observed surface cross-isobar angles becomes wider (say, −20 to 70°). Under certain conditions of strong baroclinity, α_0 may even become negative, implying that a component of the near-surface winds is directed toward the surface 'high' (Arya and Wyngaard, 1975; Arya, 1978).

Example Problem 2

The following observations are for a midlatitude ($f = 10^{-4}\ s^{-1}$) PBL during the afternoon convective conditions:

Height (m)	10	100	1000	1500
Wind speed (m s^{-1})	9.0	15.5	17.0	15.0
Wind direction (deg.):	250	253	258	260

The PBL height = 1500 m.
The surface geostrophic wind speed = 20.0 m s^{-1}.
The surface geostrophic wind direction = 275°.

(a) Calculate the average magnitudes of the geostrophic wind shear and the actual wind shear in the PBL, assuming geostrophic balance at the top of the PBL.
(b) Estimate the total wind veering and the frictional veering across the PBL, as well as the cross-isobar angle of the near-surface winds.
(c) Compute the pressure gradient, Coriolis, and friction forces per unit mass at the 10 m level and show the force balance at this level.

Solution

(a) Using the geographical coordinate system with the x-axis pointed toward the east and the y-axis toward the north, the horizontal components of geostrophic winds at the surface and the top (h = 1500 m) of the PBL are:

$$U_{g0} = G_0 \cos(270 - 275) = 19.92 \text{ m s}^{-1}$$
$$V_{g0} = G_0 \sin(270 - 275) = -1.74 \text{ m s}^{-1}$$
$$U_{gh} = G_h \cos(270 - 260) = 14.77 \text{ m s}^{-1}$$
$$V_{gh} = G_h \sin(270 - 260) = 2.60 \text{ m s}^{-1}$$

The components of the geostrophic wind shear across the PBL can, then, be calculated as

$$\Delta U_g = 14.77 - 19.92 = -5.15 \text{ m s}^{-1}$$
$$\Delta V_g = 2.60 - (-1.74) = 4.34 \text{ m s}^{-1}$$

So that, $|\Delta \mathbf{G}| = [(5.15)^2 + (4.34)^2]^{1/2} = 6.73 \text{ m s}^{-1}$

The magnitude of the geostrophic wind shear across the PBL = $|\Delta \mathbf{G}|/h \cong 4.49 \times 10^{-3}\ s^{-1}$.
The average magnitude of the actual wind shear

$$\frac{|\mathbf{V}_h|}{h} = 10^{-2}\ s^{-1}$$

(b) The total (actual) wind veering = 260 − 250 = 10°.
The geostrophic wind veering = 260 − 275 = −15°.

Frictional veering = total veering − geostrophic veering
= 10 − (−15) = 25°

The cross-isobar angle of near-surface winds = 275 − 250 = 25°, which is equal to the frictional veering across the PBL.

(c) The various forces per unit mass or accelerations are given as follows:

Type of force	Magnitude	Direction
Pressure gradient (**P**)	$f\|\mathbf{G}\|$	Normal to **G** oriented to the left of **G**
Coriolis force (**C**)	$f\|\mathbf{V}\|$	Normal to **V** oriented to the right of **V**
Friction force (**F**)	$f\|\mathbf{V} - \mathbf{G}\|$	Normal to (**V** − **G**)

In the absence of any local or advective acceleration, the above forces must balance, so that, at any height, $\mathbf{P} + \mathbf{C} + \mathbf{F} = 0$.
The magnitudes of the various forces at the 10 m level are:

$$|\mathbf{P}| = f|\mathbf{G}| = 2.0 \times 10^{-3}\ \mathrm{m\,s^{-2}}$$

$$|\mathbf{C}| = f|\mathbf{V}| = 0.9 \times 10^{-3}\ \mathrm{m\,s^{-2}}$$

$$|\mathbf{F}| = f|\mathbf{V} - \mathbf{G}| = 1.24 \times 10^{-3}\ \mathrm{m\,s^{-2}}$$

Note that $|\mathbf{V} - \mathbf{G}] \neq |\mathbf{V}| - |\mathbf{G}|$ but

$$|\mathbf{V} - \mathbf{G}] = [(U - U_g)^2 + (V - V_g)^2]^{1/2}$$

The force balance can easily be shown in a vector plot similar to Figure 6.4(b), knowing the magnitudes and directions of geostrophic and actual winds.

6.4 Stability Effects on the PBL Winds

We have discussed earlier how the diurnal cycle of heating and cooling of the surface, in conjunction with radiative, convective, and advective processes occurring within the PBL, determine the static or thermal stability of the PBL. Through its strong influence on the vertical movements of air parcels, thermal stability strongly influences the vertical exchange of momentum and, hence, the wind distribution in the PBL.

On a clear day the surface warms up relative to the air above, in response to solar heating. This gives rise to a variety of convective circulations, such as near-surface plumes, thermals (updrafts) and downdrafts, which can directly transfer momentum and heat in the vertical direction. The upward transfer of heat through the lower part of the PBL is largely responsible for convective or buoyancy-generated turbulence. The resulting vigorous mixing of momentum leads to considerable weakening and, sometimes, elimination of mean wind shears in the PBL. Thus, it is not uncommon to find nearly uniform wind speed and wind direction profiles through much of the daytime convective boundary layer (CBL). Strong wind shears are largely confined to the lower part of the surface layer and the shallow transition layer near the base of inversion which often caps the convective mixed layer. Changes in mean wind direction between the surface and the inversion base are typically less than 20°.

On a clear evening and night, on the other hand, the surface cools down in response to longwave radiation and a surface inversion begins to form and develop. Since buoyancy inhibits vertical momentum exchanges in the inversion layer, significant wind speed and direction shears develop in this layer. Wind direction changes of 30° or more across the shallow stable boundary layer (SBL), which may comprise only a part of the surface inversion layer, are not uncommon. The wind speed profile is generally characterized by a low-level jet in which winds are often supergeostrophic. The winds in the outer part of the SBL may undergo inertial oscillations in response to the inertially oscillating geostrophic flow. Internal gravity waves may also develop in such a stratified environment; such waves frequently appear mixed with turbulence.

Thermal stability also has a strong influence on the PBL height, which, in turn, affects the wind distribution. In the morning hours following sunrise, the PBL grows rapidly at first, in response to heating from below. This growth continues throughout the day, although at a progressively decreasing rate, resulting in a 1- to 2-km-deep mixed layer by mid-afternoon. Immediately following the evening transition period, when the sensible heat flux at the surface changes sign, the unstable PBL suddenly collapses and is replaced by a much shallower stable boundary layer. For a given wind speed at the top of the PBL, one would expect the average wind shear in the PBL to be an order of magnitude larger at night than that during the midday period, because of the difference in the PBL heights. Very close to the surface, however, wind shears are generally larger during the daytime than they are at night.

In the presence of wind shear, a dynamic stability parameter such as the gradient Richardson number

$$\mathrm{Ri} = \frac{g}{T_v} \frac{\partial \Theta_v}{\partial z} \left| \frac{\partial \mathbf{V}}{\partial z} \right|^{-2} \tag{6.8}$$

is considered to be a more appropriate stability parameter than the static stability parameter $s = (g/T_v)(\partial\Theta_v/\partial z)$ introduced earlier. Ri is dimensionless and has the same sign as s. It will be shown in Chapter 8 that the Richardson number is a better measure of the intensity of mixing (turbulence) and provides a simple criterion for the existence or nonexistence of turbulence in a stably stratified environment (a large positive value of $Ri > 0.25$ is indicative of weak and decaying turbulence or a completely nonturbulent environment). Therefore, the vertical distribution of Ri may be used to determine the vertical extent of the boundary layer when other, more direct, measurements of the PBL height are not available.

6.5 Observed Wind Profiles

A considerable amount of wind profile data have been collected by meteorologists in the course of a number of large and small field experiments (Garratt and Hicks, 1990; Tunick, 1999), as well as from routine upper air soundings. Observations are made using pilot balloons, rawindsondes, tethersondes, doppler radars, acoustic sounders (sodars), instrumented aircraft, and tall towers. Consequently, scientists have developed a good understanding of the effects of surface roughness and stability on wind distribution in the PBL and a fair understanding of the influence of baroclinity on the same. More complicated effects of entrainment, advection, complex topography, clouds, and precipitation have received little attention from boundary layer meteorologists and remain poorly understood. The underlying philosophy in micrometeorology has been to study simple situations first in order to have a better understanding of the basic phenomena and then include progressively more complicating factors. Unfortunately, difficulties of measuring and computing turbulence in the atmospheric boundary layer have restricted micrometeorologists for a long time to the study of the more or less idealized (horizontally homogeneous, stationary, nonentraining, dry, etc.) PBL. Only recently have efforts been made to study the effects of complicating factors which influence the real-world PBL.

First, we present a few selected wind profiles for the 'ideal' PBL, taken under different stability conditions, in order to illustrate some qualitative effects of stability. Figure 6.5 shows the pibal wind and potential temperature profiles under fairly convective conditions during day 33 of the Wangara Experiment in southern Australia. Here, U and V are the horizontal wind components in the x and y directions, respectively, with the x axis parallel to the direction of near-surface winds and the y axis normal to the same (this choice of coordinate axes is more commonly used in micrometeorology than the geographical coordinate system, which is used in other branches of meteorology).

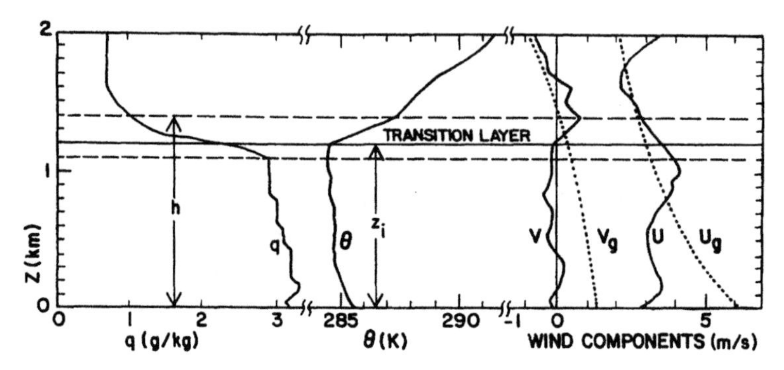

Figure 6.5 Measured wind, potential temperature and specific humidity profiles in the PBL under convective conditions on day 33 of the Wangara Experiment. [From Deardorff (1978).]

Note that despite the large geostrophic shear present on this day, wind profiles show the characteristic features of nearly uniform distributions in the convective mixed layer (the undulations in the PBL are probably caused by large eddies and organized updraft and downdraft motions whose effects are not smoothed out in pibal soundings), and strong gradients in the surface layer below and the transition layer above. Similar features have been observed in convective boundary layers over other homogeneous land and ocean surfaces (Figure 6.6a).

The observed mean wind and virtual potential temperature profiles under moderately unstable conditions in a trade wind marine PBL are shown in Figure 6.6b. At the time of these observations widely scattered shallow convective clouds were present whose bases coincided with the inversion base. Note that although the Θ_v profile is uniform throughout the subcloud layer, the wind profiles show significant shears in this layer. Still, the total wind veering across the unstable PBL remains small (about 8°), due to the combined effects of mechanical and buoyant mixing, smooth ocean surface, and low latitude.

The theoretician's ideal of a steady state, neutral, barotropic PBL is so rare in the atmosphere that relevant observations of the same do not exist. The wind profiles taken under slightly unstable and slightly stable conditions during the morning and evening transition hours differ considerably, due to the effects of nonstationarity and even slight stability or instability. An average of profiles taken during completely overcast and very windy conditions over sea might be more representative of a neutral PBL (Nicholls, 1985).

From observations of the nocturnal stable boundary layer (SBL), one can perhaps distinguish between the two broad stability regimes: (1) the moderately stable regime in which turbulent exchanges are more or less continuous in time and space through at least the lower half of the SBL; and (2) the very stable

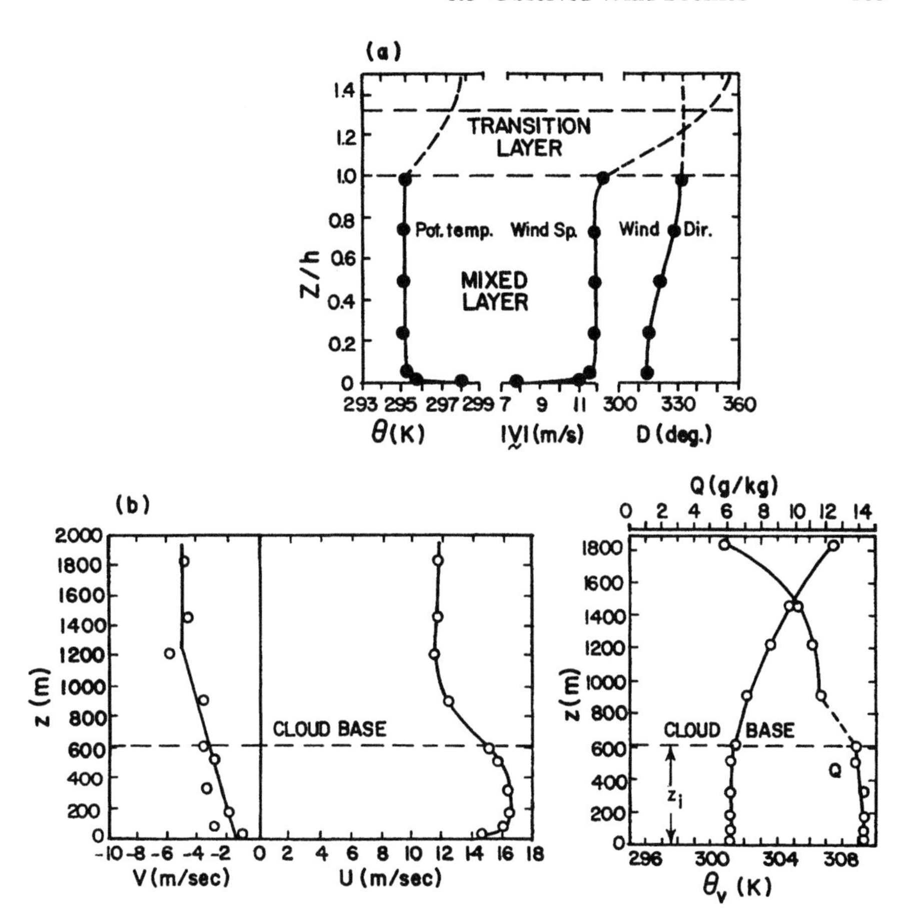

Figure 6.6 Observed vertical profiles of mean wind components, or wind speed and direction, potential temperature and specific humidity in the PBL. (a) Convective conditions overland. [After Kaimal *et al.* (1976).] (b) Unstable conditions over the ocean. [After Pennell and LeMone (1974).]

regime in which turbulent exchanges occur only intermittently in time and space through most of the SBL. The former can exist over land at night only during strong winds, and more likely at sea when the air is slightly warmer than the sea surface. The typical profiles of wind, potential temperature, and Richardson number in this moderately stable boundary layer are shown in Figure 6.7. These are based on the measurements from a tall tower near Dallas, Texas. Note that the U profile shows a pronounced maximum. The nose of the low-level jet often coincides with the top of the SBL, since it also represents the level of maximum

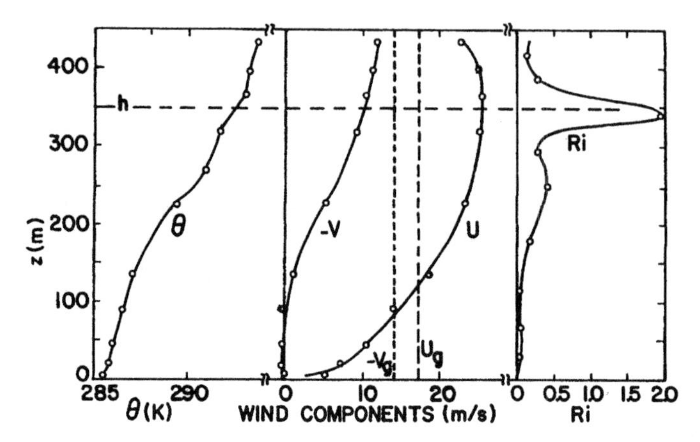

Figure 6.7 Observed vertical profiles of mean wind components and potential temperature and the calculated Ri profile in the nocturnal PBL under moderately stable conditions. [From Deardorff (1978); after Izumi and Barad (1963).]

in Ri. The Ri profile indicates that continuous turbulence may be expected only in the lower half of the SBL, where Ri < 0.25.

The sporadic turbulence regime is more common in the SBL over land. The typical observed profiles of U, V, Θ , and Ri in the very stable regime are shown in Figure 6.8. Note that both the wind components attain maximum values at low levels. The PBL depth based on the maximum in the wind speed profile is only about half of the depth of the surface inversion layer. The Ri profile indicates that the layer of possible continuous turbulence in which Ri < 0.25 is very shallow indeed and in the bulk of the SBL turbulence is likely to be intermittent, patchy, and weak.

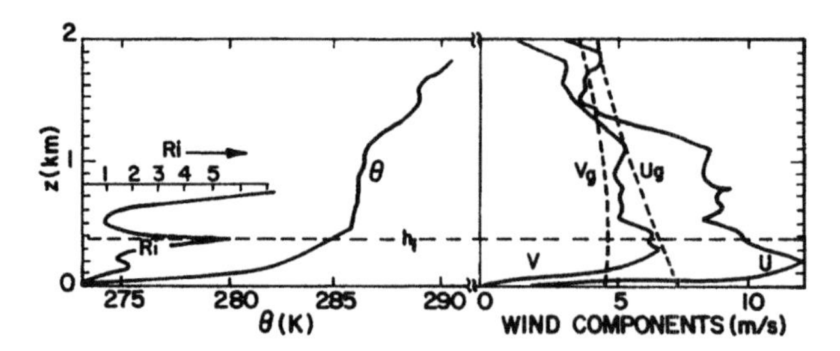

Figure 6.8 Observed wind and potential temperature profiles under very stable (sporadic turbulence) conditions at night during the Wangara Experiment. [From Deardorff (1978).]

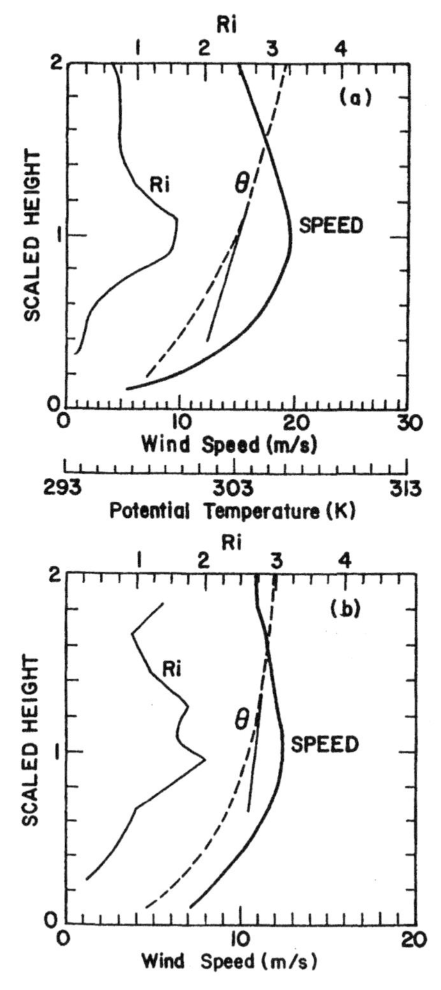

Figure 6.9 Composite profiles of wind speed, potential temperature and Richardson number scaled with respect to the low-level jet height at (a) O'Neill, Nebraska, during the Great Plains Experiment and (b) Hay, Australia, during the Wangara Experiment. [After Mahrt *et al.* Copyright © (1979) by D. Reidel Publishing Company. Reprinted by permission.]

The occurrence of a low-level jet is a common phenomenon in the SBL (Figure 6.9). The jet can become highly intensified when thermal or slope winds oppose the surface geostrophic wind. Such intensified low-level jets have frequently been observed over the Great Plains region of the United States. Figure 6.9 shows low-level jets in the composite wind speed profiles obtained from observations during the Wangara Experiment (Clark *et al.*, 1971) and the Great Plains Field Program (Lettau and Davidson, 1957).

Other characteristic features of the stably stratified PBL are large directional shear (wind veering) and small PBL depth, both of which are manifestations of the effects of stability. The veering of wind with height is better illustrated through a representation of wind components in the form of a wind hodograph, which is the locus of the end points of velocity vectors at various heights. Figure 6.10 presents such hodographs from a 32 m tower at Plateau Station, Antarctica. These are actually averages of many observed wind profiles grouped under eight stability classes (from the least stable class 1 to the most stable class 8) and three periods or seasons. The corresponding temperature profiles have already been given in Figure 5.5. Note that with increasing stability, wind veering with height also increases. For extremely stable classes changes in wind direction with height can be detected even at low levels of 1 or 2 m, so that the surface layer becomes very shallow under these conditions.

The close relationship between wind veering and stability or temperature lapse rate is further demonstrated by observations given in Figure 6.11. Note that wind veering increases with increasing stability (negative lapse rate); the negative values (backing of wind) under superadiabatic conditions are probably due to the influence of thermal winds.

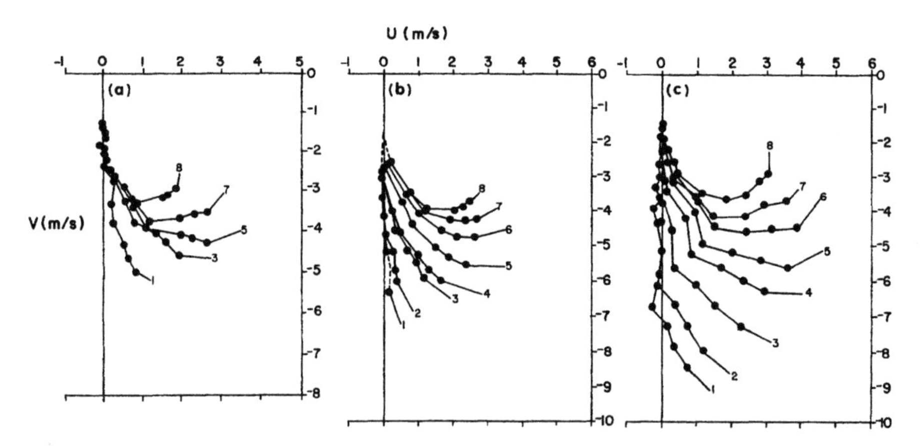

Figure 6.10 Averaged observed wind hodographs at Plateau Station, Antarctica, for (a) sunlight, (b) transitional, and (c) dark periods, grouped under different stability classes 1–8. Dots indicate ends of wind vectors at heights of 0.5, 1, 2, 4, 8, 12, 16, 20, 24, and 32 m. The components shown are in the so-called geotriptic coordinate system. [After Lettau *et al.* (1977).]

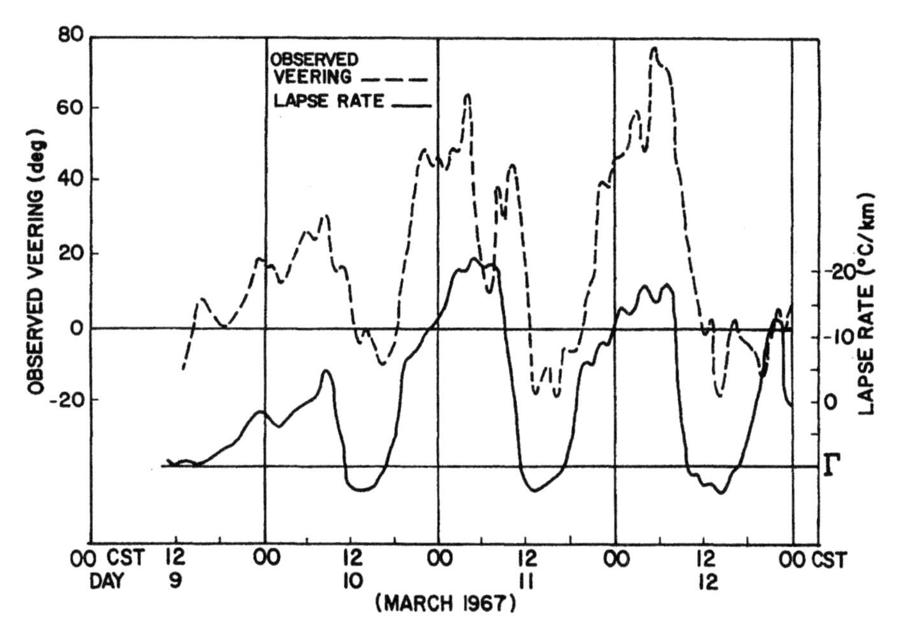

Figure 6.11 Observed relationship between wind veering (– – –) and lapse rate (——) near Oklahoma City. [After Gray and Mendenhall (1973).]

6.6 Diurnal Variations

Diurnal variations of wind speed and wind direction in the PBL have been inferred from observations collected at different locations. The evaluation of wind profiles in the course of a diurnal cycle may show large day-to-day variations due to changes in the synoptic weather situation and the surface energy balance. When averaged over long periods of time (order of a month or longer), however, the diurnal variations are better discerned.

Figure 6.12 shows the diurnal variations of mean wind speed (averaged over a period of 1 year) observed from an instrumented 500 m tower near Oklahoma City, Oklahoma. The range of times of sunrise and sunset, as well as the observation heights, are indicated in the figure. Note that the near-surface wind speed increases sharply after sunrise, attains a broad maximum in early afternoon, and decreases sharply near sunset. The diurnal wave at the higher level in the surface layer is similar, but with a reduced amplitude. The increase in the strength of surface winds following the morning inversion breakup is due to more rapid and efficient transfer of momentum from aloft through the evolving unstable or convective PBL in the daytime.

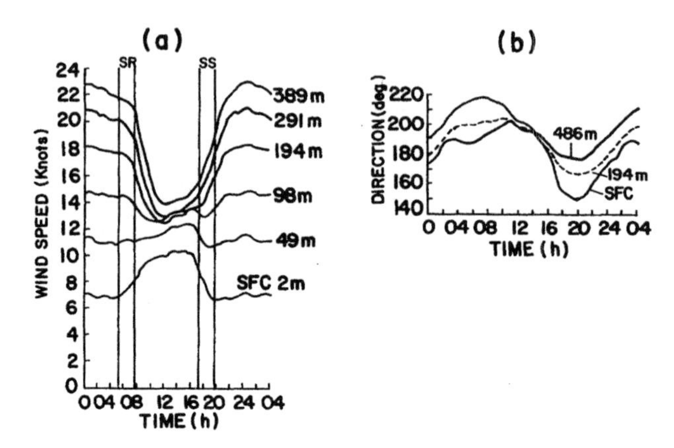

Figure 6.12 Diurnal variations of (a) mean wind speed and (b) mean wind direction, on an annual basis, for various height levels in the PBL near Oklahoma City, Oklahoma. [After Crawford and Hudson (1973).]

Above the surface layer, the diurnal wave becomes nearly 180° out of phase and wind speed decreases sharply following sunrise, attains a minimum value around noon, and increases in the afternoon and evening hours. The amplitude of the wave increases with height and attains its maximum value somewhere in the middle of the convective PBL. Although these observations did not extend to higher levels, the amplitude of diurnal variation of wind speed is expected to decrease farther up and vanish at the maximum height of the daytime PBL. Similar diurnal patterns of wind speed at various heights in the PBL are shown in Figure 6.13, which is based on 40-day averages of pibal observed winds during the Wangara expedition at a smooth rural site in Australia.

It can be inferred from the diurnal evolution of potential temperature and specific humidity profiles shown in Chapter 5 and, to some extent, from the wind profiles also, that PBL depth has a strong diurnal variation, especially during the daytime under clear skies. The unstable or convective PBL is usually capped by an inversion and the height of the inversion base z_i is considered to be a measure of the PBL depth h (actually, h is 5–30% larger than z_i, if one includes the interfacial transition layer in the former). The diurnal evolution of the height of the lowest inversion base on a typical day during the 1973 Minnesota experiment is shown in Figure 6.14. Also represented in the same figure is the diurnal variation of the surface heat flux H_0 whose cumulative input (integration with respect to time) to the PBL is primarily responsible for the growth of z_i or h.

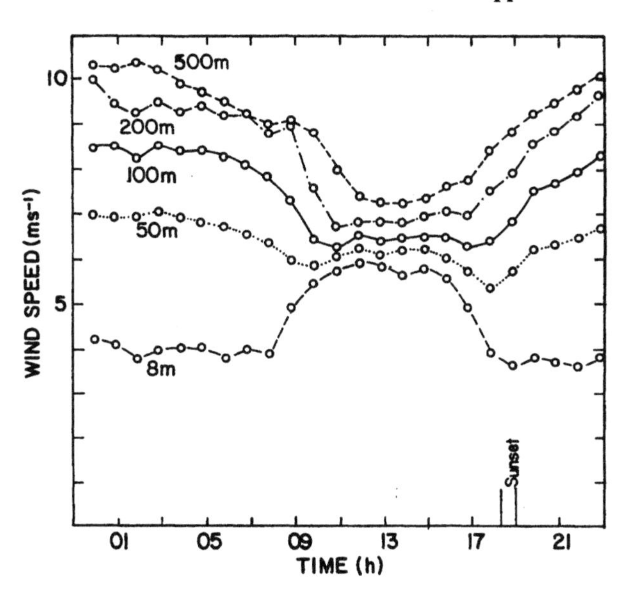

Figure 6.13 Diurnal variations of 40-day-averaged wind speeds at various heights in the PBL during the Wangara Experiment. [After Mahrt (1981).]

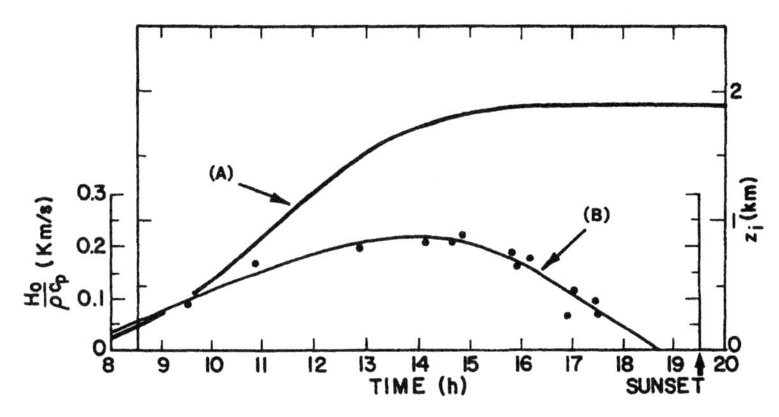

Figure 6.14 Typical variations of (A) the height of inversion base and (B) sensible heat flux during the daytime period in the Minnesota Experiment. [After Kaimal *et al.* (1976).]

6.7 Applications

The knowledge of wind distribution in the PBL has the following practical applications:

- Determining the rate of dissipation of the kinetic energy in the lower atmosphere.
- Determining local and regional transports of pollutants and air trajectories in the lower atmosphere.
- Estimating momentum flux or shear stress through the PBL and the surface drag.
- Estimating wind energy potential and designing wind power-generating systems.
- Designing tall buildings, towers, and bridges for wind loads.
- Designing wind shelters and other protective measures.

Problems and Exercises

1. Derive the thermal wind equations from the geostrophic wind relations, the hydrostatic equation, and the equation of state for the lower atmosphere, and indicate when and why the terms involving the vertical temperature gradient may be neglected.

2. Show that the net horizontal friction force per unit volume of a fluid element in the PBL is $\partial \boldsymbol{\tau}/\partial z$ and that it need not be opposite to the velocity vector **V**. How is the friction force related to the ageostrophic wind vector?

3. The following measurements of mean winds and temperatures were made from the 200 m mast at Cabauw in The Netherlands on a September night:

Height (m)	Wind speed ($m\ s^{-1}$)	Wind direction (deg)	Temperature (°C)
5	2.2	194	8.58
10	2.7	194	8.81
20	3.5	202	9.08
40	5.4	217	10.15
80	8.2	234	12.19
120	9.1	240	12.96
160	8.7	246	12.86
200	7.8	250	12.63

Geostrophic wind speed at the surface $= 5.94\ m\ s^{-1}$
Geostrophic wind direction at the surface $= 263°$
Horizontal temperature gradient toward east $= -2.69 \times 10^{-5}\ K\ m^{-1}$
Horizontal temperature gradient toward north $= 8.85 \times 10^{-6}\ K\ m^{-1}$

Boundary layer depth = 100 m
Surface pressure = 1000 mbar

(a) Plot on a graph the wind speed, wind direction, and potential temperature profiles.
(b) Plot the wind hodograph, using the geographical coordinate system, and indicate geostrophic wind vectors at the surface and the top of the PBL. Compare the actual wind with the geostrophic wind at 100 m.
(c) Calculate and plot the wind component profiles, taking the x axis along the near-surface (5 m) wind.

4.
(a) Using the wind component and potential temperature profiles of the previous problem, determine the magnitudes of wind shear, potential temperature gradient, and Richardson number at the heights of 7.5, 15, 30, 60, 100, 140, and 180 m.
(b) Plot Ri as a function of height and indicate the probable top of the PBL based on the maximum in the Ri profile.

5. The following observations are for a midlatitude ($f = 10^{-4}\ s^{-1}$) PBL during the afternoon convective conditions:

Height[a] (m)	Wind speed[b] ($m\ s^{-1}$)	Wind direction[c] (deg)
1	7.7	311
32	11.0	315
61	11.4	315
610	11.7	321
1220	12.1	331

[a] PBL height = 1220 m.
[b] Surface geostrophic wind speed = 10.0 $m\ s^{-1}$.
[c] Surface geostrophic wind direction = 295°.

(a) Calculate the average magnitudes of geostrophic wind shear and the actual wind shear in the PBL, assuming geostrophic balance at the top of the PBL.
(b) Compute the total wind veering and frictional veering across the PBL, as well as the cross-isobar angle of the near-surface winds.
(c) Compute the pressure gradient, Coriolis, and friction forces per unit volume ($\rho = 1.2\ kg\ m^{-3}$) at the 610 m level and vectorially represent (to scale) the force balance at this level.

6. Calculate the magnitudes and directions of pressure gradient, Coriolis, and friction forces per unit mass of a fluid element at the top of a nocturnal surface layer where the actual winds are 7.07 m s^{-1} from the west, the geostrophic wind speed is 10 m s^{-1}, and the cross-isobar angle is 45° (take $f = 10^{-4}$ s^{-1}).

Chapter 7 An Introduction to Viscous Flows

7.1 Inviscid and Viscous Flows

In the preceding two chapters, our treatment of temperature, humidity, and wind distributions in the PBL was largely empirical and based on rather limited observations. A better understanding of vertical profiles of wind, temperature, and humidity and their relationships to the important exchanges of momentum, heat, and moisture through the PBL has been gained through applications of mathematical, statistical, and semiempirical theories. A brief description of some of the fundamentals of viscous flows and turbulence and the associated theory will be given in this and the following chapters.

For theoretical treatments, fluid flows are commonly divided into two broad categories, namely, inviscid and viscous flows. In an inviscid or ideal fluid the effects of viscosity are completely ignored, i.e., the fluid is assumed to have no viscosity, and the flow is considered to be nonturbulent. Inviscid flows are smooth and orderly, and the adjacent fluid layers can easily slip past each other or against solid surfaces without any friction or drag. Consequently, there is no mixing and no transfer of momentum, heat, and mass across the moving layers. Such properties can only be transported along the streamlines through advection. The inviscid flow theory obviously results in some serious dilemmas and inconsistencies when applied anywhere close to solid surfaces or density interfaces. Such dilemmas can be resolved only by recognizing the presence of boundary layers or interfacial mixing layers in which the effects of viscosity or turbulence cannot be ignored. Far away from the boundaries and density interfaces, however, the fluid viscosity can be ignored and the inviscid or ideal flow model provides a very good approximation to many real fluid flows encountered in geophysical and engineering applications. Extensive applications of this are given in books on hydrodynamics and geophysical fluid dynamics (Lamb, 1932; Pedlosky, 1979).

7.1.1 Viscosity and its effects

Fluid viscosity is a molecular property which is a measure of the internal resistance of the fluid to deformation. All real fluids, whether liquids or gases, have finite viscosities associated with them. An important manifestation of the effect of viscosity is that fluid particles adhere to a solid surface as they come in contact with the latter and consequently there is no relative motion between the fluid and the solid surface. If the surface is at rest, the fluid motion right at the surface must also vanish. This is called the no-slip boundary condition, which is also applicable at the interface of two fluids with widely different densities (e.g., air and water).

Within the fluid flow, viscosity is responsible for the frictional resistance between adjacent fluid layers. The resistance force per unit area is called the shearing stress, because it is associated with the shearing motion (variation of velocity) between the layers. A simple demonstration of this is provided by the smooth, streamlined, laminar flow between two large parallel planes, one fixed and the other moving at a slow constant speed, U_h, which are separated by a small distance h (see Figure 7.1a). At sufficiently small values of U_h and h, laminar flow can be maintained and the velocity within the fluid varies linearly from zero at the fixed plane to U_h at the moving plane, so that the velocity gradient $\partial u/\partial z = U_h/h$, everywhere in the flow. From observations in such a flow, Newton found the relationship that shearing stress is proportional to the rate of strain or velocity gradient (fluids following a linear relationship between the stress and the rate of strain are known as Newtonian fluids), that is,

$$\tau = \mu(\partial u/\partial z) \tag{7.1}$$

Where the coefficient of proportionality μ is called the dynamic viscosity of the fluid. In fluid flow problems, it is more convenient to use the kinematic viscosity $\nu \equiv \mu/\rho$, which has dimensions of $L^2\,T^{-1}$.

In particular, for the gaseous fluids, such as air, a more rigorous theoretical derivation of the relationship between the shearing stress and velocity gradient can be obtained from the kinetic theory (Sutton, 1953), which clearly shows viscosity to be a molecular property which depends on the temperature and pressure in the fluid (this dependence is, however, weak and is usually ignored in most atmospheric applications).

Equation (7.1) is strictly valid only for a unidirectional flow. In most real flows, in general, spatial variations of velocity in different directions give rise to shear stresses in different directions. More general constitutive relations between the components of shearing stress and velocity gradient at any point in the fluid are (Kundu, 1990, Chapter 4)

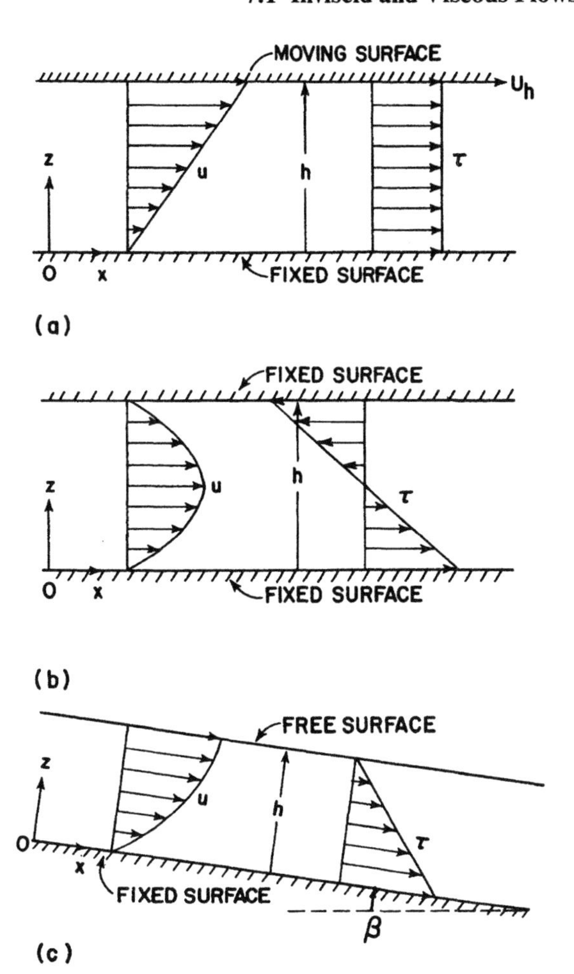

Figure 7.1 Schematics of velocity and shear stress profiles in laminar plane-parallel flows: (a) Couette flow; (b) channel flow; (c) free-surface gravity flow.

$$
\begin{aligned}
\tau_{xy} &= \tau_{yx} = \mu(\partial u/\partial y + \partial v/\partial x) \\
\tau_{xz} &= \tau_{zx} = \mu(\partial u/\partial z + \partial w/\partial x) \\
\tau_{yz} &= \tau_{zy} = \mu(\partial v/\partial z + \partial w/\partial y)
\end{aligned}
\tag{7.2}
$$

in which the first member of the subscript to τ denotes the direction normal to the plane of the shearing stress and the second member denotes the direction of the stress. Equation (7.2) implies that in Newtonian fluids, shear stresses are proportional to the applied strain rates or deformations (represented by quantities in parentheses). Here, stresses and deformations are all instantaneous quantities at a point in the flow.

Another important effect of viscosity is the dissipation of kinetic energy of the fluid motion, which is constantly converted into heat. Therefore, in order to maintain the motion, the energy has to be continuously supplied externally, or converted from potential energy, which exists in the form of pressure and density gradients in the fluid.

Although the above-mentioned effects of viscosity are felt, in varying degrees, in all real fluid flows, at all times, they are found to be particularly significant only in certain regions and in certain types of flows. Such flows are called viscous flows, as opposed to the inviscid flows mentioned earlier. Examples of such flows are boundary layers, mixing layers, jets, plumes, and wakes, which are dealt with in many books on fluid mechanics (Batchelor, 1970; Townsend, 1976; Kundu, 1990).

7.2 Laminar and Turbulent Flows

All viscous flows can broadly be classified as laminar and turbulent flows, although an intermediate category of transition between the two has also been recognized. A laminar flow is characterized by smooth, orderly, and slow motion in which adjacent layers (laminae) of fluid slide past each other with very little mixing and transfer (only at the molecular scale) of properties across the layers. The main difference between the laminar flows and the inviscid flows introduced earlier is the importance of viscous and other molecular transfers of momentum, heat, and mass in the former. In laminar flows, the flow field and the associated temperature and concentration fields are regular and predictable and vary only gradually in space and time.

In sharp contrast to laminar and inviscid flows, turbulent flows are highly irregular, almost random, three-dimensional, highly rotational, dissipative, and very diffusive (mixing) motions. In these, all the flow and scalar properties exhibit highly irregular variations (fluctuations) in both time and space, with a wide range of temporal and spatial scales. For example, turbulent fluctuations of velocity in the atmospheric boundary layer typically occur over time scales ranging from 10^{-3} to 10^4 s and the corresponding spatial scales from 10^{-3} to 10^4 m – more than a millionfold range of scales. Due to their nearly random nature (there is some order, persistence, and correlation between fluctuations in time and space), turbulent motions cannot be predicted or calculated exactly as functions of time and space; one usually deals with their average statistical properties.

Although there are many examples and applications of laminar flows in industry, in the laboratory, and in biological systems, their occurrence in natural environments, particularly the atmosphere, is rare and confined to the so-called viscous sublayers over smooth surfaces (e.g., ice, mud flats, relatively undisturbed water, and tree leaves). Most fluid flows encountered in nature and engineering applications are turbulent. In particular, the various small-scale motions in the lower atmosphere are turbulent. The turbulent mixing layer may vary in depth from a few tens of meters in clear, calm, nocturnal cooling conditions to several kilometers in highly disturbed (stormy) weather conditions involving deep penetrative convection. In the upper troposphere and stratosphere also, extensive regions of clear air turbulence are found to occur. Some meteorologists consider all atmospheric motions right up to the scale of general circulation as turbulent. But, then, one has to distinguish between the small-scale three-dimensional turbulence of the type we encounter in micrometeorology, and the large-scale 'two-dimensional turbulence.' There are fundamental differences in some of their properties and mechanisms of energy transfers up or down the scales. We will only be concerned with the former here.

Turbulent, too, are the flows in upper oceans (e.g., oceanic mixed layer), lakes, rivers, and channels; practically all flows in liquid and gas pipelines; in boundary layers over moving aircraft, missiles, ships, and boats; in wakes of structures, propeller blades, turbine blades, projectiles, bullets, and rockets; in jets of exhaust gases and liquids; and in chimney and smokestack plumes. Thus turbulence is literally all around us, particularly in the form of the atmospheric boundary layer, from which we can escape only briefly. The next and the following chapters will be devoted to fundamentals of turbulence and their application to micrometeorology. But first we discuss the basic equations for viscous motion, which are valid for both laminar and turbulent flows.

7.3 Navier–Stokes Equations of Motion

Mathematical treatments of fluid flows, including those in the atmosphere, are almost always based on the equations of motion, which are mathematical expressions of the fundamental laws of the conservation of mass, momentum, and energy. For example, considerations of mass conservation in an elemental volume of fluid leads to the equation of continuity, which for an incompressible fluid, in a Cartesian coordinate system, is given by

$$\partial u/\partial x + \partial v/\partial y + \partial w/\partial z = 0 \tag{7.3}$$

The continuity equation imposes an important constraint on the fluid motion such that the divergence of velocity must be zero at all times and at every point in the flow. The assumption of incompressible fluid or flow is quite well justified in micrometeorological applications.

The application of Newton's second law of motion, or consideration of momentum conservation in an elemental volume of fluid, leads to the so-called Navier–Stokes equations, which, in a Cartesian frame of reference tied to the surface of the rotating earth with x and y axes in the horizontal and z axis in the vertical, are

$$\begin{aligned} \frac{Du}{Dt} - fv &= -\frac{1}{\rho}\frac{\partial p}{\partial x} + \nu\nabla^2 u \\ \frac{Dv}{Dt} + fu &= -\frac{1}{\rho}\frac{\partial p}{\partial y} + \nu\nabla^2 v \\ \frac{Dw}{Dt} + g &= -\frac{1}{\rho}\frac{\partial p}{\partial z} + \nu\nabla^2 w \end{aligned} \tag{7.4}$$

Where the total derivative and Laplacian operator are defined as

$$D/Dt \equiv \partial/\partial t + u(\partial/\partial x) + v(\partial/\partial y) + w(\partial/\partial z) \tag{7.5}$$

$$\nabla^2 \equiv \partial^2/\partial x^2 + \partial^2/\partial y^2 + \partial^2/\partial z^2 \tag{7.6}$$

The derivation of the above equations of motion is outside the scope of this text and can be found elsewhere (Haltiner and Martin, 1957; Dutton, 1976; Kundu, 1990). The physical interpretation and significance of each term can be given as follows. Terms in Equation (7.4) represent accelerations or forces per unit mass of the fluid element in the x, y, and z direction, respectively. On the left-hand sides, the first terms are called the inertia terms, because they represent the inertia forces arising due to local and advective accelerations on a fluid element. The second terms in the equations of horizontal motion represent the Coriolis accelerations or forces which apparently act on the fluid element due to the earth's rotation. The rotational term is insignificant in the equation of vertical motion and has been omitted. Instead, the gravitational acceleration term appears in the equation for vertical motion. On the right-hand sides of Equation (7.4), the first terms represent the pressure gradient forces and the second terms are viscous or friction forces on the fluid element.

When fluid viscosity or friction terms can be ignored, the equations of motion reduce to the so-called Euler's equations for an inviscid or ideal fluid. The solutions of these equations for a variety of density or temperature stratifications and initial and boundary conditions now form a rich body of literature in classical hydrodynamics (Lamb, 1932) and geophysical fluid dynamics (Pedlosky, 1979). Euler's equations of motion are considered to be adequate for describing the behavior of atmosphere and oceans outside of any boundary, interfacial, and mixing layers in which the flows are likely to be turbulent.

For a mathematical description of viscous flows, one has to use the complete set of Navier–Stokes equations, which are nonlinear partial differential equations of second order and are extremely difficult (often impossible) to solve. The combination of nonlinear inertia terms and viscous terms is responsible for the extreme difficulty and, in many cases, the intractability of solutions. Analytical solutions have been found only for limited cases of very low-speed laminar or creeping flows when nonlinear terms are absent or can be simplified. Some of these 'exact' solutions are given in the following sections in order to introduce the reader to certain classical fluid flows.

7.4 Laminar Plane-Parallel Flows

The simplest viscous flows are the steady, one-dimensional, laminar flows between two infinite parallel planes. Since velocity varies only normal to the planes and there is no motion across the planes, the inertia terms in the Navier–Stokes equations are zero. For small-scale laboratory flows, the earth's rotational effects (Coriolis terms) can also be ignored. Furthermore, choosing the x axis in the direction of flow and the z axis normal to the plane boundaries, the equations of motion reduce to

$$d^2u/dz^2 = (1/\mu)(\partial p'/\partial x) \tag{7.7}$$

which can easily be solved for different flow situations (boundary conditions). Here, p' is the modified or excess pressure obtained after subtracting the hydrostatic part from the actual pressure. For a fluid of uniform density, introduction of modified pressure eliminates the influence of gravity, so that the gravity term can be dropped from the equations of motion. The procedure of subtracting the hydrostatic part from the actual pressure is also widely used for stratified geophysical flows. For these, the pressure term in the equation of vertical motion contains the deviation of the pressure from that of the reference medium at rest which is assumed to be hydrostatic. The above procedure should not be used for liquid flows with a free surface, which are essentially forced by gravity (e.g., water flow down a sloping surface).

7.4.1 Plane-Couette flow

Consider the laminar flow between two parallel boundaries, one fixed and the other moving in the x direction at a constant velocity of U_h, separated by a small distance h, as depicted in Figure 7.1a. The relevant no-slip boundary conditions are

$$\begin{aligned} u &= 0, \quad \text{at } z = 0 \\ u &= U_h, \quad \text{at } z = h \end{aligned} \tag{7.8}$$

The solution of Equation (7.7) satisfying the above boundary conditions is given by

$$u = U_h \frac{z}{h} - \frac{1}{2\mu} \frac{\partial p'}{\partial x} z(h - z) \tag{7.9a}$$

or, in the dimensionless form,

$$\frac{u}{U_h} = \frac{z}{h} - \frac{h^2}{2\mu U_h} \frac{\partial p'}{\partial x} \frac{z}{h} \left(1 - \frac{z}{h}\right) \tag{7.9b}$$

Note that the velocity profile in this so-called plane-Couette flow is a combination of linear and parabolic profiles. For the special case of zero pressure gradient ($\partial p'/\partial x = 0$), the velocity profile becomes linear, i.e.,

$$u/U_h = z/h \tag{7.10}$$

and the shear stress is equal to the surface shear stress $\tau_0 = \mu U_h/h$ everywhere in the flow. This flow is entirely forced by the moving surface, or the relative motion between the two parallel surfaces, and is depicted in Figure 7.1a.

7.4.2 Plane-Poiseuille or channel flow

When both of the parallel bounding surfaces are fixed, the appropriate boundary conditions for the unidirectional flow between them are

$$\begin{aligned} u &= 0, \quad \text{at } z = 0 \\ u &= 0, \quad \text{at } z = h \end{aligned} \tag{7.11}$$

Then, the solution of Equation (7.7) yields a parabolic velocity profile

$$u = -\frac{1}{2\mu}\frac{\partial p'}{\partial x}z(h - z) \tag{7.12}$$

which implies a linear shear stress profile

$$\tau = \mu\frac{\partial u}{\partial z} = -\frac{1}{2}\frac{\partial p'}{\partial x}(h - 2z) \tag{7.13}$$

with the maximum value at the surface $\tau_0 = -(h/2)(\partial p'/\partial x)$. Note that in order to have a steady flow through the channel, there must be a constant negative pressure gradient in the direction of the flow. The velocity and shear stress profiles in a plane channel flow are schematically shown in Figure 7.1b.

The solution of simplified equations of motion in a cylindrical coordinate system yields similar velocity and stress profiles in a circular pipe (Hagen–Poiseuille) flow.

7.4.3 Gravity flow down an inclined plane

Here we consider a unidirectional liquid flow with a free surface or other gravity flows over uniformly sloping surfaces. The relevant equation of motion with the x axis in the direction of flow (down the slope β) and the z axis normal to the surface (Figure 7.1c) is

$$d^2u/dz^2 = -(g/\nu)\sin\beta \tag{7.14}$$

in which we have included the gravity force term but have ignored the pressure gradient. The boundary conditions are

$$\begin{aligned} u = 0, &\quad \text{at } z = 0 \\ du/dz = 0, &\quad \text{at } z = h \end{aligned} \tag{7.15}$$

The upper boundary condition implies zero shear stress at the free surface, or at $z = h$, where h represents the constant thickness of the frictional shear layer.

The solution of Equation (7.14) satisfying the above boundary conditions is given by

$$u = \frac{g\sin\beta}{2\nu}z(2h - z) \tag{7.16}$$

which is again a parabolic profile with the maximum velocity at $z = h$

$$u_h = \frac{gh^2}{2\nu} \sin\beta \tag{7.17}$$

The shear stress profile is linear with the maximum value at the surface

$$\tau_0 = \rho gh \sin\beta \tag{7.18}$$

These profiles are schematically shown in Figure 7.1c. Note that both the maximum velocity and the surface shear stress are proportional to the sine of the slope angle.

Example Problem 1

Consider fully-developed laminar flow through a circular tube of diameter d. Using the simplified forms of the equations of motion in cylindrical coordinates (r, θ, x), with the x axis coinciding with the axis of the tube and r representing the radial distance from the center of the tube, obtain expressions for the velocity and shear stress distributions across the tube and compare them with those for the plane channel flow.

Solution

Due to symmetry, all the terms in the equations of motion in r and θ directions are identically zero, because there is no motion along those directions. This implies that $\partial p/\partial r = 0$ and $\partial p/\partial \theta = 0$. The x momentum equation reduces to

$$0 = -\frac{\partial p'}{\partial x} + \frac{\mu}{r}\frac{d}{dr}\left(r\frac{du}{dr}\right)$$

or,

$$\frac{1}{r}\frac{d}{dr}\left(r\frac{du}{dr}\right) = \frac{1}{\mu}\frac{\partial p'}{\partial x}$$

Since the left-hand side of the above equation can be a function only of r, and the right-hand side can be a function only of x, it follows that both terms must be constant. Therefore, the pressure gradient along the direction of flow must be constant and pressure falls linearly along the length of the tube.

Multiplying the above equation by r and integrating with respect to r, we get

$$r\frac{du}{dr} = \frac{r^2}{2\mu}\frac{\partial p'}{\partial x} + A$$

or,

$$\frac{du}{dr} = \frac{r}{2\mu}\frac{\partial p'}{\partial x} + \frac{A}{r}$$

in which the integration constant $A = 0$, because $du/dr = 0$ at $r = 0$ (velocity is maximum at the center of the tube). Integrating the above equation again gives

$$u = \frac{r^2}{4\mu}\frac{\partial p'}{\partial x} + B$$

where B is another integration constant, which can be evaluated from the no-slip boundary condition at the tube surface ($r = d/2$, $u = 0$) as $B = -(d^2/16\mu)\partial p'/\partial x$. Thus, the velocity distribution in the tube is given by

$$u = -\frac{1}{16\mu}\frac{\partial p'}{\partial x}(d^2 - 4r^2)$$

The corresponding shear stress distribution is given by

$$\tau = \mu\frac{\partial u}{\partial r} = \frac{r}{2}\frac{\partial p'}{\partial x}$$

with the maximum value at the tube surface $\tau_0 = 0.25d\,\partial p'/\partial x$.

Note that the velocity profile across the tube cross-section is parabolic and the shear stress profile is linear, similar to those for the plane channel flow given by Equations (7.12) and (7.13). The volumetric flow rate is given by

$$Q = \int_0^{d/2} 2\pi r u dr = -\frac{\pi}{8\mu}\frac{\partial p'}{\partial x}\int_0^{d/2} r(d^2 - 4r^2)dr$$

or,

$$Q = -\frac{d^4}{128\mu}\frac{\partial p'}{\partial x}$$

The average velocity through the tube can be obtained by dividing the above volume flow rate by the cross-sectional area of the tube, i.e.,

$$V = -\frac{d^2}{32\mu}\frac{\partial p'}{\partial x}$$

7.5 Laminar Ekman Layers

In this section, we consider the laminar Ekman boundary layers on rotating surfaces, particularly those of the atmosphere and oceans, aside from the conditions of their existence.

7.5.1 *Ekman layer below the sea surface*

First, we examine the problem of drift currents set up at or just below the sea surface, in response to a steady wind stress forcing. For the sake of simplicity, surface waves are being ignored and so are the horizontal pressure and density gradients in water. Then, the equations of horizontal motion [Equation (7.4)] reduce to

$$-fv = \nu \frac{d^2 u}{dz^2}; \quad fu = \nu \frac{d^2 v}{dz^2} \tag{7.19}$$

Assuming the x axis is in the direction of the applied surface stress τ_0 and the z axis is pointed vertically upward, the boundary conditions to be satisfied are

$$\begin{aligned} \mu \frac{du}{dz} = \tau_0, \quad \mu \frac{\partial v}{\partial z} = 0, \quad & \text{at } z = 0 \\ u \to 0, \quad v \to 0, \quad & \text{as } z \to -\infty \end{aligned} \tag{7.20}$$

The two equations can be combined in a single equation for the complex current $c = u + iv$, by multiplying the second part of Equation (7.19) by $i = \sqrt{-1}$ and adding it to the first. The resulting equation

$$d^2 c/dz^2 - i(f/\nu)c = 0 \tag{7.21}$$

has the solution satisfying the boundary condition at infinity

$$c = u + iv = A \exp[a(1+i)z] \tag{7.22}$$

where $a = (f/2\nu)^{1/2}$ and A is a complex constant determined from the surface boundary condition to be

$$A = (\tau_0/2a\mu)(1-i) \tag{7.23}$$

Substituting from Equation (7.23) into Equation (7.22) and separating into real and imaginary parts, one obtains

$$u = (\tau_0/\sqrt{2}a\mu)e^{az}\cos(az - \pi/4)$$
$$v = (\tau_0/\sqrt{2}a\mu)e^{az}\sin(az - \pi/4) \tag{7.24}$$

Note that according to the above solution, the current has a maximum speed of $\tau_0/\sqrt{2}a\mu$ at the surface and is directed 45° in a clockwise sense from the direction of applied stress. With increasing depth (negative z) the current speed decreases exponentially and the current direction rotates in a clockwise sense. A common method of representing the current speed and direction as a function of depth below the surface is to plot a velocity hodograph as in Figure 7.2a. Note that the velocity hodograph represented by Equation (7.24) is a spiral; it is known as the Ekman spiral, after the Swedish oceanographer V.W. Ekman who first derived the above solution.

Although the induced current theoretically disappears only at an infinite depth, in practice, the influence of surface stress forcing becomes insignificant at a depth of the order of $a^{-1} = (2\nu/f)^{1/2}$. Conventionally, the Ekman layer depth h_E is defined as the depth where the current direction becomes exactly opposite to the surface current direction. According to Equation (7.24), this happens at $z = -\pi a^{-1}$, so that $h_E = \pi(2\nu/f)^{1/2}$; at this depth the magnitude of the current has fallen to a small fraction ($e^{-\pi} \cong 0.04$) of its surface value.

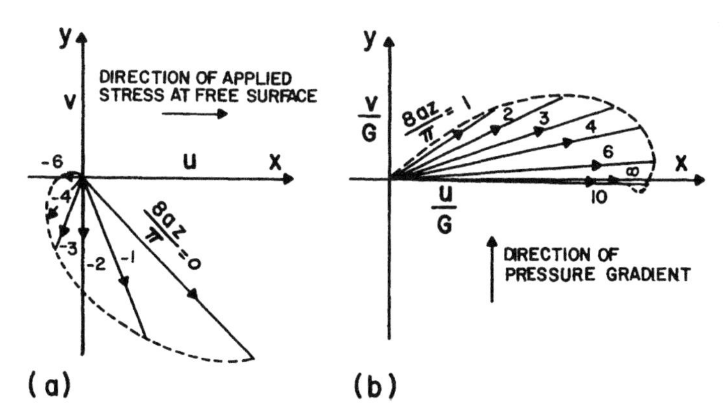

Figure 7.2 Velocity hodographs in laminar Ekman layers: (a) below free surface where tangential stress is applied; (b) above a rigid surface where a constant pressure gradient is applied. [From Batchelor (1970).]

Another parameter of interest is the net transport of water across vertical planes due to induced currents in the Ekman layer, which is given by the integral

$$\int_0^{-\infty} (u + iv)dz = -i(\tau_0/\rho f) \tag{7.25}$$

Thus, the net transport is to the right and normal to the direction of applied stress; it is proportional to the stress, but independent of fluid viscosity. In fact, the above result also follows directly from the integration of the equations of motion (7.19) after expressing the viscous terms in the form of the vertical gradients of stresses, viz., $d\tau_{zx}/dz$ and $d\tau_{zy}/dz$.

7.5.2 Ekman layer at a rigid surface

Another example of an Ekman layer is that due to a uniform pressure gradient in the atmosphere near the surface or in the ocean near the bottom boundary. For the sake of simplicity, again, the surface is assumed to be flat and uniform and the flow is considered laminar. In the absence of inertia terms, the equations of motion to be solved are

$$\begin{aligned} -fv &= -\frac{1}{\rho}\frac{\partial p}{\partial x} + \nu\frac{d^2u}{dz^2} \\ fu &= -\frac{1}{\rho}\frac{\partial p}{\partial y} + \nu\frac{d^2v}{dz^2} \end{aligned} \tag{7.26}$$

Expressing pressure gradients in terms of geostrophic velocity components using Equation (6.1), Equation (7.26) can be written as

$$\begin{aligned} -f(v - V_g) &= \nu(d^2/dz^2)(u - U_g) \\ f(u - U_g) &= \nu(d^2/dz^2)(v - V_g) \end{aligned} \tag{7.27}$$

in which U_g and V_g have been taken as height-independent parameters, ignoring any small variations of these over the small depth of the Ekman layer. The appropriate boundary conditions, assuming geostrophic balance outside the Ekman layer, are

$$\begin{aligned} u = 0 \quad v = 0, &\quad \text{at } z = 0 \\ u \to U_g, \quad v \to V_g, &\quad \text{as } z \to \infty \end{aligned} \tag{7.28}$$

The previous set of equations [Equation (7.27)] is similar to Equation (7.19), and the solution, following the earlier solution, is

$$\begin{aligned} u - U_g &= -e^{-az}[U_g \cos(az) + V_g \sin(az)] \\ v - V_g &= e^{-az}[U_g \sin(az) - V_g \cos(az)] \end{aligned} \tag{7.29}$$

This is a general solution which is independent of the orientation of the horizontal coordinate axes. A more specific and simpler solution is usually given by taking the x axis to be oriented with the geostrophic wind vector, so that $U_g = G$ and $V_g = 0$ in this geostrophic coordinate system and Equation (7.29) can be written as

$$\begin{aligned} u &= G[1 - e^{-az} \cos(az)] \\ v &= Ge^{-az} \sin(az) \end{aligned} \tag{7.30}$$

The normalized wind hodograph (Ekman spiral), according to Equation (7.30), is shown in Figure 7.2b. It shows that, in the northern hemisphere, the wind vector rotates in a clockwise sense as the height increases, and that the angle between the surface wind or stress and the geostrophic wind is 45° (this is also the cross-isobar angle of the flow near the surface), irrespective of stability. Figure 7.3a shows the profiles of the normalized velocity components u/G and v/G as functions of the normalized height az. Both the components increase approximately linearly with height above the surface, attain their maxima at different heights and then oscillate around their geostrophic values with decreasing amplitude of oscillation as height increases, reflecting the influence of the e^{-az} factor.

It should be recognized that although the spiral shape of the velocity hodograph remains invariant with the rotation of horizontal (x–y) coordinate axes, the velocity component profiles and their expressions depend on the choice of the coordinate axes. For example, Equations (7.30) are the particular expressions of velocity components in the geostrophic coordinate system. In the surface-layer coordinate system with the x axis along the near-surface wind, which is more frequently used in micrometeorology, it is found that $U_g = G/\sqrt{2}$, $V_g = -G/\sqrt{2}$, and the horizontal velocity components from Equation (7.29) are given as

$$\begin{aligned} u &= \frac{G}{\sqrt{2}} - \frac{G}{\sqrt{2}} e^{-az}[\cos(az) - \sin(az)] \\ v &= -\frac{G}{\sqrt{2}} + \frac{G}{\sqrt{2}} e^{-az}[\cos(az) + \sin(az)] \end{aligned} \tag{7.31}$$

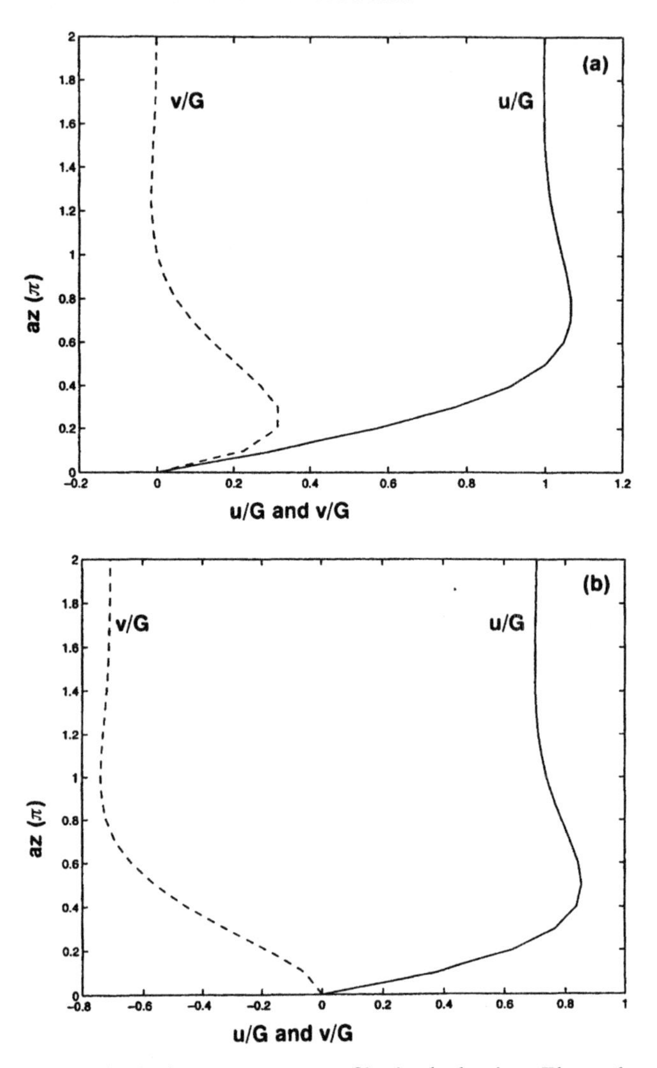

Figure 7.3 Horizontal velocity component profiles in the laminar Ekman layer in (a) the geostrophic coordinate system, and (b) the surface-layer coordinate system. Velocity components are normalized by geostrophic wind speed G, while az is the normalized height.

These component profiles (normalized by G) are represented in Figure 7.3b as functions of az and can be contrasted with those of Figure 7.3a based on the geostrophic coordinate system.

The Ekman layer depth again is given as $h_E = \pi(2\nu/f)^{1/2}$, which in middle latitudes ($f \cong 10^{-4}\ \mathrm{s}^{-1}$) is only about 1.7 m (the depth of the wind-induced oceanic Ekman spiral is about 0.5 m). There is no observational evidence for the occurrence of such shallow laminar Ekman layers in the atmosphere or oceans. Therefore, the above solutions would be of academic interest only, unless ν is replaced by a much higher effective (eddy) viscosity K to match the theoretical Ekman layer depth with that observed under a given set of conditions. Ekman (1905) originally proposed this method or procedure for estimating the 'virtual viscosity' from the observed depth, h, of frictional influence as $K = fh^2/2\pi^2$. Even then, certain features of the above solution remain inconsistent with observations. For example, the cross-isobar angle of the atmospheric flow near the surface is highly variable, as discussed in Chapter 6, and the theoretical value of $\alpha_0 = 45°$ is found more as an exception rather than the rule. Another inconsistent feature of the Ekman solution is an almost linear velocity profile near the surface, while observations indicate approximately logarithmic or log-linear velocity profiles in the surface layer. Slightly modified solutions with an appropriate choice of effective viscosity and lower boundary conditions (in the form of specified wind speed or wind direction) at the top of the surface layer have been given in the literature and largely remove the above-mentioned limitations and inconsistencies of the original Ekman solution and show better correspondence with observed wind profiles.

Example Problem 2

For the barotropic atmospheric boundary layer, the equations for the mean horizontal motion are similar to Equations (7.27) with ν replaced by an apparent or effective viscosity K. Obtain a solution to the same in a surface-layer coordinate system with the x axis oriented along the near-surface wind, using a lower boundary condition based on the observed near-surface wind U_s at $z = h_s$. Then, calculate the component velocity profiles and the boundary layer height for the following parametric values:

$$G = 10 \text{ m s}^{-1}; \quad U_s = 7 \text{ m s}^{-1}; \quad h_s = 10 \text{ m};$$
$$K = 2 \text{ m}^2 \text{ s}^{-1}; \quad f = 10^{-4} \text{ s}^{-1}$$

Solution

The equations of mean motion for the PBL over a homogeneous surface are:

$$\rho f(V - V_g) = \frac{d\tau_{zx}}{dz}$$
$$\rho f(U - U_g) = \frac{d\tau_{zy}}{dz}$$

Expressing $\tau_{zx} = \rho K\, dU/dz$ and $\tau_{zy} = \rho K\, dV/dz$, in analogy with their expressions (7.2) for laminar flows, and assuming K to be independent of height (a rather gross assumption), one can write the above equations in the form

$$\frac{d^2U}{dz^2} + \frac{f}{K}(V - V_g) = 0$$

$$\frac{d^2V}{dz^2} - \frac{f}{K}(U - U_g) = 0$$

For the barotropic PBL in which U_g and V_g are independent of height, the above two equations can be combined into one equation for the complex ageostrophic wind

$$\frac{d^2W_a}{dz^2} - \left(\frac{if}{K}\right)W_a = 0$$

where,

$$W_a = (U - U_g) + i(V - V_g)$$

The boundary conditions to be satisfied by the solution to the above equation are:

$$\text{at } z = h_s,\ U = U_s, \text{ and } V = 0$$
$$\text{as } z \to \infty,\ U \to U_g \text{ and } V \to V_g$$

For convenience, we can consider a new height variable $z' = z - h_s$ and express the ordinary differential equation and boundary conditions as follows:

$$\frac{d^2W_a}{dz'^2} - \frac{if}{K}W_a = 0$$
$$\text{at } z' = 0,\ W_a = U_s - U_g - iV_g$$
$$\text{as } z' \to \infty,\ W_a \to 0$$

The general solution to the above differential equation is

$$W_a = A\exp[-(1 + i)az'] + B\exp[(1 + i)az']$$

in which the constants A and B are determined from the boundary conditions as

$$A = U_s - U_g - iV_g;\ B = 0$$

Substituting these in the above solution and decomposing it into real and imaginary parts, we obtain

$$U = U_g - e^{-az'}[(U_g - U_s)\cos(az') + V_g \sin(az')]$$
$$V = V_g + e^{-az'}[(U_g - U_s)\sin(az') - V_g \cos(az')]$$

An important physical condition to be satisfied by the velocity distribution in the PBL is that the shear stress vector must be parallel to the wind vector near the surface. This requires that

$$\text{at } z' = 0, \ \tau_{zy} = 0, \ \text{or } \frac{\partial V}{\partial z'} = 0$$

This condition yields the following relationship between the geostrophic wind components and the actual near-surface wind:

$$U_g + V_g = U_s$$

combining it with the usual relationship

$$U_g^2 + V_g^2 = G^2$$

the geostrophic wind components can be determined as

$$U_g = \tfrac{1}{2}[U_s + (2G^2 - U_s^2)^{1/2}]$$
$$V_g = \tfrac{1}{2}[U_s - (2G^2 - U_s^2)^{1/2}]$$

after ignoring the other unphysical roots of the quadratic equations for U_g and V_g (note that, in the northern hemisphere, $U_g > -V_g > 0$).

Using the given parameter values, we can obtain from the above expressions

$$U_g = 9.64 \text{ m s}^{-1} \text{ and } V_g = -2.64 \text{ m s}^{-1}$$

which imply that the cross-isobar angle near the surface (at $z = h_s$, or $z' = 0$)

$$\alpha_o = \tan^{-1}(-V_g/U_g) \cong 15.4^\circ$$

is much smaller than 45° given by the original Ekman solution. This is a consequence of the different lower boundary condition used here.

The Ekman boundary layer height is given as

$$h = \pi \left(\frac{2K}{f} \right)^{1/2} \cong 628 \text{ m}$$

Then, using the already derived expressions for U and V, it is easy to calculate and plot the profiles of U and V as functions of z from 10 m to h or even a higher level, if desired.

7.6 Developing Laminar Boundary Layers

In the simple cases of plane-parallel flows discussed in Sections 7.4 and 7.5, inertial accelerations or forces are zero and there is a balance between the pressure gradient and friction forces, or between pressure gradient, Coriolis, and friction forces. As a result, thickness of the affected fluid layer does not change in the flow direction or x–y plane. In developing viscous flows, such as boundary layers, wakes, and jets, the thickness of the layer in which viscous or friction effects are important changes in the direction of flow, and inertial forces are as important, if not more, as the viscous forces. The ratio of the two defines the Reynolds number $\text{Re} = UL/\nu$, where U and L are the characteristic velocity and length scales (there may be more than one length scale characterizing the flow). This ratio is a very important characteristic of any viscous flow and indicates the relative importance of inertial forces as compared to viscous forces.

Atmospheric boundary layers developing over most natural surfaces are characterized by very large (10^6–10^9) Reynolds numbers. Boundary layers encountered in engineering practice also have fairly large Reynolds numbers ($\text{Re} = 10^3$–10^6, based on the boundary layer thickness as the length scale and the ambient velocity just outside the boundary layer as the velocity scale). One would expect that in such large Reynolds number flows the nonlinear inertia terms in the equations of motion will be far greater in magnitude than the viscous terms, at least in a gross sense. Still, it turns out that the viscous effects cannot be ignored in order to satisfy the no-slip boundary condition and to provide a smooth transition, through the boundary layer, from zero velocity at the surface to finite ambient velocity outside the boundary layer. Recognizing this, Prandtl (1905) first proposed an important boundary layer hypothesis which states that, under rather broad conditions, viscosity effects are significant in layers adjoining solid boundaries and in certain other layers (e.g., mixing layers and jets), the thicknesses of which approach zero as the Reynolds number of the flow approaches infinity, and are small outside these layers. This hypothesis has been applied to a variety of flow fields and is supported by many observations. The thickness of a boundary or mixing layer should be looked at in relation to the distance over which it develops. This explains the existence of relatively thick boundary layers in the atmosphere, in spite of very large Reynolds numbers characterizing the same.

The boundary layer hypothesis, for the first time, explained most of the dilemmas of inviscid flow theory, and clearly defined regions of the flow where such a theory may not be applicable and other regions where it would provide close simulations of real fluid flows. It also provided a practically useful definition of the boundary layer as the layer in which the fluid velocity makes a transition from that of the boundary (zero velocity, in the case of a fixed boundary) to that appropriate for an ambient (inviscid) flow. In theory, as well as in practice, the approach to ambient velocity is often very smooth and asymptotic, so that some arbitrariness or ambiguity is always involved in defining the boundary layer thickness. In engineering practice, the outer edge of the boundary layer is usually taken where the mean velocity has attained 99% of its ambient value. This definition is found to be quite unsatisfactory in meteorological applications, where other more useful definitions of the boundary layer thickness have been used (e.g., the Ekman layer thickness defined earlier).

Prandtl's boundary layer hypothesis is found to be valid for laminar as well as turbulent boundary layers. The fact that the boundary layer is thin compared with the distance over which it develops along a boundary allows for certain approximations to be made in the equations of motion. These boundary layer approximations, also due to Prandtl, amount to the following simplifications for the viscous diffusion terms in Equation (7.4):

$$\begin{aligned} \nu\nabla^2 u &\cong \nu(\partial^2 u/\partial z^2) \\ \nu\nabla^2 v &\cong \nu(\partial^2 v/\partial z^2) \\ \nu\nabla^2 w &\cong \nu(\partial^2 w/\partial z^2) \end{aligned} \tag{7.32}$$

which follow from the neglect of velocity gradients parallel to the boundary in comparison to those normal to it.

7.6.1 The flat-plate laminar boundary layer

Let us consider the simple case of a steady, two dimensional boundary layer developing over a thin flat plate placed in an otherwise steady, uniform stream of fluid, with streamlines of the ambient flow parallel to the plate (see Figure 7.4). This is an important classical case of fluid flow which provides a standard

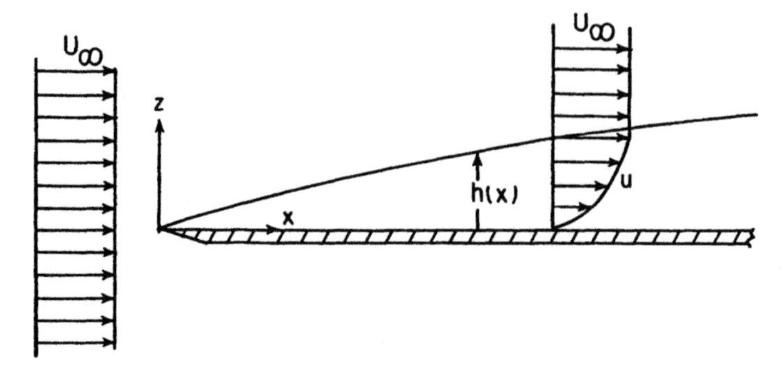

Figure 7.4 Schematic of a developing boundary layer over a flat plate.

for comparison with boundary layers developing on slender bodies, such as aircraft wings, ship hulls, and turbine blades. Except in a small region near the edge of the plate, where boundary layer approximations may not be valid, the relevant boundary layer equations to be solved are

$$\begin{gathered} u(\partial u/\partial x) + w(\partial u/\partial z) = \nu(\partial^2 u/\partial z^2) \\ \partial u/\partial x + \partial w/\partial z = 0 \end{gathered} \tag{7.33}$$

with the boundary conditions

$$\begin{gathered} u = w = 0, \quad \text{at } z = 0 \\ u \to U_\infty, \quad \text{as } z \to \infty \\ u = U_\infty, \quad \text{at } x = 0, \quad \text{for all } z \end{gathered}$$

where U_∞ is the uniform ambient velocity just outside the boundary layer.

Some idea about the growth of the boundary layer with distance downstream of the leading edge can be obtained by requiring that the inertial and viscous terms of Equation (7.32) be of the same order of magnitude in the boundary layer, i.e.,

$$u(\partial u/\partial x) \sim \nu(\partial^2 u/\partial z^2)$$

or

$$U_\infty^2/x \sim \nu U_\infty/h^2$$

or

$$h \sim (\nu x/U_\infty)^{1/2} = x/\mathrm{Re}_x^{1/2} \tag{7.34}$$

where $\mathrm{Re}_x = U_\infty x/\nu$ is a local Reynolds number. The above result indicates that the boundary thickness grows in proportion to the square root of the distance and that the ratio h/x decreases inversely proportional to the square root of the local Reynolds number. This supports the idea of a thin boundary layer in a large Reynolds number flow.

Equation (7.32) can be solved numerically. The resulting velocity profiles at various distances from the edge of the plate turn out to be similar in the sense that they collapse onto a single curve if the normalized longitudinal velocity u/U_∞ is plotted as function of the normalized distance z/h, or $z/(\nu x/U_\infty)^{1/2}$, from the surface. If, to begin with, one assumes a similarity solution of the form

$$u/U_\infty = f'(\eta) = \partial f/\partial\eta \tag{7.35}$$

where $\eta = z(U_\infty/\nu x)^{1/2}$, it is easy to show that the partial differential equations [Equation (7.33)] yield a single ordinary differential equation

$$f''' + \tfrac{1}{2}ff'' = 0 \tag{7.36}$$

where prime denotes differentiation with respect to η. The boundary conditions can be transformed to

$$\begin{aligned} f = f' = 0, &\quad \text{at } \eta = 0 \\ f' \to 1, &\quad \text{as } \eta \to \infty \end{aligned}$$

The normalized velocity profile obtained from the numerical solution of the above equations is shown in Figure 7.5. Note that the profile near the plate surface is nearly linear, similar to the velocity profiles in Ekman layers and plane-parallel flows (this linearity of profiles seems to be a common feature of all laminar flows). The shear stress on the plate surface is given by

$$\tau_0 = \mu(\partial u/\partial z)_{z=0} = 0.33\rho U_\infty^2 \,\mathrm{Re}_x^{-1/2} \tag{7.37}$$

according to which the friction or drag coefficient, $C_D \equiv \tau_0/\rho U_\infty^2$, varies inversely proportional to the square root of the local Reynolds number. The boundary layer thickness (defined as the value of z where $u = 0.99U_\infty$) is given by

$$h \cong 5(\nu x/U_\infty)^{1/2} \tag{7.38}$$

Many measurements in flat-plate laminar boundary layers have confirmed these theoretical results (Schlichting, 1960).

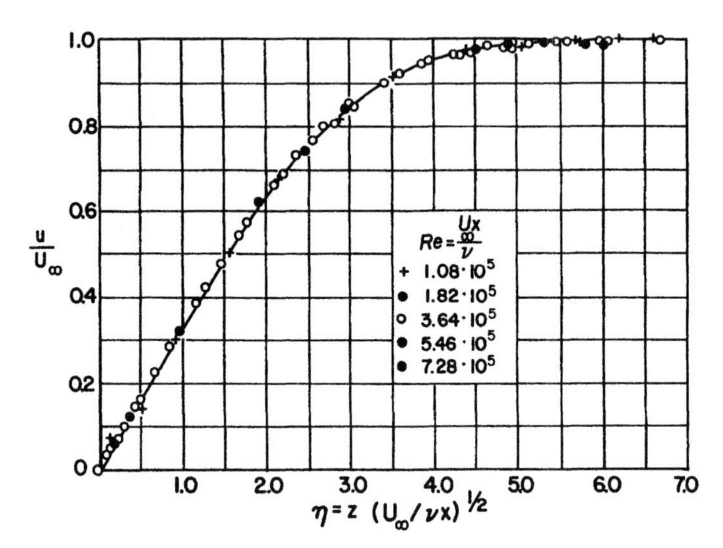

Figure 7.5 Comparison of theoretical and observed velocity profiles in the laminar flat-plate boundary layer. [From Schlichting (1960).]

7.7 Heat Transfer in Laminar Flows

In Section 4.4 we derived the one-dimensional equation of heat conduction in a solid medium and pointed out how it can be generalized to three dimensions. Heat transfer in a still fluid is no different from that in a solid medium, molecular conduction or diffusion being the only mechanism of transfer. In a moving fluid, however, heat is more efficiently transferred by fluid motions while thermal conduction is a relatively slow process.

7.7.1 Forced convection

Consideration of conservation of energy in an elementary fluid volume, in a Cartesian coordinate system, leads to the following equation of heat transfer in a low-speed, incompressible flow in which temperature or heat is considered only a passive admixture not affecting in any way the dynamics.

$$DT/Dt = \alpha_h \nabla^2 T \tag{7.39}$$

where α_h is the molecular thermal diffusivity of the fluid and the total derivative has the usual meaning as defined by Equation (7.5) and incorporates the important contribution of fluid motions.

Equation (7.39) describes the variation of temperature for a given velocity field and can be solved for the appropriate boundary conditions and flow situations. Analytical solutions are possible only for certain types of laminar flows. For two-dimensional thermal boundary layers, jets, and plumes the boundary layer approximations are used to simplify Equation (7.39) to the form

$$u(\partial T/\partial x) + w(\partial T/\partial z) = \alpha_h(\partial^2 T/\partial z^2) \tag{7.40}$$

which is similar to the simplified boundary layer equation of motion.

Analogous to the Reynolds number, one can define the Peclet number $\mathrm{Pe} = UL/\alpha_h$, which represents the ratio of the advection terms to the molecular diffusion terms in the energy equation. Instead of Pe, it is often more convenient to use the Prandtl number $\mathrm{Pr} = \nu/\alpha_h$, which is a property of the fluid only and not of the flow; note that

$$\mathrm{Pe} = \mathrm{Re} \times \mathrm{Pr} \tag{7.41}$$

For air and other diatomic gases, $\mathrm{Pr} \cong 0.7$ and is nearly independent of temperature, so that the Peclet number is of the same order of magnitude as the Reynolds number.

The difference in the temperature at a point in the flow and at the boundary is usually normalized by the temperature difference ΔT across the entire thermal layer and is represented as a function of normalized coordinates of the point, the appropriate Reynolds number, and the Prandtl number. The heat flux on a boundary is usually represented by the dimensionless ratio, called the Nusselt number

$$\mathrm{Nu} = H_0 L/k\Delta T \tag{7.42}$$

or, the heat transfer coefficient

$$C_H = H_0/\rho c_p U \Delta T \tag{7.43}$$

which is also called the Stanton number. Note that the above parameters are related (through their definitions) as

$$C_H = \mathrm{Nu}/\mathrm{RePr} = \mathrm{Nu}/\mathrm{Pe} \tag{7.44}$$

Many calculations, as well as observations of heat transfer in pipes, channels, and boundary layers have been used to establish empirical correlations between the Nusselt number or the heat transfer coefficient as a function of the Reynolds and Prandtl numbers. Some of these have found practical applications in micrometeorological problems, particularly those dealing with heat and mass transfer to or from individual leaves (Lowry, 1970; Monteith, 1973; Gates, 1980; Monteith and Unsworth, 1990).

7.7.2 *Free convection*

In free or natural convection flows, temperature cannot be considered as a passive admixture, because inhomogeneities of the temperature field in the presence of gravity give rise to significant buoyant accelerations that affect the dynamics of the flow, particularly in the vertical direction. This is usually the case in the atmosphere. The equations of motion and thermodynamic energy are slightly modified to allow for the buoyancy effects of temperature or density stratification. It is generally assumed that there is a reference state of the atmosphere at rest characterized by temperature T_0, density ρ_0, and pressure p_0, which satisfy the hydrostatic equation

$$\partial p_0/\partial z = -\rho_0 g \tag{7.45}$$

and that, in the actual atmosphere, the deviations in these properties from their reference values ($T_1 = T - T_0$, $\rho_1 = \rho - \rho_0$ and $p_1 = p - p_0$) are small compared to the values for the reference atmosphere ($T_1 \ll T_0$, $\rho_1 \ll \rho_0$, etc). These Boussinesq assumptions lead to the approximation

$$(1/\rho)(\partial p/\partial z) + g \cong (1/\rho_0)(\partial p_1 \partial z) + (g/\rho_0)\rho_1 \tag{7.46}$$

which can be used in the vertical equation of motion. In an ideal gas, such as air, changes in density are simply related to changes in temperature as

$$\rho_1 = -\beta\rho_0 T_1 \cong -(\rho_0/T_0)T_1 \tag{7.47}$$

in which the coefficient of thermal expansion $\beta = -(1/\rho_0)(\partial\rho/\partial T)\rho \cong 1/T_0$. After substituting from Equations (7.46) and (7.47) in the equation of vertical motion [Equation (7.4)], we have

$$Dw/Dt = -(1/\rho_0)(\partial p_1/\partial z) + (g/T_0)T_1 + \nu\nabla^2 w \tag{7.48}$$

Here, the second term on the right-hand side is the buoyant acceleration due to the deviation of temperature from the reference state. The equations of horizontal motion remain unchanged; alternatively, one can replace the pressure gradient terms $(1/\rho)(\partial p/\partial x)$ and $(1/\rho)(\partial p/\partial y)$ by $(1/\rho_0)(\partial p_1/\partial x)$ and

$(1/\rho_0)(\partial p_1/\partial y)$, respectively. A more appropriate form of the thermodynamic energy equation, for micrometeorological applications, is

$$D\theta/Dt = \alpha_h \nabla^2 \theta \tag{7.49}$$

in which θ is the potential temperature.

The above equations are valid for both stable and unstable stratifications. In the particular case of true free convection over a flat plate (horizontal or vertical) in which the motions are entirely generated by surface heating, the relevant dimensionless parameters on which temperature and velocity fields depend are the Prandtl number and the Grashof number

$$\mathrm{Gr} = (g/T_0)(L^3 \Delta T/\nu^2) \tag{7.50}$$

Instead of Gr, one may also use the Rayleigh number

$$\mathrm{Ra} = (g/T_0)(L^3 \Delta T/\nu\alpha_h) = \mathrm{Gr} \times \mathrm{Pr} \tag{7.51}$$

Both the Grashof and Rayleigh numbers are expected to be large in the atmosphere, although true free convection in the sense of no geostrophic wind forcing may be a rare occurrence.

7.8 Applications

Since micrometeorology deals primarily with the phenomena and processes occurring within the atmospheric boundary layer, some familiarity with the fundamentals of viscous and inviscid flows is essential. In particular, the basic differences between viscous and inviscid flows and between laminar and turbulent viscous flows should be recognized. The Navier–Stokes equations of motion and the thermodynamic energy equation and difficulties of their solution must be familiar to the students of micrometeorology. Introduction to some of the fundamentals of fluid flow and heat transfer may provide a useful link between dynamic meteorology and fluid mechanics. A direct application of this might be in micrometeorological studies of momentum, heat, and mass transfer between the atmosphere and the physical and biological elements (e.g., snow, ice, and water surfaces and plant leaves, animals, and organisms).

Problems and Exercises

1. In what regions or situations in the atmosphere may the inviscid flow theory not be applicable and why?

2. What are the basic difficulties in the solution of the Navier–Stokes equations of motion for laminar viscous flows?

3. Distinguish between the following types of environmental fluid flows, giving an example of each type from the atmosphere:
(a) inviscid and viscous flows;
(b) laminar and turbulent flows;
(c) forced and free convection.

4. In the gravity flow down an inclined plane, determine the volume flow rate Q across the plane per unit width of the plane and express the layer depth h as a function of Q and the slope angle β.

5.
(a) What are the various limitations of the classical Ekman-layer solution when compared to observations in the atmospheric boundary layer.
(b) Discuss different ways of improving the theoretical solution to the equation of motion for the PBL.

6. Using the solution to the equations of mean horizontal motion in the PBL given in the worked-out Example Problem 2, calculate and plot the following:
(a) The cross-isobar angle near the surface as a function of the ratio U_s/G.
(b) Profiles of normalized velocity components U/G and V/G as functions of height z up to 1000 m for the parametric values given in the example problem, which are typical of a neutral PBL.
(c) Velocity component profiles for a stable boundary layer of height $h = 100$ m and $U_s/G = 0.35$, using the appropriate value of K for the given h. Compare these with those for the neutral case in (b).

7.
(a) By direct substitution verify that Equation (7.29) is a solution to Equation (7.27) which satisfies the boundary conditions [Equations (7.28)].
(b) Using the above solution [Equation (7.29)], obtain the corresponding expressions for the horizontal shear stress components τ_{zx} and τ_{zy}.
(c) Show that in a coordinate system with the x axis parallel to the surface shear stress (this implies that $\tau_{zx} = \tau_0$ and $\tau_{zy} = 0$, at $z = 0$), $U_g = G/\sqrt{2}$, $V_g = -G/\sqrt{2}$, and $\tau_0 = \rho G(\nu f)^{1/2}$.
(d) In the same coordinate system write down the expressions for the normalized velocity components (u/G and v/G) and the normalized shear stress components (τ_{zx}/τ_0 and τ_{zy}/τ_0).
(e) Using the above expressions, calculate and plot the vertical profiles of the normalized velocity and shear stress components as functions of az from 0 to 2π.

(f) What conclusions can you draw from the above profiles? Comment on their applicability to the real atmosphere, if ν can be replaced by an effective viscosity K.

8. Starting from the equations of motion with boundary-layer approximations, derive Equation (7.36) for the dimensionless velocity profile in a flat-plate boundary layer.

9.

(a) What Boussinesq assumptions or approximations are commonly used to simplify the equations of motion in a thermally stratified environment.

(b) Derive the approximate expression (7.46) and discuss its usefulness and application to modeling small-scale atmospheric flows.

Chapter 8 | Fundamentals of turbulence

8.1 Instability of Flow and Transition to Turbulence

8.1.1 Types of instabilities

In Chapter 5 we introduced the concept of static (gravitational) stability or instability and showed how the parameter $s = (g/T_v)(\partial\Theta_v/\partial z \cong -(g/\rho)(\partial\rho/\partial z)$ may be used as a measure of static stability of an atmospheric (fluid) layer. This was based on the simple criterion whether vertical motions of fluid parcels were suppressed or enhanced by the buoyancy force arising from density differences between the parcel and the environment. In particular, when $s < 0$, the fluid layer is gravitationally unstable; the parcel moves farther and farther away from its equilibrium position.

Another type of flow instability is the dynamic or hydrodynamic instability. A flow is considered dynamically stable if perturbations introduced in the flow, either intentionally or inadvertently, are found to decay with time or distance in the direction of flow, and eventually are suppressed altogether. It is dynamically unstable if the perturbations grow in time or space and irreversibly alter the nature of the basic flow. A flow may be stable with respect to very small (infinitesimal) perturbations but might become unstable if large (finite) amplitude disturbances are introduced in the flow.

A simple example of dynamically unstable flow is that of two inviscid fluid layers moving initially at different velocities so that there is a sharp discontinuity at the interface. The development of this inviscid shear instability can be demonstrated by assuming a slight perturbation (e.g., in the form of a standing wave) of the interface and the resulting perturbations in pressure (in accordance with the Bernoulli equation $u^2/2 + p/\rho$ = constant) at various points across the interface. The forces induced by pressure perturbations tend to progressively increase the amplitude of the wave, no matter how small it was to begin with, thus irreversibly modifying the interface and the flow in its vicinity (Tritton, 1988, Chapter 17). Shear instability at the interface between two parallel streams of different velocities and densities, with the heavier fluid at the bottom, or of continuously stratified flows is called the Kelvin–Helmholtz (KH) instability.

The KH instability is caused by the destabilizing effect of shear, which overcomes the stabilizing effect of stratification. This type of instability is ubiquitous in the atmosphere and oceans and is considered to be a major source of internal gravity waves in these environments (Kundu, 1990, Chapter 11). In real fluids, of course, the processes leading to the development of mixing layers are more complicated, but the initiation of the Kelvin–Helmholtz instability at the interface is qualitatively similar to that in an inviscid fluid.

Laminar flows in channels, tubes, and boundary layers, introduced in Chapter 7, are found to be dynamically stable (i.e., they can be maintained as laminar flows) only under certain restrictive conditions (e.g., Reynolds number less than some critical value Re_c and a relatively disturbance-free environment), which can be realized only in carefully controlled laboratory facilities. For example, by carefully eliminating all extraneous disturbances, laminar flow in a smooth tube has been realized up to $Re \cong 10^5$, while under ordinary circumstances the critical Reynolds number (based on mean velocity and tube diameter) is only about 2000. Similarly, the critical Reynolds numbers for the instability of laminar channel and boundary layer flows are also of the same order of magnitude. For other flows, such as jets and wakes, instabilities occur at much lower Reynolds numbers. Thus, all laminar flows are expected to become dynamically unstable and, hence, cannot be maintained as such, at sufficiently large Reynolds numbers ($Re > Re_c$) and in the presence of disturbances ordinarily present in the environment (Kundu, 1990, Chapter 11).

The above Reynolds number criterion for dynamic stability does not mean that all inviscid flows are inherently unstable, because zero viscosity implies an infinite Reynolds number. Actually, many nonstratified (uniform density) inviscid flows are found to be stable with respect to small perturbations, unless their velocity profiles have discontinuities or inflexion points. The development of Kelvin–Helmholtz instability along the plane of discontinuity in the velocity profile has already been discussed. Even without a sharp discontinuity, the existence of an inflexion point in the velocity profile constitutes a necessary condition for the occurrence of instability in a neutrally stratified flow; it is also a sufficient condition for the amplification of disturbances. This inviscid instability mechanism also operates in viscous flows having points of inflexion in their velocity profiles. It is for this reason that laminar jets, wakes, mixing layers, and separated boundary layers become unstable at low Reynolds numbers. The laminar Ekman layers are also inherently unstable because their velocity profiles have inflexion points. In all such flows, both the viscous and inviscid shear instability mechanisms operate simultaneously and consequently, the stability criteria arc more complex (Kundu, 1990, Chapter 11).

Gravitational instability is another mechanism through which both the inviscid and viscous streamlined flows can become turbulent. Flows with fluid density increasing with height become dynamically unstable, particularly at large Reynolds numbers (Re > Re_c) or Rayleigh numbers (Ra > Ra_c). In the lower atmosphere, this criterion can essentially be expressed in terms of virtual potential temperature decreasing with height, because Reynolds and Rayleigh numbers are much larger than their estimated critical values. Any statically unstable layers, based on either local or nonlocal stability criteria, are also expected to be dynamically unstable and, hence, generally turbulent. The daytime unstable or convective boundary layer is a prime example of gravitational instability.

Stably stratified flows with weak negative density gradients and/or strong velocity gradients can also become dynamically unstable, if the Richardson number is less than its critical value $Ri_c = 0.25$ (this value is based on a number of theoretical as well as experimental investigations). The Richardson number criterion is found to be extremely useful for identifying possible regions of active or incipient turbulence in stably stratified atmospheric and oceanic flows. If the local Richardson number in a certain region is less than 0.25 and there is also an inflexion point in the velocity profile (e.g., in jet streams and mixing layers), the flow there is likely to be turbulent. This criterion is found to be generally valid in stratified boundary layers, as well as in free shear flows, and is often used in forecasts of clear-air turbulence in the upper troposphere and lower stratosphere for aircraft operations. Dynamic instability promotes the growth of wave-like disturbances which may break down to produce turbulence. The instability-generated wave motion may also become dynamically stable and not always result in a turbulent flow. The nighttime stable boundary layer is a prime example of the shear instability of a stratified flow.

8.1.2 Methods of investigation

The stability of a physically realizable laminar flow can be investigated experimentally and the approximate stability criteria can be determined empirically. This approach has been used by experimental fluid dynamists ever since the classical experiments in smooth glass tubes carried out by Osborne Reynolds. Systematic and carefully controlled experiments are required for this purpose (Monin and Yaglom, 1971, Chapter 1). Empirical observations to test the validity of the Richardson number criterion have also been made by meteorologists and oceanographers.

Another, more frequently used approach is theoretical, in which small-amplitude wavelike perturbations are superimposed on the original laminar or inviscid flow and the conditions of stability or instability are examined through analytical or numerical solutions of perturbation equations (Monin and Yaglom, 1971, Chapter 1; Kundu, 1990, Chapter 11). The analysis is straightforward and simple, so long as the amplitudes of perturbations are small enough for the nonlinear terms in the equations to be dropped out or linearized. It becomes extremely complicated and even intractable in the case of finite-amplitude perturbations. Thus, most theoretical investigations give only some weak criteria of the stability of flow with respect to small-amplitude (in theory, infinitesimal) perturbations. Investigations of flow stability with respect to finite-amplitude perturbations are being attempted now using new approaches to the numerical solution of nonlinear perturbation equations on large computers. Comprehensive reviews of the literature on stability analyses of viscous and inviscid flows are given by Chandrasekhar (1961), Monin and Yaglom (1971), and Drazin and Reid (1981).

The importance of stability analysis arises from the fact that any solution to the equations of motion, whether exact or approximate, cannot be considered physically realizable unless it can be shown that the flow represented by the solution is also stable with respect to small perturbations. If the basic flow is found to be unstable in a certain range of parameters, it cannot be realized under those conditions, because naturally occurring infinitesimal disturbances will cause instabilities to develop and alter the flow field irreversibly.

8.1.3 Transition to turbulence

Experiments in controlled laboratory environments have shown that several distinctive stages are involved in the transition from laminar flow (e.g., the flat-plate boundary layer) to turbulence. The initial stage is the development of primary instability which, in simple cases, may be two-dimensional. The primary instability produces secondary motions which are generally three-dimensional and become unstable themselves. The subsequent stages are the amplification of three-dimensional waves, the development of intense shear layers, and the generation of high-frequency fluctuations. Finally, 'turbulent spots' appear more or less randomly in space and time, grow rapidly, and merge with each other to form a field of well-developed turbulence. The above stages are easy to identify in a flat-plate boundary layer and other developing flows, because changes occur as a function of increasing Reynolds number with distance from the leading edge. In fully developed two-dimensional channel and pipe flows, however, transition to turbulence occurs more suddenly and explosively over the whole length of the channel or pipe as the Reynolds number is increased beyond its critical value.

Mathematically, the details of transition from initially laminar or inviscid flow to turbulence are rather poorly understood. Much of the theory is linearized, valid for small disturbances, and cannot be used beyond the initial stages. Even the most advanced nonlinear theories dealing with finite-amplitude disturbances cannot handle the later stages of transition, such as the development of 'turbulent spots.' A rigorous mathematical treatment of transition from turbulent to laminar flow is also lacking, of course, due to the lack of a generally valid and rigorous theory or model of turbulence.

Some of the turbulent flows encountered in nature and technology may not go through a transition of the type described above and are produced as such (turbulent). Flows in ordinary pipes, channels, and rivers, as well as in atmospheric and oceanic boundary layers, are some of the commonly occurring examples of such flows. Other cases, such as clear air turbulence in the upper atmosphere and patches of turbulence in the stratified ocean, are obviously the results of instability and transition processes. Transitions from laminar flow to turbulence and vice versa continuously occur in the upper parts of the stable boundary layer, throughout the night. Micrometeorologists have a deep and abiding interest in understanding these transition processes, because turbulent exchanges of momentum, heat and mass in the SBL and other stable layers occur intermittently during episodes of turbulence generation.

8.2 The Generation and Maintenance of Turbulence

It is not easy to identify the origin of turbulence. In an initially nonturbulent flow, the onset of turbulence may occur suddenly through a breakdown of streamlined flow in certain localized regions (turbulent spots). The cause of the breakdown, as pointed out in the previous section, is an instability mechanism acting upon the naturally occurring disturbances in the flow. Since turbulence is transported downstream in the manner of any other fluid property, repeated breakdowns are required to maintain a continuous supply of turbulence, and instability is an essential part of this process.

Once turbulence is generated and becomes fully developed in the sense that its statistical properties achieve a steady state, the instability mechanism is no longer required (although it may still be operating) to sustain the flow. This is particularly true in the case of shear flows, where shear provides an efficient mechanism for converting mean flow energy to turbulence kinetic energy (TKE). Similarly, in unstably stratified flows, buoyancy provides a mechanism for converting potential energy of stratification into turbulence kinetic energy (the reverse occurs in stably stratified flows). The two mechanisms for turbulence generation become more evident from the so-called shear production (S) and buoyancy production (B) terms in the TKE equation

$$d(\mathrm{TKE})/dt = S + B - D + T_r \tag{8.1}$$

which is written here in a short-form notation for the various terms; a more complete version is given later in Chapter 9. The TKE equation also shows that there is a continuous dissipation (D) of energy by viscosity in any turbulent flow and there may be transport (T_r) of energy from or to other regions of flow. Thus, in order to maintain turbulence, one or more of the generating mechanisms must be active continuously. For example, in the daytime unstable or convective boundary layer, turbulence is produced both by shear and buoyancy (the relative contribution of each depends on Richardson number). Shear is the only effective mechanism for producing turbulence in the nocturnal stable boundary layer, in low-level jets, and in regions of the upper troposphere and stratosphere where negative buoyancy actually suppresses turbulence. Near the earth's surface wind shears become particularly intense and effective, because wind speed must vanish at the surface and air flow has to go around the various surface inhomogeneities. For this reason, shear-generated turbulence is always present in the atmospheric surface layer.

Richardson (1920) originally proposed a particularly simple condition for the maintenance of turbulence in stably stratified flows, viz., that the rate of production of turbulence by shear must be equal to or greater than the rate of destruction by buoyancy. That is, the Richardson number, which turns out to be the ratio of the buoyancy destruction to shear production, must be smaller than or equal to unity. An important assumption implied in deriving this condition was that there is no viscous dissipation of energy on reaching the critical condition for the sudden decay of turbulence. On the other hand, subsequent observations of the transition from turbulent to laminar flow have indicated viscous dissipation to remain significant; they also suggest a much lower value of the critical Richardson number ($\mathrm{Ri_c}$), in the range of 0.2 to 0.5 [it has not been possible to determine $\mathrm{Ri_c}$ very precisely (see Arya, 1972)]. There is no requirement that this should be the same as the critical Richardson number ($\mathrm{Ri_c} = 1/4$) for the transition from laminar to turbulent flow, which has been determined more rigorously and precisely from theoretical stability analyses. There is some observational evidence that the critical Richardson number for the transition from turbulent to laminar flow increases with height in the SBL and may be close to one at the top of the SBL.

8.3 General Characteristics of Turbulence

Even though turbulence is a rather familiar notion, it is not easy to define precisely. In simplistic terms, we refer to very irregular and chaotic motions as turbulent. But some wave motions (e.g., on an open sea surface) can be very irregular and nearly chaotic, but they are not turbulent. Perhaps it would be more appropriate to mention some general characteristics of turbulence.

1. *Irregularity or randomness.* This makes any turbulent motion essentially unpredictable. No matter how carefully the conditions of an experiment are reproduced, each realization of the flow is different and cannot be predicted in detail. The same is true of the numerical simulations (based on Navier–Stokes equations), which are found to be highly sensitive to even minute changes in initial and boundary conditions. For this reason, a statistical description of turbulence is invariably used in practice.

2. *Three-dimensionality and rotationality.* The velocity field in any turbulent flow is three-dimensional (we are excluding here the so-called two-dimensional or geostrophic turbulence, which includes all large-scale atmospheric motions) and highly variable in time and space. Consequently, the vorticity field is also three-dimensional and flow is highly rotational.

3. *Diffusivity or ability to mix properties.* This is probably the most important property, so far as applications are concerned. It is responsible for the efficient diffusion of momentum, heat, and mass (e.g., water vapor, CO_2, and various pollutants) in turbulent flows. Macroscale diffusivity of turbulence is usually many orders of magnitude larger than the molecular diffusivity. The former is a property of the flow while the latter is a property of the fluid. Turbulent diffusivity is largely responsible for the evaporation in the atmosphere, as well as for the spread (dispersion) of pollutants released in the atmospheric boundary layer. It is also responsible for the increased frictional resistance of fluids in pipes and channels, around aircraft and ships, and on the earth's surface.

4. *Dissipativeness.* The kinetic energy of turbulent motion is continuously dissipated (converted into internal energy or heat) by viscosity. Therefore, in order to maintain turbulent motion, the energy has to be supplied continuously. If no energy is supplied, turbulence decays rapidly.

5. *Multiplicity of scales of motion.* All turbulent flows are characterized by a wide range (depending on the Reynolds number) of scales or eddies. The transfer of energy from the mean flow into turbulence occurs at the upper end of scales (large eddies), while the viscous dissipation of turbulent energy occurs at the lower end (small eddies). Consequently, there is a continuous transfer of energy from the largest to the smallest scales. Actually, it trickles down through the whole spectrum of scales or eddies in the form of a cascade process. The energy transfer processes in turbulent flows are highly nonlinear and are not well understood.

Of the above characteristics, rotationality, diffusivity, and dissipativeness are the properties which distinguish a three-dimensional random wave motion from turbulence. The wave motion is nearly irrotational, nondiffusive, and nondissipative.

8.4 Mean and Fluctuating Variables

In a turbulent flow, such as the PBL, velocity, temperature and other variables vary irregularly in time and space; it is therefore common practice to consider these variables as sums of mean and fluctuating parts, e.g.,

$$\begin{aligned}\tilde{u} &= U + u \\ \tilde{v} &= V + v \\ \tilde{w} &= W + w \\ \tilde{\theta} &= \Theta + \theta\end{aligned} \tag{8.2}$$

in which the left-hand side represents an instantaneous variable (denoted by a tilde) and the right-hand side its mean (denoted by a capital letter) and fluctuating (denoted by a lower case letter) parts. The decomposition of an instantaneous variable in terms of its mean and fluctuating parts is called Reynolds decomposition, because it was first proposed by Reynolds (1894).

There are several types of means or averages used in theory and practice. The most commonly used in the analysis of observations from fixed instruments is the time mean, which, for a continuous record (time series), can be defined as

$$F = \frac{1}{T}\int_0^T \tilde{f}(t)dt$$

where $\tilde{f}$ is any variable or a function of variables, F is the corresponding time mean, and T is the length of record or sampling time over which averaging is desired. For the digitized data, mean is simply an arithmetic average of all the digitized values of the variable during the chosen sampling period. In certain flows, such as the atmospheric boundary layer, the choice of an optimum sampling time or the averaging period T is not always clear. It should be sufficiently long to ensure stable averages and to incorporate the effects of all the significantly contributing large eddies in the flow. On the other hand, T should not be too long to mask what may be considered as real trends (e.g., the diurnal variations) in the flow. In the analysis of micrometeorological observations, the optimum averaging time may range between 10^3 and 10^4 s, depending on the height of observation, the PBL height, and stability. Time averages are frequently used for micrometeorological variables measured by instruments placed on the ground or mounted on masts, towers, or tethered balloons.

Another type of averaging used, particularly in the analyses of aircraft, radar, and sodar observations, is the spatial average of all the observations over the covered space. The optimum length, area, or volume over which the spatial averaging is performed depends on the spatial scales of significant large eddies in the flow. Horizontal line and area averages are more meaningful, considering the inhomogeneity of the PBL in the vertical due to shear and buoyancy effects.

Finally, the type of averaging which is almost always used in theory, but rarely in practice, is the ensemble or probability mean. It is an arithmatical average of a very large (approaching infinity) number of realizations of a variable or a function of variables, which are obtained by repeating the experiment over and over again under the same general conditions. It is quite obvious that the ensemble averages would be nearly impossible to obtain under the varying weather conditions in the atmosphere, over which we have little or no control. Even in a controlled laboratory environment, it would be very time consuming to repeat an experiment many times to obtain ensemble averages.

The fact that different types of averages are used in theory and experiments requires that we should know about the conditions in which time or space averages might become equivalent to ensemble averages. It has been found that the necessary and sufficient conditions for the time and ensemble means to be equal are that the process (in our case, flow) be stationary (i.e., the averages be independent of time) and the averaging period be very large ($T \to \infty$) (for a more detailed discussion of these conditions, see Monin and Yaglom, 1971). It should suffice to say that these conditions cannot be strictly satisfied in the atmosphere so that the above-mentioned equivalence of averages can only be approximate. The corresponding conditions for the equivalence of spatial and ensemble averages are spatial homogeneity (independence of averages to spatial coordinates in one or more directions) and very large averaging paths or areas. These conditions are even more restrictive and difficult to satisfy in micrometeorological applications. Quasi-stationarity over a limited period of time, or quasi-homogeneity over a limited length or area in the horizontal plane, is the best one can hope for in an idealized PBL over a homogeneous surface under undisturbed weather conditions. Under such conditions, one might expect an approximate correspondence between time or space averages used in experiments and observations and ensemble averages used in theories and mathematical models of the PBL.

Irrespective of the type of averaging used, it is obvious from Reynolds decomposition that a fluctuation is the deviation of an instantaneous variable from its mean. By definition, then, the mean of any fluctuating variable is zero ($\bar{u} = 0$, $\bar{v} = 0$, etc.), so that there are compensating negative fluctuations for positive fluctuations on the average. Here, an overbar over a variable denotes its mean.

The relative magnitudes of mean and fluctuating parts of variables in the atmospheric boundary layer depend on the type of variable, observation height relative to the PBL height, atmospheric stability, the type of surface, and other factors. In the surface layer, during undisturbed weather conditions, the magnitudes of vertical fluctuations, on the average, are much larger than the mean vertical velocity, the magnitudes of horizontal velocity fluctuations are of the same order or less than the mean horizontal velocity, and the magnitudes of the fluctuations in thermodynamic variables are at least two orders of magnitude smaller than their mean values. The relative magnitudes of turbulent fluctuations generally decrease with increasing stability and also with increasing height in the PBL. As an example, measured time series of velocity, temperature, and humidity fluctuations at a suburban site in Vancouver, Canada, during the daytime moderately unstable conditions are shown in Figure 8.1. These observations were taken at a height of 27.4 m above ground level, which falls within the unstable surface layer. The corresponding hourly averaged values of the variables were:

$$U = 3.66\ \mathrm{m\,s^{-1}};\ V = W = 0$$

$$T = 294.6\ \mathrm{K};\ Q = 8.3\ \mathrm{g\,m^{-3}}$$

Note the difference in the character of traces of horizontal and vertical velocity components, temperature, and absolute humidity fluctuations. Under unstable and convective conditions represented by these traces, buoyant plumes and thermals cause strong asymmetry between positive and negative

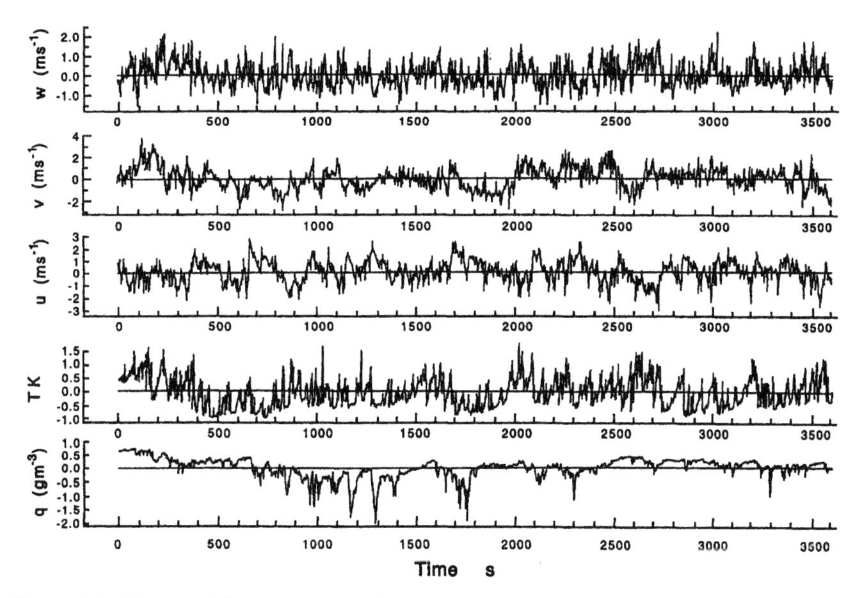

Figure 8.1 Observed time series of velocity, temperature, and absolute humidity fluctuations in the atmospheric surface layer at a suburban site in Vancouver, Canada, during moderately unstable conditions. [From Roth (1991).]

fluctuations, particularly in vertical velocity, temperature, and humidity. Under strong winds and near-neutral stability, on the other hand, fluctuations of the various variables tend to be more or less similar, with an approximate symmetry between positive and negative fluctuations.

8.5 Variances and Turbulent Fluxes

The vertical profiles of mean variables, such as mean velocity components, temperature, etc., can tell much about the mean structure of flow, but little or nothing about the turbulent structure and the various exchange processes taking place in the flow. Various statistical measures are used to study and represent the turbulence structure. These are all based on statistical analyses of turbulent fluctuations observed in the flow.

The simplest measures of fluctuation levels are the variances $\overline{u^2}$, $\overline{v^2}$, $\overline{w^2}$ [these three are often combined to define the turbulence kinetic energy per unit mass $(1/2)(\overline{u^2} + \overline{v^2} + \overline{w^2})$], $\overline{\theta^2}$, etc., and standard deviations $\sigma_u = (\overline{u^2})^{1/2}$, etc. The ratios of standard deviations of velocity fluctuations to the mean wind speed are called turbulence intensities (e.g., $i_u = \sigma_u/|\mathbf{V}|$, $i_v = \sigma_v/|\mathbf{V}|$ and $i_w = \sigma_w/|\mathbf{V}|$), which are measures of relative fluctuation levels in different directions (velocity components). Note that the mean wind speed rather than the particular component velocity, is used in the definition of turbulence intensities. When the x axis is oriented with the mean wind, i_u, i_v, and i_w are designated as longitudinal, lateral, and vertical turbulence intensities, respectively.

Observations indicate that turbulence intensities are typically less than 10% in the nocturnal boundary layer, 10–15% in a near-neutral surface layer, and greater than 15% in unstable and convective boundary layers. Turbulence intensities are generally largest near the surface and decrease with height in stable and near-neutral boundary layers. Under convective conditions, however, the secondary maxima in turbulence intensities often occur at the inversion base and sometimes in the middle of the CBL.

Even more important in turbulent flows are the covariances $\overline{uw}$, $\overline{vu}$, $\overline{\theta w}$, etc., which are directly related to and sometimes referred to as turbulent fluxes of momentum, heat, etc. As covariances, they are averages of products of two fluctuating variables and depend on the correlations between the variables involved. These can be positive, negative, or zero, depending on the type of flow and symmetry conditions. For example, if u and w are the velocity fluctuations in the x and z directions, respectively, their product uw will also be a fluctuating quantity but with a nonzero mean; $\overline{uw}$ will be positive if u and w are positively correlated, and negative if the two variables are negatively correlated. A casual inspection of u and w time series in Figure 8.1 indicates that these variables are negatively correlated, so that $\overline{uw} < 0$.

A better measure of the correlation between two variables is provided by the correlation coefficient

$$r_{uw} = \overline{uw}/\sigma_u \sigma_w \tag{8.3}$$

whose value always lies between -1 and 1, according to Schwartz's inequality $|\overline{uw}| \leqslant \sigma_u \sigma_w$. Similarly, one can define correlation coefficients for all the covariances to get an idea of how well correlated different variables are in a turbulent flow. For example, observations in Figure 8.1 indicate that temperature fluctuations are well correlated with vertical velocity fluctuations with $r_{w\theta} \cong 0.6$. There is still some order to be found in an apparently chaotic motion which is responsible for all the important exchange processes taking place in the flow.

In order to see a clear connection between covariances and turbulent fluxes, let us consider the vertical flux of a scalar with a variable concentration $\tilde{c}$ in the flow. The flux of any scalar in a given direction is defined as the amount of the scalar per unit time per unit area normal to that direction. It is quite obvious that the velocity component in the direction of flux is responsible for the transport and hence for the flux. For example, it is easy to see that in the vertical direction the flux at any instant is $\tilde{c}\tilde{w}$ and the mean flux with which we are normally concerned is $\overline{\tilde{c}\tilde{w}}$. Further writing $\tilde{c} = C + c$ and $\tilde{w} = W + w$ and following the Reynolds averaging rules, it can be shown that

$$\overline{\tilde{c}\tilde{w}} = \overline{(C + c)(W + w)} = CW + \overline{cw} \tag{8.4}$$

Thus, the total scalar flux can be represented as a sum of the mean transport (transport by mean motion) and the turbulent transport. The latter, also called the turbulent flux, is usually the dominant transport term and always of considerable importance in a turbulent flow. If the scalar under consideration is the potential temperature $\tilde{\theta}$, or enthalpy $\rho c_p \tilde{\theta}$, the corresponding turbulent flux in the vertical is $\overline{\theta w}$, or $\rho c_p \overline{\theta w}$. In this way, the covariance $\overline{\theta w}$ may be interpreted as the turbulent heat flux (in kinematic units). Similarly, $\overline{qw}$ may be considered as the vertical turbulent flux of moisture or water vapor.

Turbulent fluxes of momentum involve covariances between velocity fluctuations in different directions. For example, by symmetry, $\rho\overline{uw}$ may be considered as the vertical flux of u momentum and also as the horizontal (x direction) flux of the vertical momentum. Since the rate of change (flux) of momentum is equal to the force per unit area, or the stress, turbulent fluxes of momentum can also be interpreted as turbulent stresses, that is,

$$\bar{\tau}_{zx} = -\rho\overline{wu}, \quad \bar{\tau}_{zy} = -\rho\overline{wv}, \text{ etc.} \tag{8.5}$$

Like the viscous stresses represented in Equation (7.2), the turbulent stresses above are designated by two direction subscripts. The first subscript to $\bar{\tau}$ denotes the direction normal to the plane of the stress and the second subscript denotes the direction of the stress. There are, in all, nine stress components acting on any cubical fluid element, of which three ($\bar{\tau}_{xx}$, $\bar{\tau}_{yy}$, and $\bar{\tau}_{zz}$) are normal stresses, which are proportional to velocity variances, and six ($\bar{\tau}_{xy}$, $\bar{\tau}_{xz}$, etc.) are shearing stresses which are proportional to covariances. When expressed in terms of variances and covariances of turbulent velocity fluctuations, these are called Reynolds stresses. These turbulent stresses are found to be much larger in magnitude than the corresponding mean viscous stresses

$$T_{zx} = \mu\left(\frac{\partial U}{\partial z} + \frac{\partial W}{\partial x}\right), \quad T_{zy} = \mu\left(\frac{\partial V}{\partial z} + \frac{\partial W}{\partial y}\right), \quad \text{etc.} \tag{8.6}$$

These viscous stresses can usually be ignored, except within extremely thin viscous sublayers near smooth surfaces.

Example Problem 1

The following mean flow and turbulence measurements were made at the 22.6 m height level of the micrometeorological tower during an hour-long run of the 1968 Kansas Field Program:

Mean wind speed = 4.89 m s^{-1}.
Velocity variances: $\overline{u^2} = 0.69$, $\overline{v^2} = 1.04$, $\overline{w^2} = 0.42$ m^2 s^{-2}.
Turbulent fluxes: $\overline{uw} = -0.081$ m^2 s^{-2}, $\overline{\theta w} = 0.185$ K m s^{-1},
$\overline{\theta u} = -0.064$ K m s^{-1}.

Calculate the following:
(a) standard deviations of velocity fluctuations;
(b) turbulence intensities;
(c) correlation coefficients;
(d) turbulence kinetic energy (TKE) and the ratio TKE/$\overline{uw}$.

Solution

(a) From the definition, standard deviations are

$$\sigma_u = (\overline{u^2})^{1/2} = 0.83 \text{ m s}^{-1}$$
$$\sigma_v = (\overline{v^2})^{1/2} = 1.02 \text{ m s}^{-1}$$
$$\sigma_w = (\overline{w^2})^{1/2} = 0.64 \text{ m s}^{-1}$$

(b) The corresponding turbulence intensities can be computed as

$$i_u = \sigma_u/|\mathbf{V}| = 0.17$$
$$i_v = \sigma_v/|\mathbf{V}| = 0.21$$
$$i_w = \sigma_w/|\mathbf{V}| = 0.13$$

(c) The correlation coefficients can be calculated as

$$r_{uw} = \overline{uw}/\sigma_u\sigma_w = -0.15$$
$$r_{w\theta} = \overline{\theta w}/\sigma_\theta\sigma_w = 0.59$$
$$r_{u\theta} = \overline{\theta u}/\sigma_\theta\sigma_u = -0.16$$

which indicate that θ and w are positively well correlated, while θ and u are only weakly correlated with a negative correlation coefficient of -0.16 in the convective surface layer.

(d) Turbulence kinetic energy can be calculated as

$$\text{TKE} = \tfrac{1}{2}(\overline{u^2} + \overline{v^2} + \overline{w^2}) = 1.075 \text{ m}^2 \text{ s}^{-2}$$
$$\text{TKE}/\overline{uw} = -13.27$$

8.6 Eddies and Scales of Motion

It is a common practice to speak in terms of eddies when turbulence is described qualitatively. An eddy is by no means a clearly defined structure or feature of the flow which can be isolated and followed through, in order to study its behavior. It is rather an abstract concept used mainly for qualitative descriptions of turbulence. An eddy may be considered akin to a vortex or a whirl in common terminology. Turbulent flows are highly rotational and have all kinds of vortexlike structures (eddies) buried in them. However, eddies are not simple two-dimensional circulatory motions of the type in an isolated vortex, but are believed to be complex, three-dimensional structures. Any analogy between turbulent eddies and vortices can only be very rough and qualitative.

On the basis of flow visualization studies, statistical analyses of turbulence data, and some of the well-accepted theoretical ideas, it is believed that a turbulent flow consists of a hierarchy of eddies of a wide range of sizes (length scales), from the smallest that can survive the dissipative action of viscosity to the largest that is allowed by the flow geometry. The range of eddy sizes increases with the Reynolds number of the overall mean flow. In particular, for the ABL, the typical range of eddy sizes is 10^{-3} to 10^3 m.

Of the wide and continuous range of scales in a turbulent flow, a few have special significance and are used to characterize the flow itself. One is the characteristic large-eddy scale or macroscale of turbulence, which represents the length (l) or time scale of eddies receiving the most energy from the mean flow. Another is the characteristic small-eddy scale or microscale of turbulence, which represents the length (η) or time scale of most dissipating eddies. The ratio l/η is found to be proportional to $Re^{3/4}$ and is typically 10^5 (range: 10^4–10^6) for the atmospheric PBL.

The macroscale, or simply the scale of turbulence, is generally comparable (same order of magnitude) to the characteristic scale of the mean flow, such as the boundary layer thickness or channel depth, and does not depend on the molecular properties of the fluid. On the other hand, the microscale depends on the fluid viscosity ν, as well as on the rate of energy dissipation ε. Dimensional considerations further lead to the defining relationship

$$\eta \equiv \nu^{3/4} \varepsilon^{-1/4} \tag{8.7}$$

In stationary or steady-state conditions, the rate at which the energy is dissipated is exactly equal to the rate at which the energy is supplied from mean flow into turbulence. This leads to an inviscid estimate for the rate of energy dissipation $\varepsilon \sim u_\ell^3/l$, where u_ℓ is the characteristic velocity scale of turbulence (large eddies), which can be defined in terms of the turbulence kinetic energy as

$$u_\ell = \sqrt{\text{TKE}} \tag{8.8}$$

Some investigators define the turbulence velocity scale as the square-root of $(\overline{u^2} + \overline{v^2} + \overline{w^2})$ which is twice the turbulence kinetic energy (Tennekes and Lumley, 1972).

8.7 Fundamental Concepts, Hypotheses, and Theories

Here, we briefly describe, in qualitative terms only, some of the fundamental concepts, hypotheses, and theories of turbulence, which were mostly proposed in the first-half of the twentieth century and subsequently refined and verified by experimental data on turbulence from the laboratory as well as the atmosphere. More comprehensive reviews of theories and models of turbulence are given elsewhere (Monin and Yaglom, 1971; 1975; Hinze, 1975; Panofsky and Dutton, 1984; McComb, 1990).

8.7.1 Energy cascade hypothesis

One of the fundamental concepts of turbulence is that describing the transfer of energy among different scales or eddy sizes. It has been recognized for a long time that in large Reynolds number flows almost all of the energy is supplied by mean flow to large eddies, while almost all of it is eventually dissipated by small eddies. The transfer of energy from large (energy-containing) to small (energy-dissipating) eddies occurs through a cascade-type process involving the whole range of intermediate size eddies. This energy cascade hypothesis was originally suggested by Lewis Richardson in 1922 in the form of a parody

> Big whorls have little whorls,
> Which feed on their velocity;
> And little whorls have lesser whorls,
> And so on to viscosity.

This may be considered to be the qualitative picture of turbulence structure in a nutshell and the concept of energy transfer down the scale. The actual mechanism and quantitative aspects of energy transfer are very complicated and are outside the scope of this text. Smaller eddies are assumed to be created through an instability and breakdown of larger eddies and the energy transfer from larger to smaller eddies presumably occurs during the breakdown process. The largest eddies are produced by mean flow shear and thermal convection. Their characteristic length scale l is of the same order as the characteristic length scale (e.g., the PBL height) of the mean flow, especially for the neutral, unstable, and convective boundary layers. In the stable boundary layer, on the other hand, large-eddy size is found to be severely restricted by negative buoyancy forces, and l is usually much smaller than the PBL height h. The stronger the stability (large Ri), the smaller is the large-eddy length scale, which becomes independent of the PBL height.

If the large-eddy Reynolds number $\mathrm{Re}_\ell = u_\ell l/\nu$ is sufficiently large, these large eddies become dynamically unstable and produce eddies of somewhat smaller size, which themselves become unstable and produce eddies of still smaller size, and so on further down the scale. This cascade process is terminated when the Reynolds number based on the smallest eddy scales becomes small enough (order of one) for the smallest eddies to become stable under the influence of viscosity. This qualitative concept of energy cascade provided an underpinning for subsequent theoretical ideas, as well as experimental studies of turbulence.

8.7.2 Statistical theory of turbulence

A sophisticated mathematical and statistical theory of turbulence has been developed for the simplest type of turbulence, called homogeneous and isotropic turbulence (Hinze, 1975; Monin and Yaglom, 1975). This idealized field of turbulence has no boundaries, no mean flow and shear, and no thermal or density stratification, that may cause deviations from perfect homogeneity and isotropy. Consequently, there are no momentum, heat, and mass fluxes in any direction and the variance of velocity fluctuations is the same in all directions. In the absence of any turbulence production mechanisms (shear and buoyancy), an isotropic turbulence decays with time. Such a turbulence can be approximately realized when a uniform grid is towed through a still fluid, as in a laboratory water channel or a wind tunnel (uniform flow passing through a grid is often used as a surrogate for a homogeneous and isotropic turbulence).

Atmospheric turbulence is not homogeneous and isotropic when one considers its large-scale structure. The presence of mean wind shears and thermal stratification, at least in the vertical direction, are responsible for the generation and maintenance of turbulence. Horizontal inhomogeneities of mean flow and turbulence are often caused by variations of surface roughness, temperature, and topography. It is not easy to extend the rigorous mathematical formulation of the statistical theory to such an inhomogeneous flow and turbulence field. For the horizontally homogeneous PBL, however, approximate formulations including the effects of vertical wind shear and buoyancy have been made, but their numerical solutions still depend on many closure hypotheses and assumptions. In short, theoretical rigor and simplicity are lost, whenever significant deviations from the complete homogeneity of turbulence have to be considered.

8.7.3 Kolmogorov's local similarity theory

Kolmogorov (1941) postulated that at sufficiently large Reynolds numbers, small-scale motions in all turbulent flows have similar universal characteristics. He proposed a local similarity theory to describe these characteristics. The basic foundation of his widely accepted theory lies in the energy cascade hypothesis. Kolmogorov argued that if the characteristic Reynolds number of the mean flow or that of the most energetic large eddies is sufficiently large, there will be many steps in the energy cascade process before the energy is dissipated by small-scale motions. The large-scale motions or eddies that receive energy directly from mean flow shear or buoyancy are expected to be inhomogeneous. But the small eddies, that are formed after many successive breakdowns of large and intermediate size eddies, are likely to become homogeneous and isotropic, because they are far enough removed from the original sources (shear and buoyancy) of inhomogeneity and have no memory of large-scale processes.

Following the above reasoning, Kolmogorov (1941) proposed his local isotropy hypothesis, which states that at sufficiently large Reynolds numbers, small-scale structure is locally isotropic whether large-scale motions are isotropic or not. Here 'local' refers to the particular range of small-scale eddy motions that may be considered isotropic.

The concept of local isotropy has proved to be very useful in that it applies to all turbulent flows with large Reynolds numbers. It also made possible for the sufficiently well-developed statistical theory of isotropic turbulence to be applied to small-scale motions in most turbulent flows encountered in practice. The condition of sufficiently large Reynolds number is particularly well satisfied in the atmosphere. Consequently, the local isotropy concept or hypothesis is found to be valid over a wide range of scales and eddies in the atmosphere.

For locally isotropic, small-scale turbulence Kolmogorov (1941) proposed a local similarity theory or scaling based on his well-known equilibrium range hypothesis. This states that at sufficiently high Reynolds numbers there is a range of small scales for which turbulence structure is in a statistical equilibrium and is uniquely determined by the rate of energy dissipation (ε) and kinematic viscosity (ν). Here the statistical equilibrium refers to the stationarity of small-scale turbulence, even when the large-scale turbulence may not be stationary. The above two parameters lead to the following Kolmogorov's microscales from dimensional considerations:

$$\begin{aligned} &\text{Length scale: } \eta = \nu^{3/4}\varepsilon^{-1/4} \\ &\text{Velocity scale: } \upsilon = \nu^{1/4}\varepsilon^{1/4} \end{aligned} \tag{8.9}$$

which are used to normalize any turbulence statistics and to examine their universal similarity forms (Hinze, 1975; Monin and Yaglom, 1975; Panofsky and Dutton, 1984).

8.7.4 Taylor's frozen turbulence hypothesis

While trying to relate the spatial distribution of turbulence and related statistics to temporal statistics obtained from measurements at a fixed point, Taylor (1938) proposed the following plausible hypothesis:

> If the velocity of air stream which carries the eddies is very much greater than the turbulent velocity, one may assume that the sequence of changes in u at the fixed point are simply due to the passage of an unchanging pattern of turbulent motion over the

point, i.e., one may assume that $u(t) = u(x/U)$ where x is measured upstream at time $t = 0$ from the fixed point where u is measured.

This so-called frozen-turbulence hypothesis was originally suggested for a homogeneous field of low-intensity turbulence in a uniform mean flow. It implies a direct correspondence between spatial changes in a turbulence variable in the mean flow direction and its temporal changes at a fixed point, assuming an unchanging or frozen pattern of turbulence. The hypothesis is frequently used for converting temporal statistics to spatial statistics in the direction of mean flow. Earlier, it provided a practical and convenient way of testing the results of the statistical theory of homogeneous and isotropic turbulence. Subsequently, the frozen-turbulence hypothesis has been extended and generalized to all turbulent variables and also to nonhomogeneous shear flows.

Taylor's hypothesis has been critically examined theoretically, as well as experimentally by many investigators. Two primary limitations are found on the validity of the hypothesis in the atmospheric boundary layer. The first arises from the sufficiently large turbulence intensities, especially under unstable and convective conditions, which are found to have the apparent effect of increasing the effective advection velocity above the actual mean velocity. The validity of the hypothesis becomes particularly questionable under free convection conditions when velocity fluctuations are of the same order or larger than the mean velocity. The second limitation arises from the mean flow shear. In the presence of shear, large eddies may not be advected at the same local mean velocity which advects small eddies, because large velocity differences are likely to exist across large eddies. According to a well-known theoretical criterion, only eddies with frequencies much greater than the magnitude of mean wind shear may be assumed to be transported by the local mean velocity, as implied by Taylor's hypothesis. Aside from turbulence, Taylor's hypothesis has widespread applications in atmospheric dispersion theories and models (Arya, 1999).

8.8 Applications

A knowledge of the fundamentals of turbulence is essential to any qualitative or quantitative understanding of the turbulent exchange processes occurring in the PBL, as well as in the free atmosphere. The statistical description in terms of variances and covariances of turbulent fluctuations, as well as in terms of eddies and scales of motion, provide the bases for further quantitative analyses of observations. At the same time the basic concepts of the generation and maintenance of turbulence, of supply, transfer, and dissipation of energy, and of local similarity are the foundation stones of turbulence theory.

Problems and Exercises

1. Discuss the various types of local or small-scale instability mechanisms that might be operating in the atmosphere and their possible effects on atmospheric motions.

2.
If you have available only the upper air velocity and temperature soundings, how would you use this information to determine which layers of the atmosphere might be turbulent?

3.
(a) What are the possible mechanisms of generation and maintenance of turbulence in the atmosphere?
(b) What criterion may be used to determine if turbulence in a stably stratified layer could be maintained or if it would decay?

4. What are the special characteristics of turbulence that distinguish it from a random wave motion?

5. Compare and contrast the time and ensemble averages. Under what conditions might the two types of averages be equivalent?

6. What is the possible range of turbulence intensities in the atmospheric PBL and under what conditions might the extreme values occur?

7. Write down all the Reynolds stress components and indicate which of these have the same magnitudes.

8. The following mean flow and turbulence measurements were made at a height of 22.6 m during the 1968 Kansas Field Program under different stability conditions:

Run no.	19	20	43	54	18	17	23
Ri	−1.89	−1.15	−0.54	−0.13	0.08	0.12	0.22
U (m s^{-1})	4.89	6.20	8.06	9.66	7.49	5.50	5.02
σ_u (m s^{-1})	0.83	1.04	1.27	1.02	0.73	0.45	0.20
σ_v (m s^{-1})	1.02	1.07	1.26	0.91	0.56	0.36	0.17
σ_w (m s^{-1})	0.65	0.73	0.73	0.64	0.48	0.28	0.10
σ_θ (K)	0.48	0.61	0.56	0.25	0.14	0.15	0.18
$\overline{uw}$ (m^2 s^{-2})	−0.081	−0.111	−0.244	−0.214	−0.118	−0.043	−0.005
$\overline{\theta w}$ (m s^{-1} K)	0.185	0.273	0.188	0.072	−0.029	−0.016	−0.005
$\overline{\theta u}$ (m s^{-1} K)	−0.064	−0.081	−0.175	−0.129	0.068	0.039	0.023

(a) Calculate and plot turbulence intensities as functions of Ri.
(b) Calculate and plot the magnitudes of correlation coefficients r_{uw}, $r_{w\theta}$, and $r_{u\theta}$ as functions of Ri. Explain their different signs under unstable and stable conditions.
(c) Calculate the turbulent kinetic energy (TKE) for each run and plot the ratio $-\text{TKE}/\overline{uw}$ as a function of Ri.
(d) What conclusion can you draw about the variation of the above turbulence quantities with stability?

9.

(a) Calculate Kolmogorov's microscales of length and velocity for the following values of the rate of energy dissipation observed in the atmospheric boundary layer under different stability conditions and at different heights in the PBL:

$$\varepsilon = 10^{-4}, 10^{-3}, 10^{-2}, \text{ and } 10^{-1}\ \text{m}^2\,\text{s}^{-3}$$

(b) Comment on the possible variation of η with height in the surface layer, considering that ε generally decreases with height (it is inversely proportional to z near the surface).

Chapter 9 | Models and Theories of Turbulence

9.1 Mathematical Models of Turbulent Flows

In Chapter 7 we introduced the equations of continuity, motion, and thermodynamic energy as mathematical expressions of the conservation of mass, momentum, and heat in an elementary volume of fluid. These are applicable to laminar as well as turbulent flows. In the latter, all the variables and their temporal and spatial derivatives are highly irregular and rapidly varying functions of time and space. This essential property of turbulence makes all the terms in the conservation equations significant, so that no further simplifications of them are feasible, aside from the Boussinesq approximations introduced in Chapter 7.

In particular, the instantaneous equations to the Boussinesq approximations for a thermally stratified turbulent flow, in a rotating frame of reference tied to the earth's surface, are

$$\begin{aligned}
&\frac{\partial \tilde{u}}{\partial x} + \frac{\partial \tilde{v}}{\partial y} + \frac{\partial \tilde{w}}{\partial z} = 0 \\
&\frac{\partial \tilde{u}}{\partial t} + \tilde{u}\frac{\partial \tilde{u}}{\partial x} + \tilde{v}\frac{\partial \tilde{u}}{\partial y} + \tilde{w}\frac{\partial \tilde{u}}{\partial z} = f\tilde{v} - \frac{1}{\rho_0}\frac{\partial \tilde{p}_1}{\partial x} + \nu\nabla^2\tilde{u} \\
&\frac{\partial \tilde{v}}{\partial t} + \tilde{u}\frac{\partial \tilde{v}}{\partial x} + \tilde{v}\frac{\partial \tilde{v}}{\partial y} + \tilde{w}\frac{\partial \tilde{v}}{\partial z} = -f\tilde{u} - \frac{1}{\rho_0}\frac{\partial \tilde{p}_1}{\partial y} + \nu\nabla^2\tilde{v} \qquad (9.1) \\
&\frac{\partial \tilde{w}}{\partial t} + \tilde{u}\frac{\partial \tilde{w}}{\partial x} + \tilde{v}\frac{\partial \tilde{w}}{\partial y} + \tilde{w}\frac{\partial \tilde{w}}{\partial z} = \frac{g}{T_0}\tilde{T}_1 - \frac{1}{\rho_0}\frac{\partial \tilde{p}_1}{\partial z} + \nu\nabla^2\tilde{w} \\
&\frac{\partial \tilde{\theta}}{\partial t} + \tilde{u}\frac{\partial \tilde{\theta}}{\partial x} + \tilde{v}\frac{\partial \tilde{\theta}}{\partial y} + \tilde{w}\frac{\partial \tilde{\theta}}{\partial z} = \alpha_h\nabla^2\tilde{\theta}
\end{aligned}$$

Here, a tilde denotes an instantaneous variable which is the sum of its mean and fluctuating parts.

No general solution to this highly nonlinear system of equations is known and there appears to be no hope of finding one through purely analytical methods. However, approximate numerical solutions through the use of supercomputers have been obtained for a variety of low-Reynolds-number turbulent flows. For large-Reynolds-number flows, such as those encountered in the atmosphere including the PBL, only the Reynolds-averaged equations obtained from a grid-volume averaging (spatial filtering) or an ensemble averaging of the instantaneous equations can be solved. Their numerical solutions including the various finite-difference and finite-element methods constitute the rapidly growing field of computational fluid dynamics.

9.1.1 Direct numerical simulation

In this age of superfast computers, it is natural to attempt direct numerical solutions of the instantaneous Navier–Stokes equations for the various turbulent flows of interest using the brute force method of numerical integration on supercomputers or networks of small computers or work stations. This method has been called full-turbulence simulation or direct numerical simulation (DNS); so far it has been limited to only low-Reynolds-number turbulent flows.

Since the instantaneous equations describe all the details of the turbulent fluctuating motion, their direct numerical solution with appropriate boundary conditions should yield every detail of its complex behavior in time and space. Numerical methods are available for solving Equations (9.1) in principle, but there is a serious problem of a numerical model's ability to resolve the whole wide range of scales encountered in turbulent flow. In any direct numerical simulation of turbulence, the flow domain is discretized into three-dimensional grid elements. For an appropriate resolution of small-scale motions, which are mostly responsible for the dissipation of turbulence kinetic energy, the grid size must be less than the microscale η, the characteristic length scale of such motions. For the proper resolution of large energy-containing eddies, the model domain should be larger than the largest scales (l) of interest. Thus, the number of three-dimensional grid elements or grid points at which instantaneous equations are solved at each small time step (this is also determined by the microscale) is of the order of $(l/\eta)^3$. This number for the atmospheric boundary layer can be estimated to fall in the range of 10^{15}–10^{20}. The present supercomputers cannot handle more than 10^9 grid points for direct numerical simulation of turbulence. Thus, so far, DNS has been limited to low-Reynolds-number flows with $l/\eta < 10^3$. Apart from the very large storage requirements, much faster computer speeds will be needed before DNS becomes practically feasible for atmospheric applications.

9.1.2 Large-eddy simulation

A computationally more feasible but less fundamental approach which has been used in micrometeorology and engineering fluid mechanics is large-eddy simulation (LES). It attempts to faithfully simulate only the scales of motion in a certain range of scales between the smallest grid size and the largest dimension of simulated flow domain. The smaller subgrid scales are not resolved, but their important contributions to energy dissipation and minor contributions to turbulent transports are usually parameterized through the use of simpler subgrid scale (SGS) models (Mason, 1994). The origins of the LES technique lie in the early global weather prediction and general circulation models in which the permissible grid spacing could hardly resolve large-scale atmospheric structures.

Large-eddy simulations of turbulent flows, including the PBL, started with the pioneering work of Deardorff (1970a, b, 1972a, b, 1973). The soundness of this modeling approach in simulating neutral and unstable PBLs, even in the presence of moist convection and clouds, has been amply demonstrated by subsequent studies of Deardorff and others (Mason, 1994). A pressing need has been the extension of the LES to the nocturnal stable boundary layer, including the morning and evening transition periods, which are poorly understood. The main difficulty in simulating the stably stratified boundary layer is that the characteristic large-eddy scale (l) becomes small (say, < 10 m) under very stable conditions that are typically encountered in the SBL over land surfaces and most of the energy transfer and other exchange processes are overly influenced or dominated by subgrid scale motions. Therefore, recent LES studies of the SBL have been limited to conditions of mild or moderate stability, strong wind shears, and continuous turbulence across the SBL (Andren, 1995; Ding *et al.*, 2000b; Kosovic and Curry, 2000). Satisfactory large-eddy simulations of the transition from the late afternoon, unstable or convective boundary layer to moderately stable boundary layer in the early evening hours have also been accomplished, especially during the early period of continuous but decaying turbulence. Similar LES studies of very stable boundary layers with frequent episodes of turbulence generation by shear and destruction by buoyancy should become possible in the near future as more powerful next-generation computers become available.

A more detailed description of the large-eddy simulation approach and its applications to the PBL will be given in Chapter 13. Here, it should suffice to say that LES is viewed as the most promising tool for future research in turbulence and is expected to provide a better understanding of the transport phenomena, well beyond the reach of semiempirical theories and most routine observations. Since it requires an investment of huge computer resources (there are also a few other limitations), LES will most likely remain a research tool. It would be very useful, of course, for generating numerical turbulence data for studying large-eddy structures, developing parameterizations for simpler models, and designing physical experiments (Wyngaard and Moeng, 1993).

9.1.3 Ensemble-averaged turbulence models

Direct numerical simulation of turbulent flows is based on the 'primitive' Navier–Stokes equations of motion. The large-eddy simulation utilizes the same equations, but averaged over the finite grid volume. Numerical integrations of both systems yield highly irregular, turbulent flow variables as functions of time and space. Average statistics, such as means, variances, and higher moments, are then calculated from this 'simulated turbulence' data. In most applications, however, only the average statistics are needed and how they come about and other details of turbulent motions are usually of little or no concern. One may then ask, 'Why not use the equations of mean motion to compute the desired average statistics?' This question was first addressed by Osborne Reynolds toward the end of the nineteenth century. He also suggested some simple averaging rules or conditions and derived what came to be known as the Reynolds-averaged equations.

If f and g are two independent variables or functions of variables with mean values $\bar{f}$ and $\bar{g}$, and c is a constant, the Reynolds-averaging conditions are

$$\begin{aligned} &\overline{f+g} = \bar{f} + \bar{g} \\ &\overline{cf} = c\bar{f};\ \overline{f\bar{g}} = \bar{f}\bar{g} \\ &\overline{\partial f/\partial s} = \partial \bar{f}/\partial s;\ \overline{\int f ds} = \int \bar{f} ds \end{aligned} \tag{9.2}$$

where $s = x$, y, z, or t. These conditions are used in deriving the equations for mean variables from those of the instantaneous variables. Since these are strictly satisfied only by ensemble averaging, this type of averaging is generally implied in theory. The time and space averages often used in practice can satisfy the Reynolds-averaging rules only under certain idealized conditions (e.g., stationarity and homogeneity of flow), which were discussed in Chapter 8.

The usual procedure for deriving Reynolds-averaged equations is to substitute in Equation (9.1) $\tilde{u} = U + u$, $\tilde{v} = V + v$, etc., and take their averages using the Reynolds-averaging conditions [Equation (9.2)]. Consider first the continuity equation

$$\partial(U + u)/\partial x + \partial(V + v)/\partial y + \partial(W + w)/\partial z = 0 \tag{9.3}$$

which, after averaging, gives

$$\partial U/\partial x + \partial V/\partial y + \partial W/\partial z = 0 \tag{9.4}$$

Subtracting Equation (9.4) from (9.3), one obtains the continuity equation for the fluctuating motion

$$\partial u/\partial x + \partial v/\partial y + \partial w/\partial z = 0 \tag{9.5}$$

Thus, the form of continuity equation remains the same for instantaneous, mean, and fluctuating motions. This is not the case, however, for the equations of conservation of momentum and heat, because of the presence of nonlinear advection terms in those equations.

Let us consider, for example, the advection terms in the instantaneous equation of the conservation of heat

$$\tilde{a}_\theta = \tilde{u}(\partial\tilde{\theta}/\partial x) + \tilde{v}(\partial\tilde{\theta}/\partial y) + \tilde{w}(\partial\tilde{\theta}/\partial z) \tag{9.6a}$$

which, after utilizing the continuity equation, can also be written in the form

$$\tilde{a}_\theta = (\partial/\partial x)(\tilde{u}\tilde{\theta}) + (\partial/\partial y)(\tilde{v}\tilde{\theta}) + (\partial/\partial z)(\tilde{w}\tilde{\theta}) \tag{9.6b}$$

Expressing variables as sums of their mean and fluctuating parts in Equation (9.6b) and averaging, one obtains the Reynolds-averaged advection terms

$$\begin{aligned} A_\theta = {} & \frac{\partial}{\partial x}(U\Theta) + \frac{\partial}{\partial y}(V\Theta) + \frac{\partial}{\partial z}(W\Theta) \\ & + \frac{\partial}{\partial x}(\overline{u\theta}) + \frac{\partial}{\partial y}(\overline{v\theta}) + \frac{\partial}{\partial z}(\overline{w\theta}) \end{aligned} \tag{9.7a}$$

or, after using the mean continuity equation,

$$A_\theta = U\frac{\partial\Theta}{\partial x} + V\frac{\partial\Theta}{\partial y} + W\frac{\partial\Theta}{\partial z} + \frac{\partial}{\partial x}(\overline{u\theta}) + \frac{\partial}{\partial y}(\overline{v\theta}) + \frac{\partial}{\partial z}(\overline{w\theta}) \tag{9.7b}$$

Thus, upon averaging, the nonlinear advection terms yield not only the terms which may be interpreted as advection or transport by mean flow, but also several additional terms involving covariances or turbulent fluxes. These latter terms are the spatial gradients (divergence) of turbulent transports.

Following the above procedure with each component of Equation (9.1), one obtains the Reynolds-averaged equations for the conservation of mass, momentum, and heat:

$$
\begin{aligned}
&\frac{\partial U}{\partial x}+\frac{\partial V}{\partial y}+\frac{\partial W}{\partial z}=0 \\
&\frac{\partial U}{\partial t}+U\frac{\partial U}{\partial x}+V\frac{\partial U}{\partial y}+W\frac{\partial U}{\partial z}=fV-\frac{1}{\rho_0}\frac{\partial P_1}{\partial x}+\nu\nabla^2 U \\
&\qquad -\left(\frac{\partial \overline{u^2}}{\partial x}+\frac{\partial \overline{uv}}{\partial y}+\frac{\partial \overline{uw}}{\partial z}\right) \\
&\frac{\partial V}{\partial t}+U\frac{\partial V}{\partial x}+V\frac{\partial V}{\partial y}+W\frac{\partial V}{\partial z}=-fU-\frac{1}{\rho_0}\frac{\partial P_1}{\partial y}+\nu\nabla^2 V \\
&\qquad -\left(\frac{\partial \overline{uv}}{\partial x}+\frac{\partial \overline{v^2}}{\partial y}+\frac{\partial \overline{vw}}{\partial z}\right) \\
&\frac{\partial W}{\partial t}+U\frac{\partial W}{\partial x}+V\frac{\partial W}{\partial y}+W\frac{\partial W}{\partial z}=\frac{g}{T_0}T_1-\frac{1}{\rho_0}\frac{\partial P_1}{\partial z}+\nu\nabla^2 W \\
&\qquad -\left(\frac{\partial \overline{wu}}{\partial x}+\frac{\partial \overline{wv}}{\partial y}+\frac{\partial \overline{w^2}}{\partial z}\right) \\
&\frac{\partial \Theta}{\partial t}+U\frac{\partial \Theta}{\partial x}+V\frac{\partial \Theta}{\partial y}+W\frac{\partial \Theta}{\partial z}=\alpha_h\nabla^2\Theta-\left(\frac{\partial \overline{u\theta}}{\partial x}+\frac{\partial \overline{v\theta}}{\partial y}+\frac{\partial \overline{w\theta}}{\partial z}\right)
\end{aligned}
\tag{9.8}
$$

When these equations are compared with the corresponding instantaneous equations [Equation (9.1)], one finds that most of the terms (except for the turbulent transport terms) are similar and can be interpreted in the same way. However, there are several fundamental differences between these two sets of equations, which forces us to treat them in entirely different ways. While Equation (9.1) deals with instantaneous variables varying rapidly and irregularly in time and space, Equation (9.8) deals with mean variables which are comparatively well behaved and vary only slowly and smoothly. While all the terms in the former equation set may be significant and cannot be ignored *a priori*, the mean flow equations can be greatly simplified by neglecting the molecular diffusion terms outside of possible viscous sublayers and also other terms on the basis of certain boundary layer approximations and considerations of stationarity and horizontal homogeneity, whenever applicable. For example, it is easy to show that for a horizontally homogeneous and stationary PBL the equations of mean flow reduce to

$$
\begin{aligned}
-f(V - V_g) &= -\partial \overline{uw}/\partial z \\
f(U - U_g) &= -\partial \overline{vw}/\partial z
\end{aligned}
\tag{9.9}
$$

An important consequence of averaging the equations of motion is the appearance of turbulent flux-divergence terms which contain unknown variances and covariances. Note that in Equation (9.8) there are many more unknowns than the number of equations. While the number of extra unknowns may be reduced to one or two in certain simple flow situations, the Reynolds-averaged equations remain essentially unclosed and, hence, unsolvable. This so-called closure problem of turbulence has been a major stumbling block in developing a rigorous and general theory of turbulence. It is a consequence of the nonlinearity of the original (instantaneous) equations of motion. Many semiempirical theories and turbulence closure models have been proposed to get around the closure problem, but none of them has proved to be entirely satisfactory. The simplest gradient-transport theories, hypotheses, or relations, which are still widely used in micrometeorology, form the basis of the so-called first-order closure models; these are based on the equations of mean motion. More sophisticated higher-order closure models are based on the equations of mean motion, turbulence variances, covariances, and even higher moments.

Derivation of the dynamical equations for the various turbulence statistics is outside the scope of this book. It would suffice to say that the fundamental closure problem remains, or even gets worse, as one includes second and higher-order moments in a turbulence closure model. The addition of the turbulence kinetic energy equation to the equations of motion forms an optimal set of dynamical equations on which most frequently used turbulence models in computational fluid dynamics, boundary-layer meteorology, and mesoscale meteorology are based.

9.1.4 Turbulence kinetic energy equation

An approximate form of the turbulence kinetic energy equation for gradually developing atmospheric boundary layer flows is

$$
\frac{DE}{Dt} = -\overline{uw}\frac{\partial U}{\partial z} - \overline{vw}\frac{\partial V}{\partial z} + \frac{g}{T_{v0}}\overline{w\theta_v} - \frac{\partial}{\partial z}\left(\overline{we} + \frac{\overline{wp}}{\rho_0}\right) - \varepsilon \tag{9.10}
$$

where E and e are mean and fluctuating turbulence kinetic energies per unit mass

$$
E = \tfrac{1}{2}(\overline{u^2} + \overline{v^2} + \overline{w^2}) \tag{9.11}
$$

$$
e = \tfrac{1}{2}(u^2 + v^2 + w^2) \tag{9.12}
$$

$\overline{w\theta_v}$ is the virtual heat flux, and ε is the rate of energy dissipation. The left-hand side of Equation (9.10) represents local and advective changes of E, while the various terms on the right-hand side represent shear production, buoyancy production or destruction, turbulent transport including that due to pressure fluctuations, and the rate of dissipation due to viscosity.

The relative importance of shear generation and buoyancy production/destruction terms for the maintenance of turbulence in the atmospheric boundary layer has already been discussed in Section 8.2. Their ratio is used to define the flux Richardson number

$$\mathrm{Rf} = \frac{g\overline{w\theta_v}}{T_{v0}} \bigg/ \left(\overline{uw}\frac{\partial U}{\partial z} + \overline{vw}\frac{\partial V}{\partial z}\right) \qquad (9.13)$$

which is related to the gradient Richardson number as

$$\mathrm{Rf} = (K_h/K_m)\mathrm{Ri} \qquad (9.14)$$

For the horizontally homogeneous PBL, the left-hand side of Equation (9.10) simplifies to $\partial E/\partial t$, and any imbalance between the various source (e.g., shear production) and sink (e.g., dissipation and buoyancy destruction) terms will determine whether turbulence kinetic energy increases, remains constant, or decreases with time.

In the unstable surface layer, both shear and buoyancy mechanisms act in concert to produce vigorous turbulence, part of which is transported upward to the mixed layer. Due to much reduced wind shears in the convective mixed layer, however, turbulence there is produced primarily by buoyancy. Although turbulence kinetic energy increases with time in response to increasing heat flux and increasing PBL height in the morning hours, $\partial E/\partial t$ is found to be an order of magnitude smaller than the individual production and dissipation terms in the TKE equation. Thus, there is an approximate balance between production, dissipation, and transport terms of that equation.

In the stable boundary layer, on the other hand, the buoyancy acts in opposition to the shear production, resulting in rather weak and, sometimes, decaying turbulence. Based on their relative magnitudes, Richardson (1920) arrived at his well-known criterion of $\mathrm{Rf} < \mathrm{Rf_c} = 1$ for the maintenance of turbulence. However, he did not include the important dissipation term in his theoretical considerations. Subsequent theoretical and experimental studies of turbulence in the SBL have yielded much lower values of $\mathrm{Rf_c} = 0.2$–0.5 (Arya, 1972; Stull, 1988). The criterion is better formulated in terms of the critical flux Richardson number ($\mathrm{Rf_c}$), rather than $\mathrm{Ri_c}$.

9.2 Gradient-transport Theories

In order to close the set of equations given by Equation (9.8) or its simplified version for a given flow situation, the variances and covariances must either be specified in terms of other variables or additional equations must be developed. In the latter approach, the closure problem is only shifted to a higher level in the hierarchy of equations that can be developed. We will not discuss here the so-called higher-order closure schemes or models that have been developed in recent years and that have been found to have their own limitations and problems. The older and more widely used approach has been based on the assumed (hypothetical) analogy between molecular and turbulent transfers. It is called the gradient-transport approach, because turbulent transports or fluxes are sought to be related to the appropriate gradients of mean variables (velocity, temperature, etc.). Several different hypotheses have been used in developing such relationships.

9.2.1 Eddy viscosity (diffusivity) hypothesis

In analogy with Newton's law of molecular viscosity [see Equation (7.1), Chapter 7], J. Boussinesq in 1877 proposed that the turbulent shear stress in the direction of flow may be expressed as

$$\tau = \rho K_m(\partial U/\partial z) \tag{9.15}$$

in which K_m is called the eddy exchange coefficient of momentum or simply the eddy viscosity, which is analogous to the molecular kinematic viscosity ν. In analogy with the more general constitutive relations [Equation (7.2)], one can also generalize Equation (9.15) to express the various Reynolds stress components in terms of mean gradients. In particular, when the mean gradients in the x and y directions can be neglected in comparison to those in the z direction (the usual boundary layer approximation), we have the simpler eddy viscosity relations for the vertical fluxes of momentum

$$\begin{aligned} \overline{uw} &= -K_m(\partial U/\partial z) \\ \overline{vw} &= -K_m(\partial V/\partial z) \end{aligned} \tag{9.16}$$

Similar relations have been proposed for the turbulent fluxes of heat, water vapor, and other transferable constituents (e.g., pollutants), which are analogous to Fourier's and Fick's laws of molecular diffusion of heat and mass.

Those frequently used in micrometeorology are the relations for the vertical fluxes of heat ($\overline{\theta w}$) and water vapor ($\overline{qw}$)

$$\begin{aligned}\overline{\theta w} &= -K_{\mathrm{h}}(\partial\Theta/\partial z)\\ \overline{qw} &= -K_{\mathrm{w}}(\partial Q/\partial z)\end{aligned} \tag{9.17}$$

in which K_h and K_w are called the eddy exchange coefficients or eddy diffusivities of heat and water vapor, respectively, and Q and q denote the mean and fluctuating parts of specific humidity.

It should be recognized that the above gradient-transport relations are not the expressions of any sound physical laws in the same sense that their molecular counterparts are. These are not based on any rigorous theory, but only on an intuitive assumption of similarity or analogy between molecular and turbulent transfers. Under ordinary circumstances, one would expect heat to flow from warmer to colder regions, roughly in proportion to the temperature gradient. Similarly, momentum and mass transfers may be expected to be proportional to and down the mean gradients. However, these expectations are not always borne out by experimental data in turbulent flows, including the atmospheric boundary layer.

The analogy between molecular and turbulent transfers has subsequently been found to be very weak and qualitative only. Eddy diffusivities, as determined from their defining relations, Equations (9.16) and (9.17), are usually several orders of magnitude larger than their molecular counterparts, indicating the dominance of turbulent mixing over molecular exchanges. More importantly, eddy diffusivities cannot simply be regarded as fluid properties; these are actually turbulence or flow properties, which can vary widely from one flow to another and from one region to another in the same flow. Eddy diffusivities show no apparent dependence on molecular properties, such as mass density, temperature, etc., and have nothing in common with molecular diffusivities, except for the same dimensions.

There are other limitations of the K theory which we have not mentioned so far. The basic notion of down-gradient transport implied in the theory may be questioned. There are practical situations when turbulent fluxes are in no way related to the local gradients. For example, in a convective mixed layer the potential temperature gradient becomes near zero or slightly positive, while the heat is transported upward in significant amounts. This would imply infinite or even negative values of K_h, indicating that K theory becomes invalid in this case. Even in other situations, the specification of eddy diffusivities in a rational manner is not always easy, but quite difficult.

In spite of the above limitations of the implied analogy between molecular and turbulent diffusion, Equations (9.16) and (9.17) need not be restrictive, since they replace only one set of unknowns (fluxes) for another (eddy diffusivities). Some restrictions are imposed, however, when one assumes that eddy diffusivities depend on the coordinates and flow parameters in some definite manner. This, then, constitutes a semiempirical theory that is based on a hypothesis and is subject to experimental verification. The simplest assumption, which Boussinesq proposed originally, is that eddy diffusivities are constants for the whole flow. It turns out that this assumption works well in free turbulent flows such as jets, wakes, and mixing layers, away from any boundaries, and is often used in the free atmosphere. But when applied to boundary layers and channel flows, it leads to incorrect results. In general, the assumption of constant eddy diffusivity is not applicable near a rigid surface. But here other reasonable hypotheses can be made regarding the variation of eddy diffusivity with distance from the surface. For example, a linear distribution of K_{m} in the neutral surface layer works quite well. Suggested modifications of the K distribution in thermally stratified conditions are usually based on other theoretical considerations and empirical data, which will be discussed later. In any case, the K theory is quite useful and is widely used in micrometeorology.

9.2.2 Mixing-length hypothesis

In an attempt to specify eddy viscosity as a function of geometry and flow parameters, L. Prandtl in 1925 further extended the molecular analogy by ascribing a hypothetical mechanism for turbulent mixing. According to the kinetic theory of gases, momentum and other properties are transferred when molecules collide with each other. The theory leads to an expression of molecular viscosity as a product of mean molecular velocity and the mean free path length (the average distance traveled by molecules before collision). Prandtl hypothesized a similar mechanism of transfer in turbulent flows by assuming that eddies or 'blobs' of fluid (analogous to molecules) break away from the main body of the fluid and travel a certain distance, called the mixing length (analogous to free path length), before they mix suddenly with the new environment. If the velocity, temperature, and other properties of a blob or parcel are different from those of the environment with which the parcel mixes, fluctuations in these properties would be expected to occur as a result of the exchanges of momentum, heat, etc. If such eddy motions occur more or less randomly in all directions, it can be easily shown that net (average) exchanges of momentum, heat, etc., will occur only in the direction of decreasing velocity, temperature, etc.

In order to illustrate the above mechanism for the generation of turbulent fluctuations and their covariances (fluxes), we consider the usual case of increasing mean velocity with height in the lower atmosphere (Figure 9.1).

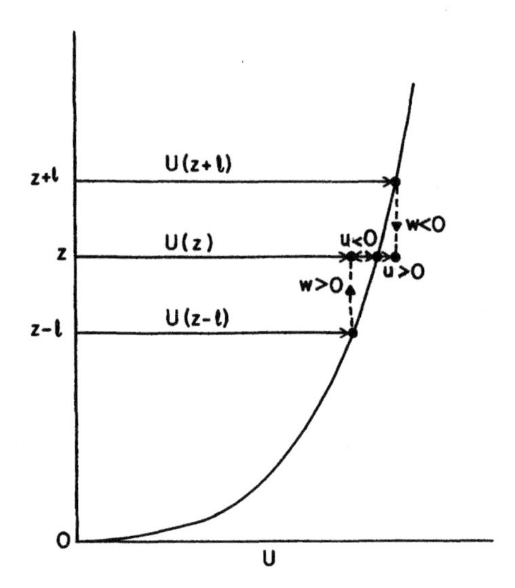

Figure 9.1 Schematic of mean velocity in the lower atmosphere and expected correlations between the longitudinal and vertical velocity fluctuations due to fluid parcels coming from above or below mixing with the surrounding fluid.

The longitudinal velocity fluctuations at a level z may be assumed to occur as a result of mixing with the environment, at this level, of fluid parcels arriving from different levels above and below. For example, a parcel arriving from below (level $z - l$) will give rise to a negative fluctuation at level z of magnitude

$$u = U(z - l) - U(z) \sim -l(\partial U/\partial z) \tag{9.18}$$

associated with its positive vertical velocity (fluctuation) w. The last approximation is based on the assumption of a linear velocity profile over the mixing length l, which is considered here a fluctuating quantity with positive values for the upward motions and negative for the downward motions of the parcel. Considering the action of many parcels arriving at z and taking the average, we obtain an expression for the momentum flux

$$\overline{uw} = -\overline{lw}(\partial U/\partial z) \tag{9.19}$$

which is not very helpful, because there is no way of measuring l. Another mixing-length expression for $\overline{uw}$ was obtained by Prandtl, after assuming that in a turbulent flow the velocity fluctuations in all directions are of the same order of magnitude and are related to each other, so that

$$w \sim -u \cong l(\partial U/\partial z) \tag{9.20}$$

From Equations (9.18) and (9.20) one obtains

$$\overline{uw} \sim -\overline{l^2}(\partial U/\partial z)^2$$

or Prandtl's original mixing-length relation

$$\overline{uw} = -l_m^2|\partial U/\partial z|(\partial U/\partial z) \tag{9.21}$$

in which l_m is a mean mixing length, which is proportional to the root-mean-square value of the fluctuating length l, and the absolute value of the velocity gradient is introduced to ensure that the momentum flux is down the gradient.

Note that Equations (9.20) and (9.21) are derived for the unidirectional mean flow in the x direction. These can be generalized to the PBL flow with U and V profiles as

$$w \sim l|\partial \mathbf{V}/\partial z| \tag{9.22}$$

$$\overline{uw} = -l_m^2|\partial \mathbf{V}/\partial z|(\partial U/\partial z) \tag{9.23}$$

$$\overline{vw} = -l_m^2|\partial \mathbf{V}/\partial z|(\partial V/\partial z) \tag{9.24}$$

Further, extensions of the mixing-length hypothesis to the vertical transfers of heat and water vapor lead to the following analogous relations for their fluxes:

$$\overline{\theta w} = -l_m l_h|\partial \mathbf{V}/\partial z|(\partial \Theta/\partial z) \tag{9.25}$$

$$\overline{qw} = -l_m l_w|\partial \mathbf{V}/\partial z|(\partial Q/\partial z) \tag{9.26}$$

where l_h and l_w are the mixing lengths for heat and water vapor transfers, which might differ from that of momentum.

Equations (9.23)–(9.26) constitute closure relations if the various mixing lengths can be prescribed as functions of the flow geometry and, possibly, other flow properties. If large eddies are mainly responsible for momentum and other exchanges in a turbulent flow, it can be argued that l_m should be directly related to the characteristic large-eddy length scale. Some investigators make no distinction between the two, although, strictly speaking, a proportionality coefficient of the order of one is usually involved. In free turbulent flows and in the outer parts of boundary layer and channel flows, where large-eddy size may not have significant spatial variations, the assumption of constant mixing length ($l_m = l_0$) is found to be reasonable. In surface or wall layers, the large-eddy size, at least normal to the surface, varies roughly in proportion to the distance from the surface, so that mixing length is also expected to be proportional to the distance ($l_m \sim z$). In the atmospheric boundary layer, the mixing-length distribution is further expected to depend on thermal stratification and boundary layer thickness, and will be considered later.

The mixing-length hypothesis may be used to specify eddy viscosity. From a comparison of Equations (9.16) and (9.23), it is obvious that

$$K_{\mathrm{m}} = l_{\mathrm{m}}^2 |\partial \mathbf{V}/\partial z| \tag{9.27}$$

An alternative specification is obtained from a comparison of Equations (9.16) and (9.19)

$$K_{\mathrm{m}} = c_{\mathrm{m}} l_{\mathrm{m}} \sigma_{\mathrm{w}} \tag{9.28}$$

where c_{m} is a constant. Other parametric relations of eddy viscosity involving the product of a characteristic length scale and a turbulence velocity scale have been proposed in the literature. Similar relations are given for K_{h} and K_{w}; e.g.,

$$K_{\mathrm{h}} = l_{\mathrm{m}} l_{\mathrm{h}} |\partial \mathbf{V}/\partial z| \tag{9.29}$$

or, alternatively,

$$K_{\mathrm{h}} = c_{\mathrm{h}} l_{\mathrm{h}} \sigma_{\mathrm{w}} \tag{9.30}$$

Example Problem 1

In the neutral surface layer, eddy viscosity and mixing length can be expressed as $K_{\mathrm{m}} = kzu_*$ and $l_{\mathrm{m}} = kz$. Derive an expression for wind distribution, assuming that the friction velocity $u_* = (-\overline{uw})^{1/2}$ is independent of height in this constant-flux, neutral surface layer.

Solution

For the neutral surface layer, the eddy viscosity relation (9.16) can be expressed as

$$u_*^2 = K_{\mathrm{m}} \frac{\partial U}{\partial z} = kzu_* \frac{\partial U}{\partial z}$$

so that,

$$\frac{\partial U}{\partial z} = \frac{u_*}{kz}$$

whose integration with respect to z yields

$$U = \frac{u_*}{k} \ln z + A$$

where A is a constant of integration. In order to determine A we use the lower boundary condition at $z = z_0$, $U = 0$ (here, one cannot use the boundary condition right at the surface, i.e., at $z = 0$, for various reasons). This boundary condition gives $A = -(u_*/k) \ln z_0$ and the wind profile as

$$U = \frac{u_*}{k} \ln(z/z_0)$$

which is the well known logarithmic wind profile law, in which z_0 is called the roughness length.

The mixing-length relation (9.21) leads to the same result as

$$u_*^2 = l_\mathrm{m}^2 \left(\frac{\partial U}{\partial z}\right)^2 = k^2 z^2 \left(\frac{\partial U}{\partial z}\right)^2$$

or,

$$\frac{\partial U}{\partial z} = \frac{u_*}{kz}$$

9.3 Dimensional Analysis and Similarity Theories

Dimensional analysis and similarity considerations are extensively used in micrometeorology, as well as in other areas of science and engineering. Therefore, these analytical methods are briefly discussed in this section, while their applications in micrometeorology will be given in later chapters.

9.3.1 Dimensional analysis

Dimensional analysis is a simple but powerful method of investigating a variety of scientific phenomena and establishing useful relationships between the various quantities or parameters, based on their dimensions. One can define a set of fundamental dimensions, such as length [L], time [T], mass [M], etc., and express the dimensions of all the quantities involved in terms of these fundamental dimensions. A representation of the dimensions of a quantity or a parameter in terms of fundamental dimensions constitutes a dimensional formula, e.g., the dimensional formula for fluid viscosity is $[\mu] = [\mathrm{ML}^{-1}\mathrm{T}^{-1}]$. If the exponents in the dimensional formula are all zero, the parameter under consideration is dimensionless. One can form dimensionless parameters from appropriate combinations of dimensional quantities; e.g., the Reynolds number

$\text{Re} = VL\rho/\mu$ is a dimensionless combination of fluid velocity V, the characteristic length scale L, density ρ, and viscosity μ.

Dimensionless groups or parameters are of special significance in any dimensional analysis in which the main objective is to seek certain functional relationships between the various dimensionless parameters. There are several reasons for considering dimensionless groups instead of dimensional quantities or variables. First, mathematical expressions of fundamental physical laws are dimensionally homogeneous (i.e., all the terms in an expression or equation have the same dimensions) and can be written in dimensionless forms simply by an appropriate choice of scales for normalizing the various quantities. Second, dimensionless relations represented in mathematical or graphical form are independent of the system of units used and they facilitate comparisons between data obtained by different investigators at different locations and times. Third, and, perhaps, the most important reason for working with dimensionless parameters is that nondimensionalization always reduces the number of parameters that are involved in a functional relationship. This follows from the well-known Buckingham Pi theorem, which states that if m quantities $(Q_1, Q_2, \ldots, Q_m)$, involving n fundamental dimensions, form a dimensionally homogeneous equation, the relationship can always be expressed in terms of $m-n$ independent dimensionless groups $(\Pi_1, \Pi_2, \ldots, \Pi_{m-n})$ made of the original m quantities. Thus, the dimensional functional relationship

$$f(Q_1, Q_2, \ldots, Q_m) = 0 \tag{9.31}$$

is equivalent to the dimensionless relation

$$F(\Pi_1, \Pi_2, \ldots, \Pi_{m-n}) = 0 \tag{9.32}$$

or, alternatively,

$$\Pi_1 = F_1(\Pi_2, \Pi_3, \ldots, \Pi_{m-n}) \tag{9.33}$$

In particular, when only one dimensionless group can be formed out of all the quantities, i.e., when $m - n = 1$, that group must be a constant, since it cannot be a function of any other parameters. If there are two Π-groups, one must be a unique function of the other, and so on. Dimensional analysis does not give actual forms of the functions F, F_1, etc., or values of any dimensionless constants that might result from the analysis. This must be obtained by other means, such as further theoretical considerations and experimental observations. It is common practice to follow dimensional analysis by a systematic experimental study of the phenomenon to be investigated.

9.3.2 Similarity theory

The Buckingham Pi theorem and dimensional analysis discussed above are merely mathematical formalisms and do not deal with the physics of the problem. The actual formulation of a similarity theory involves several steps, some of which require physical intuition, other theoretical considerations, prior observational information, and possibly new experiments designed to test the theory. The five steps involved in developing and testing a similarity theory are:

(1) Define the scope of the theory with all the restrictive assumptions clearly stated.
(2) Select an optimal set of relevant independent variables on which one or more variable of interest may depend. This constitutes a similarity hypothesis about the functional dependence between the various variables. Only one dependent variable is considered at a time in such a functional relationship, but one can have several functional relationships (e.g., one for each dependent variable).
(3) Perform dimensional analysis after determining the number of possible independent dimensionless groups and organize the variables into dimensionless groups.
(4) Express functional relationships between dimensionless groups, one of which should contain the dependent variable. These constitute the similarity relations or similarity theory predictions (one for each dependent variable).
(5) Gather relevant data from previous experiments that satisfy the restrictive assumptions of the similarity theory, or perform a new experiment to test the initial similarity hypothesis and similarity theory predictions. Experimental data will tell us whether the original similarity hypothesis is correct or not. If the theory is verified by experimental data, the latter can also be used to determine empirical forms of the various similarity functions by appropriate curve fitting through data plots.

The ultimate result of this five-step procedure is a set of empirical equations or fitted curves through plots of experimental data, all involving dimensionless similarity parameters. For a successful similarity theory, verified by experiments, the empirical similarity relations are expected to be universal and can be used at other locations with different surface characteristics and under different meteorological conditions.

In the first step, certain restrictive assumptions are made in order to reduce the number of independent variables involved in a similarity hypothesis, so that one can reduce the number of dimensionless parameters to a minimum that is consistent with the physics. The fewer the dimensionless parameters, the more powerful are the similarity theory predictions, and the easier it is to verify them and to determine empirical similarity relations or constants from carefully conducted experiments and observations. For example, the most commonly used simplifying assumptions in all the proposed PBL similarity theories are: (1) mean flow is steady or stationary; (2) mean flow is horizontally homogeneous, implying a flat and homogeneous surface; (3) viscosity and other molecular diffusivities are not relevant in the bulk of the PBL flow outside molecular sublayers; and (4) near-surface canopy variables can be ignored in the formulation of a similarity theory for the horizontally homogeneous part of the PBL. Additional assumptions are sometimes made depending on the type of the PBL (e.g., for a barotropic PBL, geostrophic winds are independent of height) and its stability regime (e.g., under convective conditions, shear effects and the Coriolis parameter are ignored). In surface-layer similarity theories, the Coriolis parameter, geostrophic winds and shears, and the PBL height are all considered irrelevant.

The second step, involving the selection of independent variables in the formulation of a similarity hypothesis, is the most crucial step in the development of a successful similarity theory. On the one hand, one cannot ignore any of the important variables or parameters on which the dependent variable really depends, because this might lead to a completely wrong or unphysical relationship. On the other hand, if unnecessary and irrelevant variables are included in the original similarity hypothesis, they will unnecessarily complicate the analysis and make empirical determination of the various functional relationships extremely difficult, if not impossible. In principle, experimental data should tell us if there are irrelevant variables or similarity parameters, which can be dropped from the similarity theory without any significant loss of generality. It is always desirable to keep the number of independent variables to a minimum, consistent with physics. Sometimes, it may become necessary to break down the domain of the problem or phenomenon under investigation into several small subdomains, so that simpler similarity hypotheses can be formulated for each of them separately. For example, the atmospheric boundary layer is usually divided into a surface layer and an outer layer or a mixed layer for dimensional analysis and similarity considerations.

The third step is straightforward, once the number of independent dimensionless groups is determined by the Buckingham's Pi theorem. But, there is always some flexibility and choice in formulating dimensionless parameters. For this, it is often convenient to first determine the appropriate scales of length, velocity, etc., from the independent variables. Then, dimensionless parameters can be determined merely by inspection. If there are more than one length or velocity scales, their ratio forms a dimensionless similarity parameter.

The similarity relations or predictions in the fourth step are simply expressions of dimensionless groups containing dependent variables as unspecified functions of other dimensionless (similarity) parameters. If there is no similarity parameter that can be formed by independent variables only, the dependent parameter (Π-group) must simply be a constant.

Finally, a thorough experimental verification of the similarity hypothesis and resulting similarity relations is necessary before a proposed similarity theory becomes widely accepted and successful. For this, experiments are conducted at different locations, under more or less idealized conditions implied in the theory. However, the whole wide range of parameters should be covered. Sometimes, experimental data can verify some of the similarity relations, but not others, in which case the theory is considered only partially successful with applicability to only specified variables.

9.3.3 Examples of similarity theories

In order to illustrate the method and usefulness of dimensional analysis and similarity, let us consider the possible relationship for the mean potential temperature gradient ($\partial\Theta/\partial z$) as a function of the height (z) above a uniform heated surface, the surface heat flux (H_0), the buoyancy parameter (g/T_0) which appears in the expressions for static stability and buoyant acceleration, and the relevant fluid properties (ρ and c_p) in the near-surface layer when free convection dominates any mechanical mixing (this latter condition permits dropping of all shear-related parameters from consideration). To establish a functional relationship in the dimensional form

$$f(\partial\Theta/\partial z, H_0, g/T_0, z, \rho, c_p) = 0 \tag{9.34}$$

would require extensive observations of temperature as a function of height and surface heat flux at different times and locations (to represent different types of surfaces and radiative regimes). If the relationship is to be further generalized to other fluids, laboratory experiments using different fluids will also be necessary. Considerable simplification can be achieved, however, if ρ and c_p are combined with H_0 into what may be called kinematic heat flux $H_0/\rho c_p$, so that Equation (9.34) can be written as

$$F(\partial\Theta/\partial z, H_0/\rho c_p, g/T_0, z) = 0 \tag{9.35}$$

If we now use the method of dimensional analysis, realizing that only one dimensionless group can be formed from the above quantities, we obtain

$$(\partial\Theta/\partial z)(H_0/\rho c_p)^{-2/3}(g/T_0)^{1/3}z^{4/3} = C \tag{9.36}$$

in which the left-hand side is the dimensionless group that is predicted to be a constant. The value of the constant C can be determined from only one carefully conducted experiment, although a thorough experimental verification of the above relationship might require more extensive observations.

The desired dimensionless group from a given number of quantities can often be formed merely by inspection. A more formal and general approach would be to write and solve a system of algebraic equations for the exponents of various quantities involved in the dimensionless group. For example, the dimensionless group formed out of all the parameters in Equation (9.35) may be assumed as

$$\Pi_1 = (\partial\Theta/\partial z)(H_0/\rho c_p)^a (g/T_0)^b z^c \tag{9.37}$$

in which we have arbitrarily assigned a value of unity to one of the indices (here, the exponent of $\partial\Theta/\partial z$), because any arbitrary power of a dimensionless quantity is also dimensionless. Writing Equation (9.37) in terms of our chosen fundamental dimensions (length, time, and temperature), we have

$$[\mathrm{L}^0\mathrm{T}^0\mathrm{K}^0] = [\mathrm{KL}^{-1}][\mathrm{KLT}^{-1}]^a[\mathrm{LT}^{-2}\,\mathrm{K}^{-1}]^b[\mathrm{L}]^c$$

from which we obtain the algebraic equations

$$\begin{aligned} 0 &= -1 + a + b + c \\ 0 &= -a - 2b \\ 0 &= 1 + a - b \end{aligned} \tag{9.38}$$

whose solution gives $a = -2/3$, $b = 1/3$, and $c = 4/3$. Substituting these values in Equation (9.37) and equating the only dimensionless group to a constant, then, yields Equation (9.36).

Another approach is to first formulate the characteristic scales of length, velocity, etc., from combinations of independent variables and then use these scales to normalize the dependent variables. In the case of multiple scales, their ratios form the independent dimensionless groups. In the above example of temperature distribution over a heated surface, with $\partial\Theta/\partial z$ as the dependent variable and the remaining quantities as independent variables, the following scales can be formulated out of the latter:

$$
\begin{aligned}
&\text{length:} && z \\
&\text{temperature:} && \theta_{\mathrm{f}} = \left(\frac{H_0}{\rho c_{\mathrm{p}}}\right)^{2/3} \left(\frac{g}{T_0}\right)^{-1/3} z^{-1/3} \\
&\text{velocity:} && u_{\mathrm{f}} = \left(\frac{H_0}{\rho c_{\mathrm{p}}} \frac{g}{T_0} z\right)^{1/3}
\end{aligned}
\tag{9.39}
$$

Then, the appropriate dimensionless group involving the dependent variable is $(z/\theta_{\mathrm{f}})(\partial\Theta/\partial z)$, which must be a constant, since no other independent dimensionless groups can be formed from independent variables. This procedure also leads to Equation (9.36); it is found to be more convenient to use when a host of dependent variables are functions of the same set of independent variables. For example, standard deviations of temperature and vertical velocity fluctuations in the free convective surface layer are given by

$$
\begin{aligned}
\sigma_\theta/\theta_{\mathrm{f}} &= C_\theta \\
\sigma_w/u_{\mathrm{f}} &= C_{\mathrm{w}}
\end{aligned}
\tag{9.40}
$$

or, after substituting from Equation (9.39),

$$
\begin{aligned}
\sigma_\theta &= C_\theta (H_0/\rho c_{\mathrm{p}})^{2/3} (g/T_0)^{-1/3} z^{-1/3} \\
\sigma_w &= C_w[(H_0/\rho c_{\mathrm{p}})(g/T_0)z]^{1/3}
\end{aligned}
\tag{9.41}
$$

Equations (9.41) have proved to be quite useful for the atmospheric surface layer under daytime unstable conditions and are supported by many observations (Monin and Yaglom, 1971, Chapter 5; Wyngaard, 1973) from which $c_\theta \cong 1.3$ and $C_{\mathrm{w}} \cong 1.4$.

The above similarity theory was originally proposed by Obukhov and is more commonly known as the local free convection similarity theory. Local free convection similarity and scaling are not found to be applicable to horizontal velocity fluctuations, so that σ_u and σ_v are not simply proportional to u_{f}. Instead, they are found to be strongly influenced by large-eddy motions (convective updrafts and downdrafts) which extend through the whole depth of the CBL. Since the PBL height is ignored in the local free convection similarity theory, its applicability is limited to vertical velocity and temperature fluctuations.

If the boundary layer height (h) is also added to the list in Equation (9.35), we would have, according to the pi theorem, two independent dimensionless groups and the predicted functional relationship would be

$$\frac{\partial \Theta}{\partial z}\left(\frac{H_0}{\rho c_p}\right)^{-2/3}\left(\frac{g}{T_0}\right)^{1/3} z^{4/3} = F(z/h) \tag{9.42}$$

Similarly, σ_w/u_f and σ_θ/θ_f would be predicted to be functions of z/h. However, subsequent analysis of experimental data might indicate that the dependence of these on z/h is very weak or nonexistent and, hence, the irrelevance of h in the original hypothesis. The inclusion of h would be quite justified and even necessary, however, if we were to investigate the turbulence structure of the mixed layer, or σ_u and σ_v even in the surface layer. For example, the mixed-layer similarity hypothesis proposed by Deardorff (1970b) states that turbulence structure in the convective mixed layer depends on z, g/T_0, $H_0/\rho c_p$, and h. Then, the relevant mixed-layer similarity scales are:

$$\begin{aligned} &\text{length:} \quad h \\ &\text{temperature:} \quad T_* = \left(\frac{H_0}{\rho c_p}\right)^{2/3}\left(\frac{gh}{T_0}\right)^{-1/3} \\ &\text{velocity:} \quad W_* = \left(\frac{H_0}{\rho c_p}\frac{g}{T_0}h\right)^{1/3} \end{aligned} \tag{9.43}$$

The corresponding similarity predictions are that the dimensionless structure parameters σ_u/W_*, σ_w/W_*, etc., must be some unique function of z/h. This mixed-layer similarity theory has proved to be very useful in describing turbulence and diffusion in the convective boundary layer (Arya, 1999).

As the number of independent variables and parameters is increased in a similarity hypothesis, not only the number of independent dimensionless groups increases, but also the possible combinations of variables in forming such groups become large. The possibility of experimentally determining their functional relationships becomes increasingly remote as the number of dimensionless groups increases beyond two or three. A generalized PBL similarity theory will have to include all the possible factors influencing the PBL under the whole range of conditions encountered and, hence, might be too unwieldy for practical use. Simpler PBL similarity theories and scaling will be described in Chapter 13, while the surface layer similarity theory/scaling will be discussed in more detail in Chapters 10–12.

Example Problem 2

In a convective boundary layer during the midday period when $T_0 = 300$ K, $H_0 = 500$ W m^{-2}, and $h = 1500$ m, calculate and compare the standard deviations of vertical velocity and temperature fluctuations at the 10 m and 100 m height levels.

Solution

Using the local free-convection similarity relations (9.41) with $C_\theta = 1.3$, $C_w = 1.4$, and $\rho c_p = 1200\ \mathrm{J\ K^{-1}\ m^{-3}}$, the computed values of σ_w and σ_θ at the two heights in the convective surface layer ($z < 0.1h = 150$ m) are as follows:

Height, z (m)	σ_w (m s^{-1})	σ_θ (K)
10	0.72	1.05
100	1.55	0.49

Note that σ_w increases and σ_θ decreases by a factor of $10^{1/3} \cong 2.15$ as the height increases from 10 to 100 m.

9.4 Applications

The theories and models of turbulence discussed in this chapter are widely used in micrometeorology. Some of the specific applications are as follows:

- Calculating the mean structure (e.g., vertical profiles of mean velocity and temperature) of the PBL.
- Calculating the turbulence structure (e.g., profiles of fluxes, variances of fluctuations, and scales of turbulence) of the PBL.
- Providing plausible theoretical explanations for turbulent exchange and mixing processes in the PBL.
- Providing suitable frameworks for analyzing and comparing micrometeorological data from different sites.
- Suggesting simple methods of estimating turbulent fluxes from mean profile observations.

Problems and Exercises

1. Compare and contrast the instantaneous and the Reynolds-averaged equations of motion for the PBL.

2.

(a) What boundary layer approximations are often used for simplifying the Reynolds-averaged equations of motion for the PBL?

(b) What are the other simplifying assumptions commonly used in micrometeorology and the conditions in which they may not be valid?

3. Show, step by step, how Equation (9.8) is reduced to Equation (9.9) for a horizontally homogeneous and stationary PBL.

4.

(a) Write down the mean thermodynamic energy equation for a horizontally homogeneous PBL and discuss the rationale for retaining the time-tendency term in the same.

(b) Describe a method for estimating the diurnal variation of surface heat flux from hourly soundings of temperature in the PBL.

5. Show that with the assumption of a constant eddy viscosity the equations of mean motion in the PBL become similar to those for the laminar Ekman layer and, hence, discuss some of the limitations of the above assumption.

6. If the surface layer turbulence under neutral stability conditions is characterized by a large-eddy length scale l, which is proportional to height, and a velocity scale u_*, which is independent of height, suggest the plausible expressions of mixing length and eddy viscosity in this layer.

7. A similarity hypothesis proposed by von Karman states that mixing length in a turbulent shear flow depends only on the first and second spatial derivatives of mean velocity ($\partial U/\partial z$, $\partial^2 U/\partial z^2$) in the direction of shear. Suggest an expression for mixing length which is consistent with the above similarity hypothesis.

8.

(a) In a neutral, barotropic PBL, ageostrophic velocity components ($U - U_g$ and $V - V_g$) may be assumed to depend only on the height z above the surface, the friction velocity scale u_*, and the Coriolis parameter f. On the basis of dimensional analysis suggest the appropriate similarity relations for velocity distribution and the PBL height h.

(b) If the PBL height in Problem 8(a) above was determined by a low-level inversion, with its base at z_i, what should be the corresponding form of similarity relations?

9.

(a) In a convective boundary layer during the midday period when $T_0 = 310$ K, $H_0 = 300$ W m^{-2}, and $h = 1000$ m, calculate and compare the standard deviations of velocity and temperature fluctuations at the heights of 10, 50, and 100 m.

(b) Show that, in the convective surface layer, the local free convection similarity relations can also be expressed in terms of mixed-layer similarity scales as

$$\sigma_\theta / T_* = C_\theta (z/h)^{-1/3}; \; \sigma_w / W_* = C_w (z/h)^{1/3}$$

Chapter 10 Near-neutral Boundary Layers

10.1 Velocity-profile Laws

Strictly neutral stability conditions are rarely encountered in the atmosphere. However, during overcast skies and strong surface geostrophic winds, the atmospheric boundary layer may be considered near-neutral, and simpler theoretical and semiempirical approaches developed for neutral boundary layers by fluid dynamists and engineers can be used in micrometeorology. One must recognize, however, that, unlike in unidirectional flat-plate boundary layer and channel flows, the wind direction in the PBL changes with height in response to the Coriolis force due to the earth's rotation. Therefore, the wind distribution in the PBL is expressed in terms of either wind speed and direction, or the two horizontal components of velocity (see Figures 6.5–6.8).

10.1.1 The power-law profile

Measured velocity distributions in flat-plate boundary layer and channel flows can be represented approximately by a power-law expression

$$U/U_h = (z/h)^m \tag{10.1}$$

which was originally suggested by L. Prandtl with an exponent $m = 1/7$ for smooth surfaces. Here, h is the boundary layer thickness or half-channel depth. Since, wind speed does not increase monotonically with height up to the top of the PBL, a slightly modified version of Equation (10.1) is used in micrometeorology:

$$U/U_r = (z/z_r)^m \tag{10.2}$$

where U_r is the wind speed at a reference height z_r, which is smaller than or equal to the height of wind speed maximum; a standard reference height of 10 m is commonly used.

The power-law profile does not have a sound theoretical basis, but frequently it provides a reasonable fit to the observed velocity profiles in the lower part of the PBL, as shown in Figure 10.1. The exponent m is found to depend on both the surface roughness and stability. Under near-neutral conditions, values of m range from 0.10 for smooth water, snow and ice surfaces to about 0.40 for well-developed urban areas. Figure 10.2 shows the dependence of m on the roughness length or parameter z_0, which will be defined later. The exponent m also increases with increasing stability and approaches one (corresponding to a linear profile) under very stable conditions. The value of the exponent may also depend, to some extent, on the height range over which the power law is fitted to the observed profile.

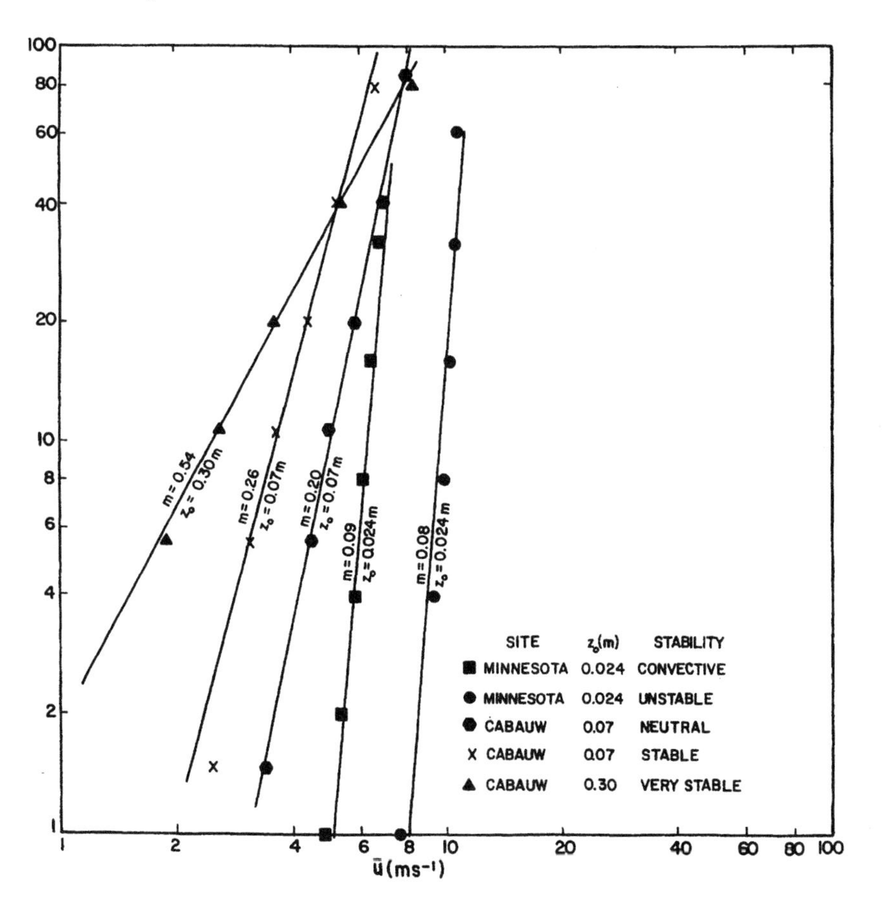

Figure 10.1 Comparison of observed wind speed profiles at different sites (z_0 is a measure of the surface roughness) under different stability conditions with the power-law profile. [Data from Izumi and Caughey (1976).]

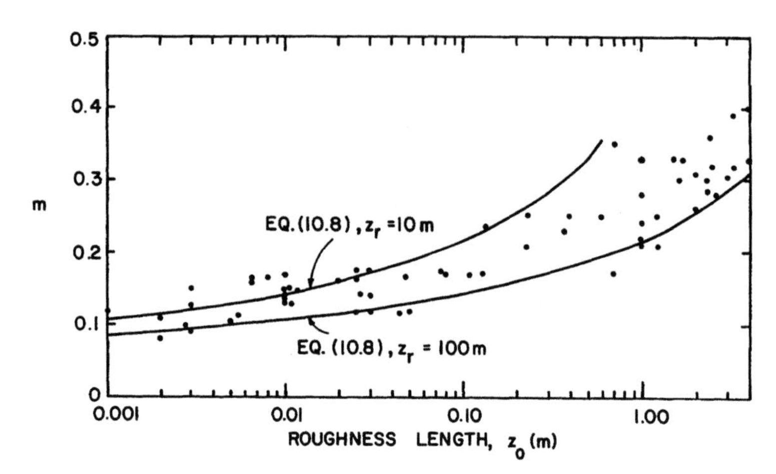

Figure 10.2 Variations of the power-law exponent with the roughness length for near-neutral conditions. [Data from Counihan (1975). Reprinted with permission from *Atmospheric Environment*. Copyright © (1975), Pergamon Journals Ltd.]

Note that the power-law wind profile would be appropriate only for the lower part of the PBL in which wind speed increases monotonically with height. At higher levels, wind speed may be specified differently (e.g., constant or linearly varying with height), depending on stability and other conditions (Arya, 1999, Chapter 4).

The power-law velocity profile implies a power-law eddy viscosity (K_m) distribution in the lower part of the boundary layer, in which the momentum flux may be assumed to remain nearly constant with height, i.e., in the constant stress layer. It is easy to show that

$$K_m/K_{mr} = (z/z_r)^n \tag{10.3}$$

with the exponent $n = 1 - m$, is consistent with Equation (10.2) in the surface layer. Equations (10.2) and (10.3) are called conjugate power laws and have been used extensively in theoretical formulations of atmospheric diffusion, including transfers of heat and water vapor from extensive uniform surfaces (Sutton, 1953). In such formulations eddy diffusivities of heat and mass are assumed to be equal or proportional to eddy viscosity, and thermodynamic energy and diffusion equations are solved for prescribed velocity and eddy diffusivity profiles in the above manner. When Equations (10.2) and (10.3) are used above the constant stress layer, however, the conjugate relationship $n = 1 - m$ need not be satisfied; this relationship may be too restrictive even in the surface layer.

10.1.2 The logarithmic profile law

Let us focus our attention now to the neutral surface layer over a flat and uniform surface in which Coriolis force can be ignored and the momentum flux may be considered constant, independent of height. Furthermore, we also exclude from our consideration any viscous sublayer that may exist over a smooth surface, or the canopy or roughness sublayer in which the flow is very likely to be disturbed by individual roughness elements. For the remaining fully turbulent, horizontally homogeneous surface layer, a simple similarity hypothesis can be used to obtain the velocity distribution, namely, the mean wind shear $\partial U/\partial z$ is dependent only on the height z above the surface (more appropriately, above a suitable reference plane near the surface), the surface drag, and the fluid density, i.e.,

$$\partial U/\partial z = f(z, \tau_0, \rho) = f(z, \tau_0/\rho) \tag{10.4}$$

An implied assumption in this similarity hypothesis is that the influence of other possible parameters, such as the surface roughness, horizontal pressure gradients (geostrophic winds), and the PBL height, is fully accounted for in τ_0, which then determines the velocity gradients in the surface layer. Note that it is appropriate to combine τ_0 and ρ into their ratio τ_0/ρ, which represents the kinematic momentum flux, because the dependent variable does not contain any mass dimension.

The only characteristic velocity scale given by the above surface layer similarity hypothesis is the so-called friction velocity $u_* \equiv (\tau_0/\rho)^{1/2}$, and the only characteristic length scale is z. Then, from dimensional analysis it follows that the dimensionless wind shear

$$(z/u_*)(\partial U/\partial z) = \text{const.} = 1/k \tag{10.5}$$

where k is called von Karman's constant.

The above similarity relation has been verified by many observed velocity profiles in laboratory boundary layer, channel, and pipe flows, as well as in the near-neutral atmospheric surface layer. The von Karman constant is presumably a universal constant for all surface or wall layers. However, it is an empirical constant with a value of about 0.40; it has not been possible to determine it with an accuracy better than 5% (Hogstrom, 1985).

Note that Equation (10.5) also follows from the mixing length and eddy viscosity hypotheses, if one assumes that $l = kz$ and/or $K_m = kzu_*$ in the constant-flux surface layer. Integration of Equation (10.5) with respect to z gives the well-known logarithmic velocity profile law,

$$U/u_* = (1/k)\ln(z/z_0) \tag{10.6}$$

in which z_0 has been introduced as a dimensional constant of integration and is commonly referred to as the roughness parameter or roughness length. The physical significance of z_0 and its relationship to surface roughness characteristics will be discussed later.

For a comparison of the power-law and logarithmic wind profiles, Equation (10.6) can be written as

$$U/U_r = 1 + \ln(z/z_r)/\ln(z_r/z_0) \tag{10.7}$$

Figure 10.3 compares the plots of U/U_r versus z/z_r, according to Equations (10.2) and (10.7), for different values of m and z_r/z_0, respectively. There is no exact correspondence between these two profile parameters, because the two profile shapes are different. An approximate relationship can be obtained by matching the velocity gradients, in addition to the mean velocities, at the reference height z_r:

$$m = \frac{d(\ln U)}{d(\ln z)} = \frac{z}{U}\frac{\partial U}{\partial z}\bigg|_{z=z_r} = \left(\frac{z_r}{z_0}\right)^{-1} \tag{10.8}$$

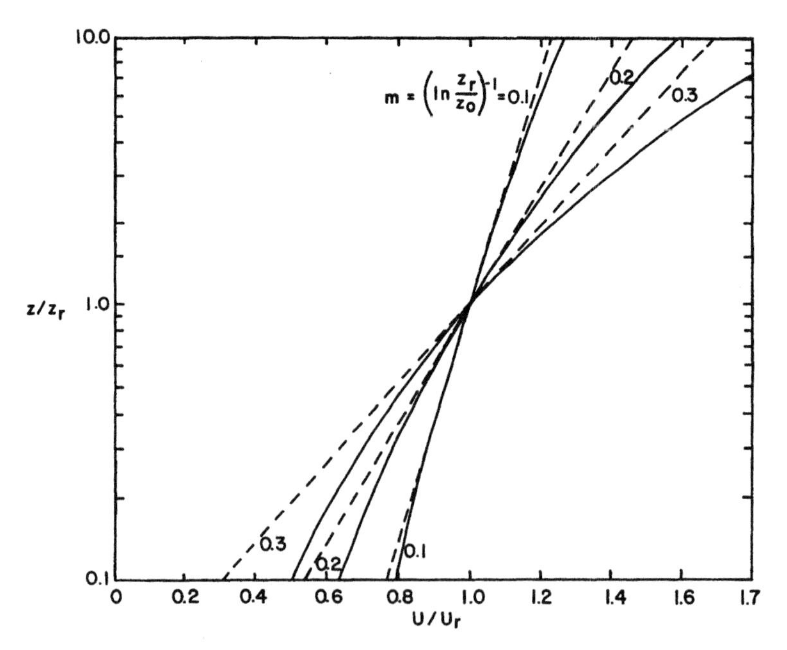

Figure 10.3 Comparison of hypothetical wind profiles following the power-law Equation (10.2) (—) and the log-law Equation (10.7) (– – –) with their parameters related by Equation (10.8).

Even with this type of matching, the two profiles deviate more and more as z deviates farther and farther from the matching height z_r. The theoretical relationship given by Equation (10.8) between the power-law exponent m and the surface roughness parameter is also compared with observations in Figure 10.2. For different data sets represented in that figure, z_r lies in the range between 10 and 100 m.

The logarithmic profile law has a sounder theoretical and physical basis than the power-law profile, especially within the neutral surface layer. The former implies linear variations of eddy viscosity and mixing length with height, irrespective of the surface roughness, while the power-law profile implies the incorrect and unphysical, nonlinear behavior of K_m and l_m, even close to the surface.

Many observed wind profiles near the surface under near-neutral conditions have confirmed the validity of Equation (10.6) up to heights of 20–200 m, depending on the PBL height. Figure 10.4 illustrates some of the observed wind

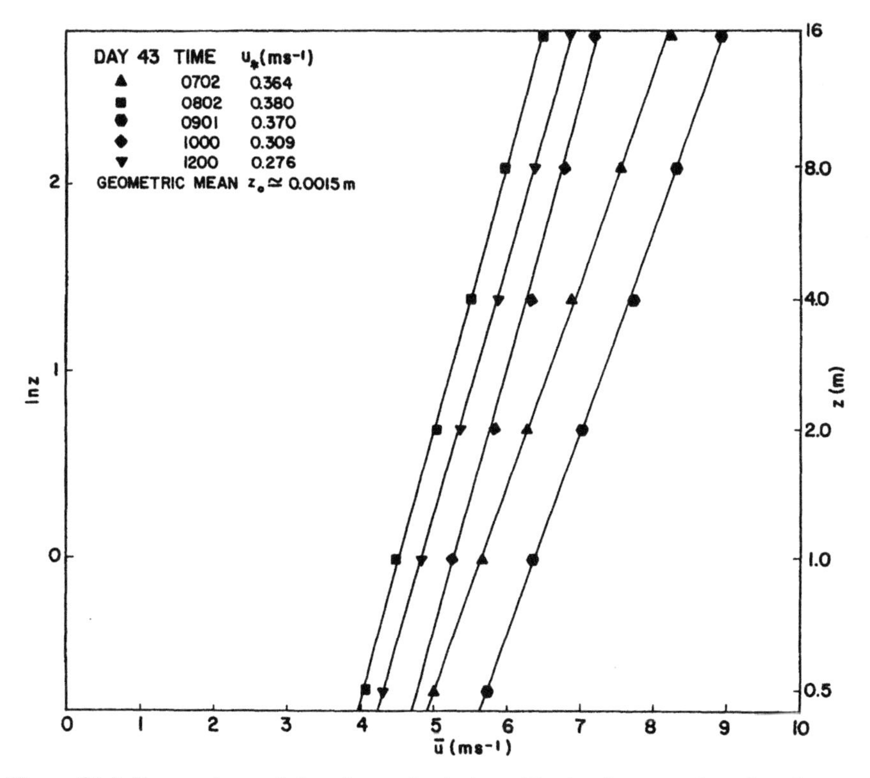

Figure 10.4 Comparison of the observed wind profiles in the neutral surface layer of Wangara Day 43 with the log law [Equation (10.6)] (solid lines). [Data from Clarke *et al.* (1971).]

profiles during the Wangara Experiment which was conducted over a low grass surface in southern Australia. These plots of $\ln z$ versus U also illustrate how the best-fitted straight line through observed data points can be used to estimate the surface-layer parameters u_* and z_0. Note that, according to Equation (10.6), the slope of the line must be equal to k/u_*, while its intercept on the $\ln z$ axis where $U = 0$ must be equal to $\ln z_0$. Thus estimates of slope and intercept of the best-fitted line yield u_* and z_0. If there are several wind profiles available at the same site, one can determine z_0 as the geometric mean of the values obtained from individual profiles.

In the engineering fluid mechanics literature, Equation (10.6) is referred to as the law of the wall, which has been found to hold in all kinds of pipe and channel flows, as well as in boundary layers. There, the distinction is made between aerodynamically smooth and rough surfaces. A surface is considered aerodynamically smooth if the small-scale surface protuberances or irregularities are sufficiently small to allow the formation of a laminar or viscous sublayer in which surface protuberances are completely submerged. If small-scale surface irregularities are large enough to prevent the formation of a viscous sublayer, the surface is considered aerodynamically rough. It is found from laboratory measurements that the thickness of the laminar sublayer in which the velocity profile is linear is about $5\nu/u_*$. For atmospheric flows the sublayer thickness would be of the order of 1 mm or less, while surface protuberances are generally larger than 1 mm. Therefore, almost all natural surfaces are aerodynamically rough; the only exceptions might be smooth ice, snow, and mud flats, as well as water surfaces under weak winds ($U_{10\,\mathrm{m}} < 3\ \mathrm{m\ s^{-1}}$). The law of the wall (velocity-profile law) for an aerodynamically smooth surface is usually expressed in the form (Monin and Yaglom, 1971, Chapter 3)

$$\frac{U}{u_*} = \frac{1}{k}\ln\frac{u_* z}{\nu} + 5.1 \cong \frac{1}{k}\ln\left(\frac{u_* z}{0.13\nu}\right) \qquad (10.9)$$

in which the characteristic height scale of the viscous sublayer (ν/u_*) is used in place of z_0 for a rough surface, and the numerical constant is evaluated from laboratory measurements in smooth flat-plate boundary layer and channel flows.

A comparison of Equations (10.6) and (10.9) shows that, if the former is used to represent wind profiles over a smooth surface, one should expect the roughness parameter to decrease with increasing u_*, according to the relation $z_0 \cong 0.13\nu/u_*$. For a rigid, aerodynamically rough surface, however, z_0 is expected to remain constant, independent of u_*. Some surfaces are neither smooth nor completely rough, but fall into a transitional roughness regime. In nature, lake, sea and ocean surfaces may fall into this transition between smooth and rough surfaces at moderate wind speeds ($2.5 < U_{10} < 7.5$ m s^{-1}) (Kitaigorodski, 1970).

The transfer of momentum from the atmosphere to the ocean or sea surface is of considerable importance to meteorologists, as well as to physical oceanographers. A part of this momentum goes into the generation of surface waves, while the remaining portion is responsible for setting up drift currents and turbulence in the upper layers of the ocean. The relative partitioning of momentum between these different modes of motion in water depends on such factors as the stage of wave growth, duration and fetch of winds, etc., and is not easy to determine. But the total momentum exchange can be determined from measurements in the atmospheric surface layer, well above the level of highest wave crests.

When one considers extending the 'law of the wall' similarity to air flow over a moving water surface, the effects of both the surface drift currents and surface waves should be taken into account. The surface current, which is typically 3–5% of the wind speed at 10 m, but can be much stronger in certain regions (e.g., the Gulf Stream, Kuroshio, and other ocean currents), can easily be taken into account by considering the air motion relative to that of surface water ($U = U_a - U_w$). The influence of moving surface waves on mean wind profile in the surface layer is much more difficult to discern.

10.1.3 The velocity-defect law

Above the surface layer, the Coriolis effects become significant and it is more appropriate to consider the deviations of actual velocity from the geostrophic velocity. For flat-plate boundary layer and channel flows, similarity considerations for the deviations of velocity from the ambient velocity lead to the so-called velocity-defect law (Monin and Yaglom, 1971, Chapter 3)

$$(U - U_\infty)/u_* = F(z/h) \tag{10.10}$$

The analogous geostrophic departure laws proposed for the neutral barotropic PBL are (Tennekes, 1982):

$$\begin{aligned} (U - U_g)/u_* &= F_u(fz/u_*) \\ (V - V_g)/u_* &= F_v(fz/u_*) \end{aligned} \tag{10.11}$$

These are based on the similarity hypothesis for the barotropic, neutral PBL that departures in mean velocity components from their geostrophic values are functions of z, u_* and f only. This similarity hypothesis was originally proposed by Kazanski and Monin (1961). It also implies that the neutral PBL height must be proportional to u_*/f, i.e., $h = cu_*/|f|$ where c is an empirical constant with estimated values between 0.2 and 0.3.

A satisfactory confirmation of Equation (10.11) and empirical determination of the similarity functions F_u and F_v has been made from atmospheric observations under near-neutral stability conditions. A strictly neutral, stationary, and barotropic PBL, which is not constrained by any inversion from above, is so rare in the atmosphere that relevant observations of the same do not exist. The averages of wind profiles measured under slightly unstable and slightly stable conditions appear to follow the above similarity scaling (Clarke, 1970; Nicholls, 1985). Some laboratory simulations of the neutral Ekman layer in a rotating wind tunnel flow have also confirmed the same. Of course, there is plenty of experimental support for the validity of the velocity-defect law [Equation (10.10)] in nonrotating flows.

Example Problem 1

The following mean wind speeds were measured over a short grass surface under near-neutral stability conditions:

z (m):	0.5	1	2	4	8	16
U(m s^{-1}):	7.82	8.66	9.54	10.33	11.22	12.01

(a) Determine the roughness length and friction velocity from a plot of $\ln z$ versus U.

(b) Using appropriate similarity relations, estimate eddy viscosity and mixing length at 10 and 100 m.

Solution

(a) Plotting $\ln z$ versus U and drawing the best-fitted regression line through the data points, or simply regressing $\ln z$ against U in a programmable calculator, one obtains

$$\ln z = 0.824U - 7.15$$

Comparing this with Equation (10.6), which can be rewritten as

$$\ln z = \frac{k}{u_*} U - \ln z_0$$

we obtain k/u_* = slope of the line = 0.824 s m^{-1}, or

$$u_* = 0.40/0.824 = 0.485 \text{ m s}^{-1},$$

and $\ln z_0 = -7.15$, so that

$$z_0 \cong 7.9 \times 10^{-4}\ \text{m} = 0.079\ \text{mm}.$$

(b) In the neutral surface layer, eddy viscosity and mixing length are given by

$$K_m = kzu_*;\ l_m = kz$$

from which the following values can be estimated at the desired height levels (10 and 100 m):

z (m)	K_m ($m^2\ s^{-1}$)	l_m (m)
10	1.94	4.0
100	19.40	40

10.2 Surface Roughness Parameters

The aerodynamic roughness of a flat and uniform surface may be characterized by the average height (h_0) of the various roughness elements, their areal density, characteristic shapes, and dynamic response characteristics (e.g., flexibility and mobility). All the above characteristics would be important if one were interested in the complex flow field within the roughness or canopy layer. There is not much hope for a generalized and, at the same time, simple theoretical description of such a three-dimensional flow field in which turbulence dominates over the mean motion. In theoretical and experimental investigations of the fully developed surface layer, however, the surface roughness is characterized by only one or two roughness characteristics which can be empirically determined from wind-profile observations.

10.2.1 Roughness length

The roughness length parameter z_0 introduced in Equation (10.6) is one such characteristic. In practice, z_0 is determined from the least-square fitting of Equation (10.6) through the wind-profile data, or by graphically plotting $\ln z$ versus U and extrapolating the best-fitted straight line down to the level where $U = 0$; its intercept on the ordinate axis must be $\ln z_0$. One should note, however, that this is only a mathematical or graphical procedure for estimating z_0 and that Equation (10.6) is not expected to describe the actual wind profile below the tops of roughness elements. An assumption implied in the derivation of (10.6) is that $z_0 \ll z$.

Empirical estimates of the roughness parameter for various natural surfaces can be ordered according to the type of terrain (Figure 10.5) or the average height of the roughness elements (Figure 10.6). Although z_0 varies over five orders of magnitude (from 10^{-5} m for very smooth water surfaces to several meters for forests and urban areas), the ratio z_0/h_0 falls within a much narrower range (0.03–0.25) and increases gradually with increasing height of roughness elements. For uniform sand surfaces Prandtl and others have suggested a value of $z_0/h_0 \cong 1/30$. An average value of $z_0/h_0 = 0.15$ has been determined for various crops and grasslands; the same may also be used for many other natural surfaces (Plate, 1971). Figure 10.5 gives some typical values of z_0 for different types of surfaces and terrains. Figure 10.6 relates z_0 to h_0 for different types of crop canopies.

Wind-profile measurements from fixed towers in shallow coastal waters and from stable floating buoys (ships are not suitable for this, as they present too much of an obstruction to the air flow) over open oceans also follow the logarithmic law [Equation (10.6)] under neutral stability conditions. However, the roughness parameter (z_0) determined from this is found to be related to both the wind and wave fields in a rather complex manner and can vary over a very wide range (say, 10^{-5}–10^{-2} m).

There have been a number of semiempirical and theoretical attempts at relating z_0 to the friction velocity, as well as to the average height, the phase speed, and the stage of development of surface waves. The simplest and still the most widely used relationship is that proposed by Charnock (1955) on the basis of dimensional arguments

$$z_0 = a(u_*^2/g) \tag{10.12}$$

where a is an empirical constant. The basic assumptions implied in Charnock's hypothesis (z_0 is uniquely determined by u_* and g) are that the winds are blowing steadily and long enough for the wave field to be in complete equilibrium with the wind field, independent of fetch, and that the surface is aerodynamically rough.

Individual estimates of z_0 plotted as a function of u_*^2/g invariably show large scatter, because they usually pertain to different fetches and different stages of wave development and also because of large errors in the experimental determination of z_0. However, when the available data sets are block averaged, a definite trend of z_0 increasing in proportion to u_*^2/g does emerge. Figure 10.7 shows such block-averaged data compiled from some 33 experimental sets of wind profile and drag data (Wu, 1980). Note that Charnock's relation is verified, particularly for large values of z_0 and u_*^2/g, and $a \cong 0.018$. Empirical

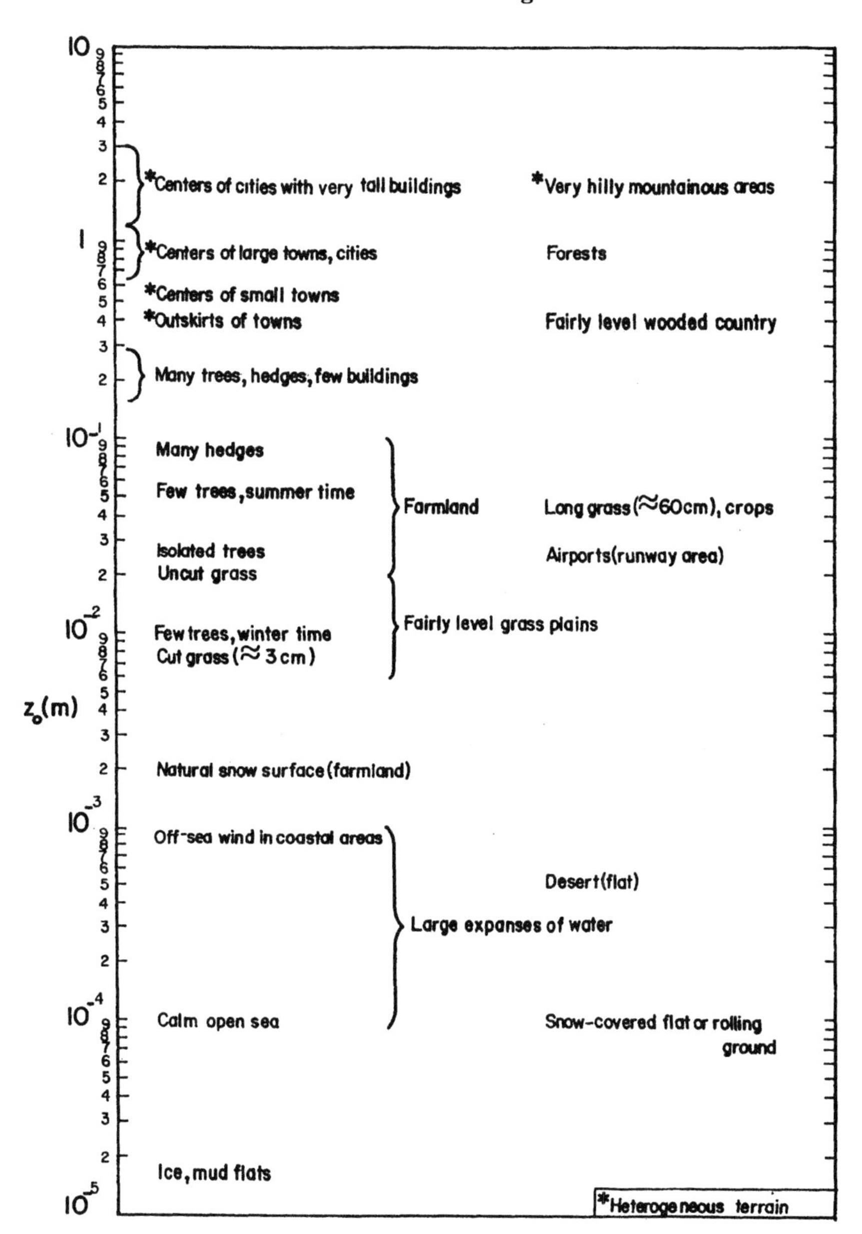

Figure 10.5 Typical values, or range of values, of the surface roughness parameter for different types of terrain. [From tables by the Royal Aeronautical Society (1972).]

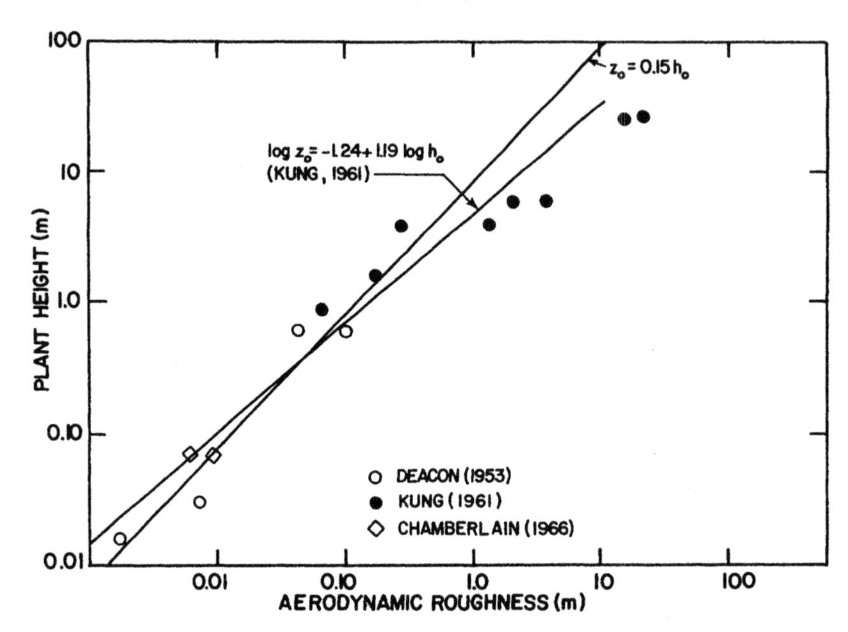

Figure 10.6 Relationship between the aerodynamic roughness parameter and the average vegetation height. [From Plate (1971).]

estimates of a by other investigators have ranged between 0.01 and 0.035 (Garratt, 1992).

The simple state of affairs assumed in Charnock's hypothesis rarely exists over lakes and oceans. More frequently, the wave field is not in equilibrium with local winds, but depends on the strength, fetch, and time history of the wind field. At low wind speeds, the surface can also become aerodynamically smooth, or be in a transitional regime between smooth and rough. A comprehensive treatment of sea-surface roughness at different development stages of wind-generated waves is given by Kitaigorodski (1970). The fundamental parameter indicating the stage of wave development, in his theory, is the ratio C_0/u_*, where C_0 is the phase speed of the most dominant (corresponding to the peak in the wave-height spectrum) waves. It is shown that only in the equilibrium stage of wave development is z_0 proportional to u_*^2/g. Furthermore, on the basis of analogy with the aerodynamic roughness of a rigid surface, z_0 is assumed proportional to h_0 for a completely rough surface; proportional to ν/u_* for an aerodynamically smooth surface; and a function of both h_0 and ν/u_* in the transitionally smooth or rough regime. Kitaigorodski has suggested plausible expressions for the roughness parameter for the various combinations of roughness and wave regimes, some of them involving still unknown empirical

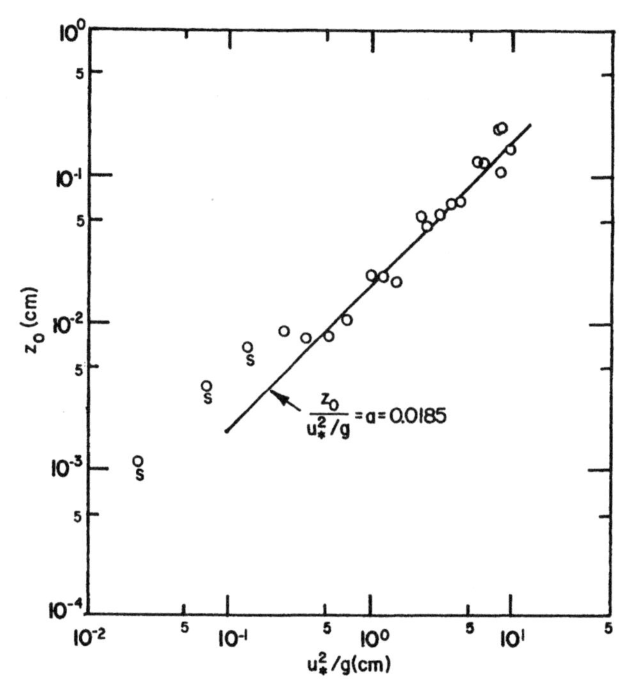

Figure 10.7 Observed roughness parameter of sea surface as a function of u_*^2/g compared with Charnock's (1955) formula. [After Wu (1980).]

constants or functions. For the fully-developed wave field, a simple interpolation relation

$$z_0 = a\frac{u_*^2}{g} + 0.13\frac{\nu}{u_*} \tag{10.13}$$

has been suggested (Smith *et al.*, 1996).

At low wind speeds (say $U_{10} < 7\ \text{m s}^{-1}$) and in the early stages of wave development, the sea surface appears to be remarkably smooth, before waves start to break. In fact, the aerodynamic roughness of even a fully developed sea under a tropical storm is observed to be less than that of mown grass. This is largely due to the fact that waves are generated by and move along with the wind and do not offer as much resistance to the flow as would an immobile surface with similar shape. There is also strong evidence suggesting that sea-surface drag is primarily due to small ripples and wavelets with phase speeds less than the near-surface wind speed, and that large waves make only a minor contribution to the drag.

Different surface roughness regimes mentioned earlier are found to occur over oceans. In the absence of swell, the sea surface is found to be aerodynamically smooth during calm and weak winds ($U_{10} < 2.5\ \mathrm{m\ s^{-1}}$), transitionally rough in moderate winds ($2.5 \leq U_{10} \leq 7.5\ \mathrm{m\ s^{-1}}$), and completely rough in strong winds ($U_{10} > 7.5\ \mathrm{m\ s^{-1}}$). The above criteria should be considered only approximate and limited to fully developed (equilibrium) wind-generated waves. The data points corresponding to small values of u_*^2/g in Figure 10.7 probably belong to the transitional roughness regime.

10.2.2 Displacement height

For very rough and undulating surfaces, the soil–air or water–air interface may not be the most appropriate reference datum for measuring heights in the surface layer. The air flow above the tops of roughness elements is dynamically influenced by the ground surface as well as by individual roughness elements. Therefore, it may be argued that the appropriate reference datum should lie somewhere between the actual ground level and the tops of roughness elements. In practice, the reference datum is determined empirically from wind-profile measurements in the surface layer under near-neutral stability conditions. The modified logarithmic wind-profile law used for this purpose is

$$U/u_* = (1/k)\ln[(z' - d_0)/z_0] \tag{10.14}$$

in which d_0 is called the zero-plane displacement or displacement height and z' is the height measured above the ground level.

For a plane surface, d_0 is expected to lie between zero and h_0, depending on the areal density of roughness elements. The displacement height may be expected to increase with increasing roughness density and approach a value close to h_0 for very dense canopies in which the flow within the canopy might become stagnant, or independent of the air flow above the canopy. These expectations are indeed borne out by empirical estimates of d_0 for vegetative canopies (see Figure 10.8) which indicate that the appropriate datum for wind-profile measurements over most vegetative canopies is displaced above the ground level by 70–80% of the average height of vegetation. The zero-plane displacement in an urban boundary layer can similarly be expected to be a large fraction of the average building height.

After the mean height of roughness elements, the next important characteristic of roughness morphology is the ratio $\lambda = A_f/A_h$, where A_f is the total frontal area of roughness elements and A_h is the horizontal area covered by them. As the frontal area density (λ) increases, the ratio d_0/h_0 may be expected to increase monotonically to its maximum value of one. The ratio z_0/h_0 is also

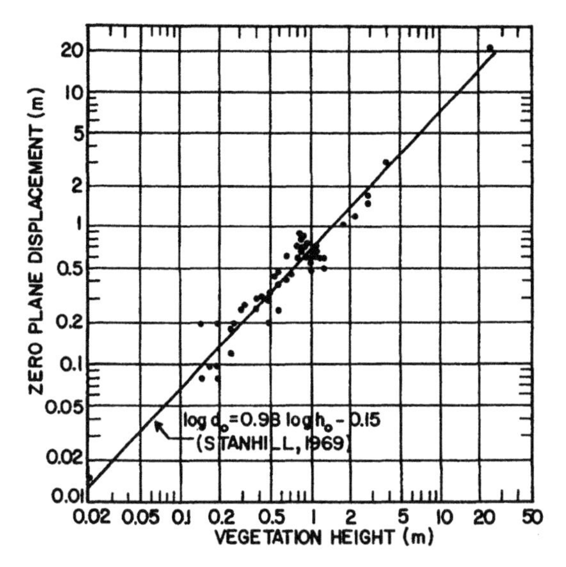

Figure 10.8 Relationship between the zero-plane displacement and average vegetation height for different types of vegetation. [After Stanhill (1969).]

expected to increase and attain its maximum value at some optimum value of $\lambda < 1$. Wind tunnel experiments with regular arrays of simulated roughness elements have confirmed these expectations (McDonald *et al.*, 1998). Methods of determining surface roughness parameters from an analysis of surface morphometry of urban areas have also been suggested (Grimmond and Oke, 1999).

Figure 10.9 shows a verification of Equation (10.14) against observed wind profiles over bare and vegetative surfaces, using appropriate estimates of z_0 and d_0 for each surface.

10.3 Surface Drag and Resistance Laws

A direct measurement of the wind drag or stress on a relatively smooth surface can be made with an appropriate drag plate of a nominal diameter of 1–2 m and surface representative of the terrain around it (Bradley, 1968). The larger the surface roughness, the more difficult it becomes for the drag plate to represent (simulate) such a roughness. Therefore, this method of drag measurement is limited to flat and uniform surfaces which are bare or have low vegetation. In most other cases, and also when drag-plate measurements cannot be made, the surface stress is determined indirectly from wind measurements in the surface layer.

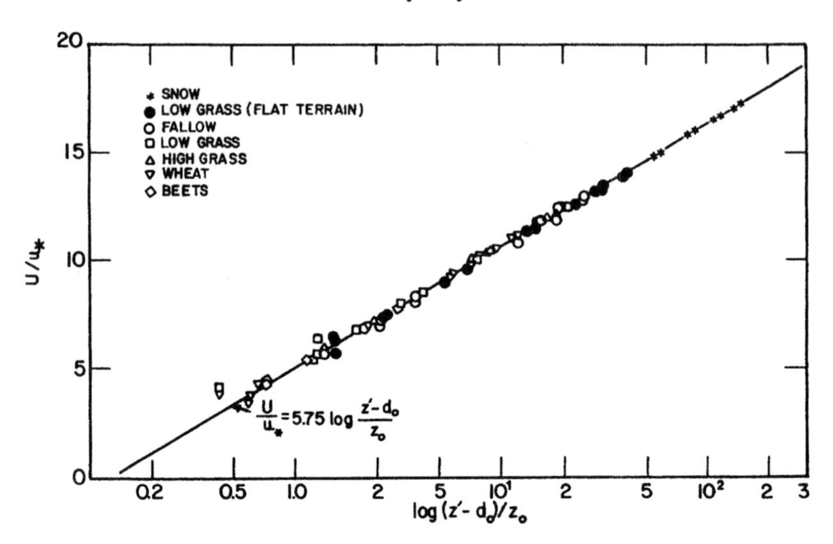

Figure 10.9 Comparison of observed velocity profiles over crops with the log law [Equation (10.14)]. [From Plate (1971).]

10.3.1 Surface drag law

If mean winds are observed at a standard reference height z_r, the surface stress can be determined using a drag relation

$$\tau_0 = \rho C_D U_r^2 \tag{10.15}$$

in which C_D is a dimensionless drag coefficient which depends on the surface roughness (more appropriately, on the ratio z_r/z_0) and atmospheric stability near the surface. In particular, for near-neutral stability, the drag coefficient is given by

$$C_{DN} = k^2/[\ln(z_r/z_0)]^2 \tag{10.16}$$

which follows from Equations (10.6) and (10.15). Note that for a given z_r, C_{DN} increases with increasing surface roughness. For a standard reference height of 10 m, values of C_{DN} range from 1.0×10^{-3} for large lake and ocean surfaces at moderate wind speeds ($U_{10} < 6$ m s^{-1}) to 7.5×10^{-3} for tall crops and moderately rough ($z_0 < 0.1$ m) surfaces. For very rough surfaces (e.g., forests and urban areas), however, a reference height of 10 m would not be adequate; it must be at least 1.5 times the height of roughness elements.

There has been considerable interest in knowing the dependence of C_{DN} over ocean surfaces on wind speed, sea state, and other factors. To this end, many experimental determinations of C_{DN} have been made by different investigators, using different measurement techniques. Early determinations were largely based on rather crude estimates of the tilt of water surface and the geostrophic departure of flow in the PBL, and are not considered very reliable. Later, the sea-surface drag was determined by using the more reliable eddy correlation, gradient, and profile methods. A comprehensive review of the existing data on C_D up to 1975 was made by Garratt (1977). After applying the appropriate corrections for atmospheric stability and actual observation height, the neutral drag coefficient C_{DN}, referred to the standard height of 10 m, was obtained as a function of wind speed at 10 m (U_{10}).

Figure 10.10(a) shows the block-averaged values of $C_{DN} \pm 1$ standard deviation (indicated by vertical bars) for intervals of U_{10} of 1 m s^{-1}, based on the eddy correlation method (closed circles) and wind-profile method (open circles). The number of data points used for averaging in each 1 m s^{-1} interval are also indicated above the abscissa axis (the top line refers to closed circles and the bottom line to open circles). Note that there are too few data points to give reliable estimates of C_{DN} at high wind speeds ($U_{10} > 15$ m s^{-1}). However, some investigators have inferred the surface drag in hurricanes, using the geostrophic departure method. Others have measured C_{DN} at high wind speeds in laboratory wind flume experiments. Figure 10.10(b) shows that there is a remarkable similarity between the hurricane data and wind flume data (usually referred to a height of 10 cm, assuming a scaling factor of 100:1), despite the large expected errors in the former and dissimilarity of wave fields in the two cases.

Considering the overall block-averaged data for the whole wide range of wind speeds represented in Figure 10.10, there is no doubt that C_{DN} increases with wind speed. The trend appears to be quite consistent with Charnock's formula [Equation (10.12)], which, in conjunction with Equations (10.15) and (10.16) yields

$$\ln C_{DN} + kC_{DN}^{-1/2} = \ln(gz/aU_r^2) \tag{10.17}$$

After determining the best-fitting value of $a \cong 0.0144$ with $k = 0.41$ (for the more commonly used value of $k = 0.4$, one would obtain $a \cong 0.017$, which is closer to the value indicated in Figure 10.7), from a regression of $\ln C_{DN} + kC_{DN}^{-1/2}$ on $\ln U_{10}$, Equation (10.17) is represented in Figures 10.10a and b. In the completely rough regime ($U_{10} > 7$ m s^{-1}), in particular, Charnock's relation is in good agreement with the experimental data. The variation of C_{DN} with U_{10} can also be approximated by a simple linear relation (Garratt, 1977)

$$C_{DN} = (0.75 + 0.067U_{10}) \times 10^{-3} \tag{10.18}$$

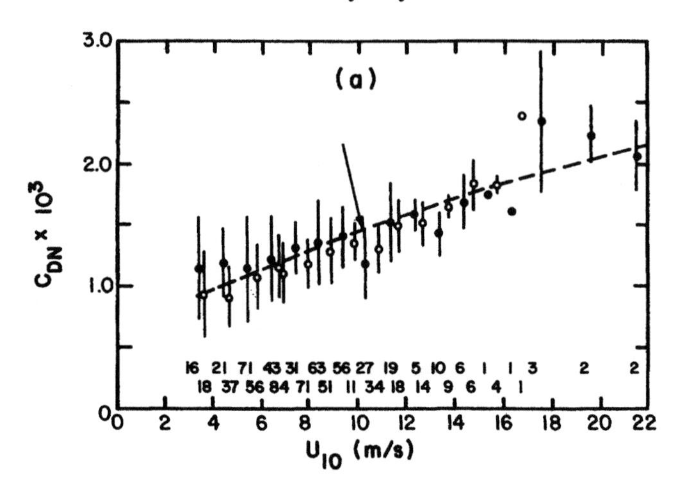

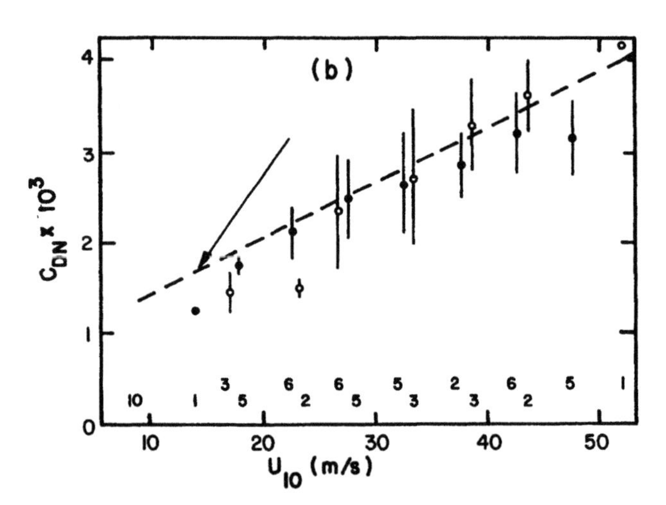

Figure 10.10 Neutral drag coefficient as a function of wind speed at 10 m height compared with Charnock's formula [Equation (10.17), indicated by arrows in (a) and (b)] with $a = 0.0144$. Block-averaged values are shown for (a) 1 m s^{-1} intervals, based on eddy correlation and profile methods, and (b) 5 m s^{-1} intervals, based on geostrophic departure method and wind flume simulation experiments. [After Garratt (1977).]

in which U_{10} is to be expressed in meters per second. A linear relationship, similar to Equation (10.18), is also suggested by other measurements of drag in gale-force winds (Smith, 1980).

There is some controversy over the variation of C_{DN} with U_{10} at low and moderate wind speeds (say, $U_{10} < 7.5 \text{ m s}^{-1}$). At very low wind speeds ($U_{10} < 2.5 \text{ m s}^{-1}$), observations from laboratory flume experiments and from over lakes and oceans indicate a slight tendency for C_{DN} to decrease with the increase in wind speed, as would be expected for an aerodynamic smooth surface. But sea surface is rarely completely smooth; ripples and wavelets generated by weak to moderate winds give it a moderate roughness, which is transitional between smooth and completely rough regimes. Some investigators have argued that the dependence of C_{DN} on U_{10} should be weak, if any, in this transitional regime ($2.5 < U_{10} < 7.5 \text{ m s}^{-1}$). The data of Figure 10.10 do suggest a weaker dependence of C_{DN} on U_{10} in this range. Other oceanic data have even indicated a nearly constant value of $C_{DN} \cong 1.2 \times 10^{-3}$ in the moderate range of wind speeds (Kraus, 1972).

Few investigators have sought to investigate the dependence of C_{DN} on wind direction, fetch, and sea-state parameters, such as C_0/u_* (Kitaigorodski, 1970; SethuRaman, 1978; Smith, 1980). There is only a slight tendency for the drag coefficient to decrease with increasing fetch. It has also been argued that the coefficient a in Charnock's formula should decrease with increasing C_0/u_* and may reach a constant asymptotic value only at the equilibrium stage of wave development. This would imply that C_{DN} should also decrease with increasing C_0/u_*. An observational evidence in support of such a trend is given by SethuRaman (1978), who also shows dramatic changes in C_{DN} in response to a rapid shift in wind direction.

10.3.2 Geostrophic drag relations

Alternative parameterizations of the surface drag use the surface geostrophic wind speed (G_0), or the actual wind speed at the top of the PBL, instead of the near-surface wind in expressing the surface stress, e.g.,

$$\tau_0 = \rho C_D G_0^2 \tag{10.19}$$

in which the value of C_D is expected to be smaller than that in Equation (10.15). The ratio $u_*/G_0 = C_D^{1/2}$ is often called the geostrophic drag coefficient and denoted by c_g. For the neutral PBL, c_g can be expressed as a function of the surface Rossby number, G_0/fz_0, using the so-called geostrophic resistance laws (Tennekes, 1982):

$$U_g/u_* = \frac{1}{k}[\ln(u_*/fz_0) - A]$$
$$V_g/u_* = -B/k \tag{10.20}$$

where A and B are similarity constants.

An elegant derivation of the above resistance laws from the asymptotic matching of the neutral surface layer and the PBL similarity wind-profile relations (10.6) and (10.11) is given by Tennekes (1982). The similarity constants have been empirically determined from the observed wind profile and geostrophic winds under near-neutral conditions. Earlier estimates from observations over land surfaces showed fairly large scatter, probably due to the large sensitivity of A and B to the stability and baroclinicity of the PBL (Sorbjan, 1989). More recently, Nicholls (1985) used the mean flow and turbulence data obtained over the ocean during the Joint Air–Sea Interaction Experiment (JASIN) to estimate $A \cong 1.4$ and $B \cong 4.2$. From Equations (10.19) and (10.20), the geostrophic drag coefficient can be expressed as

$$c_g = k[(\ln c_g + \ln \mathrm{Ro} - A)^2 + B^2]^{-1/2} \tag{10.21}$$

where $\mathrm{Ro} = G_0/fz_0$ is the surface Rossby number. Thus, c_g is a decreasing function of Ro; its evaluation and plot is left as an exercise for the reader.

10.4 Turbulence

Turbulence in a neutrally stratified boundary layer is entirely of mechanical origin and depends on the surface friction and vertical distribution of wind shear. From similarity considerations discussed in Section 10.1, the primary velocity scale is u_* and normalized standard deviations of velocity fluctuations (σ_u/u_*, σ_v/u_*, and σ_w/u_*) must be constants in the surface layer and some functions of the normalized height fz/u_* in the outer layer. Surface layer observations from different sites indicate that under near-neutral conditions $\sigma_u/u_* \cong 2.4$, $\sigma_v/u_* \cong 1.9$, and $\sigma_w/u_* \cong 1.3$ (Panofsky and Dutton, 1984, Chapter 7). Observed values of turbulence intensity (σ_u/U) are shown in Figure 10.11 as a function of the roughness parameter. These are compared with the theoretical relation $\sigma_u/U = l/\ln(z/z_0)$, which is based on $\sigma_u/u_* = 2.5$ and the logarithmic wind-profile law. The above theoretical relation seems to overestimate turbulence intensities over very rough surfaces. Counihan (1975) has suggested an

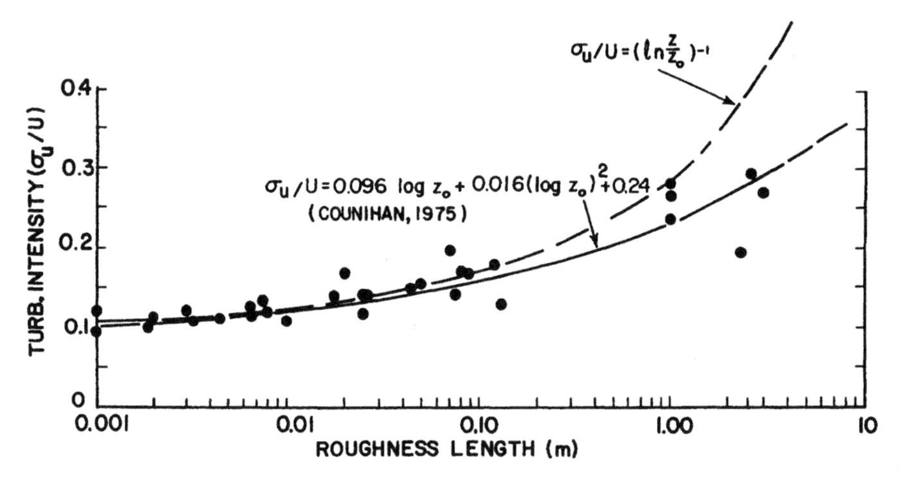

Figure 10.11 Variation of turbulence intensity in the near-neutral surface layer with the roughness length. [After Counihan (1975). Reprinted with permission from *Atmospheric Environment*. Copyright © (1975), Pergamon Journals Ltd.]

alternative empirical relationship between σ_u/U and $\log z_0$, which provides a better fit to the observed data in Figure 10.11.

Measurements of turbulence in the upper part of the near-neutral PBL are limited to few experiments in the marine PBL (Pennell and LeMone, 1974; Nicholls, 1985) but numerical model results indicate approximately exponential decrease of σ_u/u_*, etc., with height. For rough estimates of turbulent fluctuations, we suggest the following relations:

$$\begin{aligned} \sigma_u/u_* &= 2.4 \exp|-afz/u_*| \\ \sigma_v/u_* &= 1.9 \exp|-afz/u_*| \\ \sigma_w/u_* &= 1.3 \exp|-afz/u_*| \end{aligned} \tag{10.22}$$

in which the constant a may depend on the empirical constant c in the PBL height relationship $h = cu_*/|f|$. If we define the PBL height as the level where turbulent kinetic energy or momentum flux reduces to 10% of its value at the surface, then, we find that $a = 1.15/c$ (e.g., for $c = 0.25$, $a = 4.6$). Note that the above relations (10.22) imply the following expression for the TKE:

$$E = 5.5u_*^2 \exp|-2afz/u_*| \tag{10.23}$$

Example Problem 2

For the wind profile data given in the Example Problem 1 for a site in southern Australia (latitude = 34.5°S), estimate the following turbulence parameters at the indicated heights in the neutral PBL:

(a) the drag coefficient for $z_r = 10$ m;
(b) the vertical turbulence intensities at 10 and 100 m;
(c) turbulence kinetic energy (TKE) near the surface;
(d) TKE at 100, 200, and 500 m.

Solution

(a) We can estimate the wind speed at 10 m

$$U_{10} = \frac{u_*}{k} \ln \frac{z}{z_0} = \frac{0.485}{0.40} (\ln 10 + 7.15) = 11.46 \text{ m s}^{-1}$$

Then, $C_D = (u_*/U_{10})^2 = 1.79 \times 10^{-3}$.

(b) In the surface layer,

$$\sigma_w = 1.3u_* = 0.63 \text{ m s}^{-1}$$

The vertical turbulence intensity is defined as

$$i_w = \sigma_w/U$$

and depends on the wind speed at a given height.
At $z = 10$ m,

$$i_w = \frac{\sigma_w}{U_{10}} = 0.055$$

Similarly, at $z = 100$ m,

$$i_w = \frac{\sigma_w}{U_{100}} = 0.044$$

(c) Turbulence kinetic energy near the surface is given by

$$\begin{aligned} E &= \tfrac{1}{2}(\overline{u^2} + \overline{v^2} + \overline{w^2}) \\ &= \tfrac{1}{2}[(2.4)^2 + (1.9)^2 + (1.3)^2]u_*^2 \\ &\cong 5.5u_*^2 = 1.29 \text{ m}^2 \text{ s}^{-2} \end{aligned}$$

(d) For estimating TKE at higher levels in the PBL, we use Equation (10.23) with $a = 4.6$. For the given latitude $\phi = -34.5°$

$$f = 2\Omega \sin \phi = -0.826 \times 10^{-4} \text{ s}^{-1}$$

Then, the estimated values of TKE at various heights are:

z (m):	100	200	500
E (m^2 s^{-2}):	1.11	0.95	0.59

The estimated PBL height $h = 0.25u_*/|f| \cong 1468$ m.

10.5 Applications

Semiempirical wind-profile laws and other relations discussed in this chapter may have the following practical applications:

- Determining the wind energy potential of a site.
- Estimating wind loads on tall buildings and other structures.
- Calculating dispersion of pollutants in very windy conditions.
- Characterizing aerodynamic surface roughness.
- Parameterizing the surface drag in large-scale atmospheric models, as well as in wave height and storm surge formulations.

Problems and Exercises

1. Show that in the constant-stress surface layer, the power-law wind profile [Equation (10.2)] implies a power-law eddy viscosity profile [Equation (10.3)] and that the two exponents are related as $n = 1 - m$.

2. Derive the logarithmic wind-profile law on the basis of von Karman's mixing-length hypothesis, which implies

$$l_m = k\frac{\partial U}{\partial z} \Big/ \frac{\partial^2 U}{\partial z^2}$$

in the constant-stress surface layer.

3. Using the logarithmic wind-profile law, plot the expected wind profiles (using a linear height scale) to a height of 100 m for the following combinations of the roughness parameter (z_0) and the mean wind speed at 100 m:
(a) $z_0 = 0.01$ m; $U_{100} = 5$, 10, 15, and 20 m s^{-1};
(b) $U_{100} = 10$ m s^{-1}; $z_0 = 10^{-3}$, 10^{-2}, 10^{-1}, and 1 m.
Also calculate for each profile the friction velocity u_* and the drag coefficient C_D for a reference height of 10 m, and show their values on the graphs.

4. The following mean velocity profiles were measured during the Wangara Experiment conducted over a uniform short-grass surface under near-neutral stability conditions:

Height (m):	0.5	1	2	4	8	16
U (m s^{-1}):	4.91	5.44	6.06	6.64	7.17	7.71

(a) Determine the roughness parameter and the surface stress for the above observation period. Take $\rho = 1.25$ kg m^{-3}.
(b) Estimate eddy viscosity, mixing length, and turbulence intensities at 5 and 50 m heights.

5. If the sea surface behaves like a smooth surface at low wind speeds ($U_{10} < 2.5$ m s^{-1}), like a completely rough surface at high wind speeds ($U_{10} > 7.5$ m s^{-1}), and like a transitionally rough surface at speeds in between, express and plot the drag coefficient C_D as a function of U_{10} in the range $0.5 \leq U_{10}$ m s$^{-1} \leq 50$, assuming the following:
(a) $z_0 = 0.13\,\nu/u_*$ in the aerodynamically smooth regime;
(b) $z_0 = 0.02\,u_*^2/g$ in the aerodynamically rough regime;
(c) $z_0 = 2 \times 10^{-4}$ m in the transitional regime.

6.
(a) Derive the geostrophic drag relations (10.20) for the neutral PBL from the matching of the geostrophic departure laws (10.11) with the logarithmic wind profile law. Note that both the mean velocity and its vertical gradient should be matched in the surface layer.
(b) Using Equation (10.21) with $A = 1.4$ and $B = 4.2$ for the neutral barotropic PBL, calculate and plot the geostrophic drag coefficient (c_g) as a function of the surface Rossby number (Ro) for the wide range of expected values of the latter (say, Ro $= 10^4$–10^9).

7.
(a) Plot σ_u/u_*, σ_v/u_*, σ_w/u_*, and E/u_*^2 as functions of the normalized height fz/u_* in the neutral PBL, using $a = 4.6$.
(b) After comparing the computed profiles of E/u_*^2 versus fz/u_* for $a = 3.5$ and 5.5, discuss the sensitivity of the TKE profile to the empirical constant c in the PBL height relation $h = c|u_*/f|$. In order to remove the dependence on a or c, express E/u_*^2 as an exponential function of z/h.

Chapter 11 | Thermally Stratified Surface Layer

11.1 The Monin–Obukhov Similarity Theory

In the previous chapter, we showed that the atmospheric surface layer under neutral conditions is characterized by a logarithmic wind profile and nearly uniform (with respect to height) profiles of momentum flux and standard deviations of turbulent velocity fluctuations. It was also mentioned that the neutral stability condition is an exception rather than the rule in the lower atmosphere. More often, the turbulent exchange of heat between the surface and the atmosphere leads to thermal stratification of the surface layer and, to some extent, of the whole PBL, as shown in Chapter 5. It has been of considerable interest to micrometeorologists to find a suitable theoretical or semiempirical framework for a quantitative description of the mean and turbulence structure of the stratified surface layer. The Monin–Obukhov similarity theory has provided the most suitable and acceptable framework for organizing and presenting micrometeorological data, as well as for extrapolating and predicting certain micrometeorological information where direct measurements are not available.

11.1.1 The similarity hypothesis

The basic similarity hypothesis first proposed by Monin and Obukhov (1954) is that in a horizontally homogeneous surface layer the mean flow and turbulence characteristics depend only on the four independent variables: the height above the surface z, the surface drag τ_0/ρ, the surface kinematic heat flux $H_0/\rho c_p$, and the buoyancy variable g/T_0 (this appears in the expressions for buoyant acceleration and static stability given earlier). This is an extension of the earlier hypothesis for the neutral surface layer. The simplifying assumptions implied in this similarity hypothesis are that the flow is horizontally homogeneous and quasistationary, the turbulent fluxes of momentum and heat are constant (independent of height), the molecular exchanges are insignificant in comparison with turbulent exchanges, the rotational effects can be ignored in the surface layer, and the influence of surface roughness, boundary layer height, and geostrophic winds is fully accounted for through τ_0/ρ.

Because the four independent variables in the M–O similarity hypothesis involve three fundamental dimensions (length, time, and temperature), according to Buckingham's theorem, one can formulate only one independent dimensionless combination out of them. The combination traditionally chosen in the Monin–Obukhov similarity theory is the buoyancy parameter

$$\zeta = z/L$$

where

$$L = -u_*^3/[k(g/T_0)(H_0/\rho c_p)] \tag{11.1}$$

is an important buoyancy length scale, known as the Obukhov length after its originator. In his 1946 article in an obscure Russian journal (for the English translation, see Obukhov, 1946/1971), Obukhov introduced L as 'the characteristic height (scale) of the sublayer of dynamic turbulence'; he also generalized the semiempirical theory of turbulence to the stratified atmospheric surface layer and used this approach to describe theoretically the mean wind and temperature profiles in the surface layer in terms of the fundamental stability parameter z/L.

One may wonder about the physical significance of the Obukhov length (L) and the possible range of its values. From the definition, it is clear that the values of L may range from $-\infty$ to ∞, the extreme values corresponding to the limits of the heat flux approaching zero from the positive (unstable) and the negative (stable) side, respectively. A more practical range of $|L|$, corresponding to fairly wide ranges of values of u_* and $|H_0|$ encountered in the atmosphere, is shown in Figure 11.1. Here, we have assumed a typical value of $kg/T_0 = 0.013$ m s^{-2} K^{-1}, which may not differ from the actual value of this by more than 15%. In magnitude $|L|$ represents the thickness of the layer of dynamic influence near the surface in which shear or friction effects are always important. The wind shear effects usually dominate and the buoyancy effects essentially remain insignificant in the lowest layer close to the surface ($z \ll |L|$). On the other hand, buoyancy effects may dominate over shear-generated turbulence for $z \gg |L|$. Thus, the ratio z/L is an important parameter measuring the relative importance of buoyancy versus shear effects in the stratified surface layer, similar to the Richardson number (Ri) introduced earlier. The negative sign in the definition of L is introduced so that the ratio z/L has the same sign as Ri. Later on, it will be shown that Ri and z/L are intimately related to each other, even though they have different distributions with respect to height in the surface layer (z/L obviously varies linearly with height, showing the increasing importance of buoyancy with height above the surface).

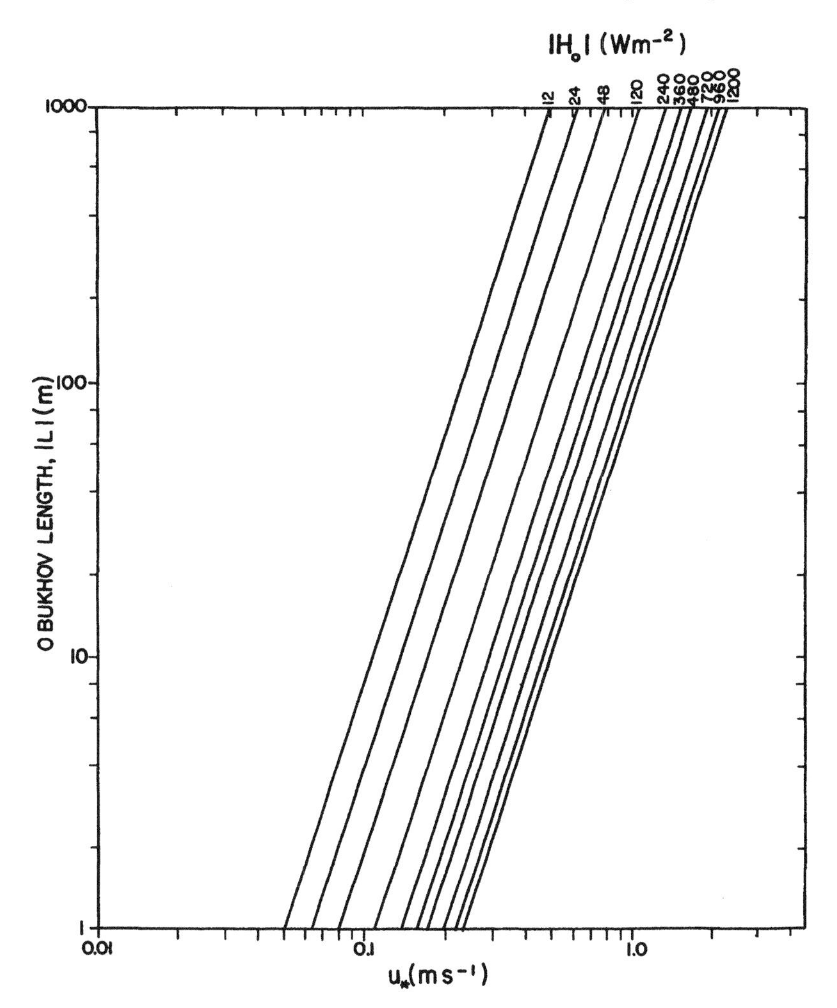

Figure 11.1 Obukhov's buoyancy length as a function of the friction velocity and the surface heat flux.

11.1.2 The M–O similarity relations

The following characteristic scales of length, velocity, and temperature are used to form dimensionless groups in the Monin–Obukhov similarity theory:

Length scales:	z and L
Velocity scale:	u_*
Temperature scale:	$\theta_* = -H_0/\rho c_p u_*$

The similarity prediction that follows from the M–O hypothesis is that any mean flow or average turbulence quantity in the surface layer, when normalized by an appropriate combination of the above-mentioned scales, must be a unique function of z/L only. Thus, a number of similarity relations can be written for the various quantities (dependent variables) of interest. For example, with the x axis oriented parallel to the surface stress or wind (the appropriate surface layer coordinate system), the dimensionless wind shear and potential temperature gradient are usually expressed as

$$\begin{aligned}(kz/u_*)(\partial U/\partial z) &= \phi_m(\zeta) \\ (kz/\theta_*)(\partial \Theta/\partial z) &= \phi_h(\zeta)\end{aligned} \tag{11.2}$$

in which the von Karman constant k is introduced only for the sake of convenience, so that $\phi_m(0) = 1$, and $\phi_m(\zeta)$ and $\phi_h(\zeta)$ are the basic universal similarity functions which relate the constant fluxes

$$\begin{aligned}\tau &= \tau_0 = \rho u_*^2 \\ H &= H_0 = -\rho c_p u_* \theta_*\end{aligned} \tag{11.3}$$

to the mean gradients in the surface layer.

It is easy to show from the definition of Richardson number and Equations (11.1)–(11.2) that

$$\mathrm{Ri} = \zeta\phi_h(\zeta)/\phi_m^2(\zeta) \tag{11.4}$$

which relates Ri to the basic stability parameter $\zeta = z/L$ of the M–O similarity theory. The inverse of Equation (11.4), namely, $\zeta = f(\mathrm{Ri})$ is often used to determine ζ and, hence, the Obukhov length L from easily measured gradients of velocity and temperature at one or more heights in the surface layer. Then, Equations (11.2) and (11.3) can be used to determine the fluxes of momentum and heat, knowing the empirical forms of the similarity functions $\phi_m(\zeta)$ and $\phi_h(\zeta)$ from carefully conducted experiments.

Other properties of flow which relate turbulent fluxes to the local mean gradients, such as the exchange coefficients of momentum and heat and the corresponding mixing lengths, can also be expressed in terms of $\phi_m(\zeta)$ and $\phi_h(\zeta)$. For example, it is easy to show that

$$\begin{aligned}K_m/(kzu_*) &= \phi_m^{-1}(\zeta) \\ K_h/(kzu_*) &= \phi_h^{-1}(\zeta) \\ K_h/K_m &= \phi_m(\zeta)/\phi_h(\zeta)\end{aligned} \tag{11.5}$$

which can be used to prescribe the eddy diffusivities or their ratio K_h/K_m in the stratified surface layer.

11.2 Empirical Forms of Similarity Functions

As mentioned earlier, a similarity theory based on a particular similarity hypothesis and dimensional analysis can only suggest plausible functional relationships between certain dimensionless parameters. It does not tell anything about the forms of those functions, which must be determined empirically from accurate observations during experiments specifically designed for this purpose. Experimental verification is also required, of course, for the original similarity hypothesis or its consequences (e.g., predicted similarity relations).

Following the proposed similarity theory by Monin and Obukhov (1954), a number of micrometeorological experiments have been conducted at different locations (ideally, flat and homogeneous terrain with uniform and low roughness elements) and under fair-weather conditions (to satisfy the conditions of quasistationarity and horizontal homogeneity) for the expressed purpose of verifying the M–O similarity theory and for accurately determining the forms of the various similarity functions. Early experiments lacked direct and accurate measurements of turbulent fluxes, which could be estimated only after making some *a priori* assumptions about flux-profile relations. The 1953 Great Plains Experiment at O'Neill, Nebraska, was one of the earliest concerted efforts in observing the PBL in which micrometeorological measurements were made by several different groups, using different instrumentation and techniques (Lettau and Davidson, 1957). But, here too, as in subsequent Australian and Russian experiments in the 1960s, the flux measurements suffered from severe instrument-response problems and were not very reliable. Still, these experiments could verify the general validity of M–O similarity theory and give the approximate forms of the similarity functions $\phi_m(\zeta)$ and $\phi_h(\zeta)$. Empirical evaluations of the M–O similarity functions based on several micrometeorological experiments conducted in southern Australia have been reported by Dyer (1967), Dyer and Hicks (1970), Webb (1970), and Hicks (1976). Garratt and Hicks (1990) presented an overview of these experiments and their results.

Perhaps the best micrometeorological experiment conducted so far, for the specific purpose of determining the M–O similarity functions, was the 1968 Kansas Field Program (Izumi, 1971). A 32 m tower located in the center of a 1 mile^2 field of wheat stubble ($h_0 \approx 0.18$ m) was instrumented with fast-response cup anemometers, thermistors, resistance thermometers, and three-dimensional sonic anemometers at various levels. These were used to determine mean velocity and temperature gradients, as well as the momentum and heat fluxes using the eddy correlation method. In addition, two large drag plates installed nearby were used to measure the surface stress directly. Both heat and momentum fluxes were found to be constant with height, within the limits of experimental accuracy ($\pm 20\%$, for fluxes). The flux-profile relations derived from the Kansas Experiment are discussed in detail by Businger *et al.* (1971).

The generally accepted forms of $\phi_m(\zeta)$ and $\phi_h(\zeta)$ on the basis of the various micrometeorological experiments are

$$\phi_m = \begin{cases} (1 - \gamma_1\zeta)^{-1/4} & \text{for } \zeta < 0 \text{ (unstable)} \\ 1 + \beta_1\zeta, & \text{for } \zeta \geq 0 \text{ (stable)} \end{cases} \quad (11.6)$$

$$\phi_h = \begin{cases} \alpha(1 - \gamma_2\zeta)^{-1/2} & \text{for } \zeta < 0 \text{ (unstable)} \\ \alpha + \beta_2\zeta, & \text{for } \zeta \geq 0 \text{ (stable)} \end{cases} \quad (11.7)$$

There remain some differences, however, in the estimated values of the constants, α, β_1, β_2, γ_1, and γ_2 in the above expressions as obtained by different investigators. The main causes of these differences are the unavoidable measurement errors and deviations from the ideal conditions assumed in the theory (Yaglom, 1977; Hogstrom, 1988). The originally estimated values from the Kansas Experiment were (Businger *et al.*, 1971):

$$\alpha = 0.74; \quad \beta_1 = \beta_2 = 4.7; \quad \gamma_1 = 15; \quad \gamma_2 = 9 \quad (11.8)$$

The corresponding formulas [Equations (11.6) and (11.7) for ϕ_m and ϕ_h with the above empirical values of constants are shown in Figures 11.2 and 11.3 together with the observed Kansas data. These data are characterized by relatively small scatter, compared with other micrometeorological data sets used for evaluating the M–O similarity functions (Hogstrom, 1988). Still, they were probably subjected to significant errors due to distortion of air flow around instrument boxes mounted on the tower. In particular, unusually low values of $k = 0.35$ and $\alpha = 0.74$, estimated from the Kansas data, have been attributed to likely flow distortion errors. The low value of the von Karman constant, compared to its traditional value of 0.40, generated some controversial hypotheses or explanations on the possible dependence of k on roughness Reynolds number. But more definitive experimental determinations of the same in the 1980s did not find any such dependence and have confirmed its traditional value (Hogstrom, 1988). More accurate estimates of α from both the field and laboratory experiments have also indicated that α may not be significantly different from one (Hogstrom, 1988). Still, the precise forms of the M–O similarity functions $\phi_m(\zeta)$ and $\phi_h(\zeta)$ remain far from being settled (Hogstrom, 1996).

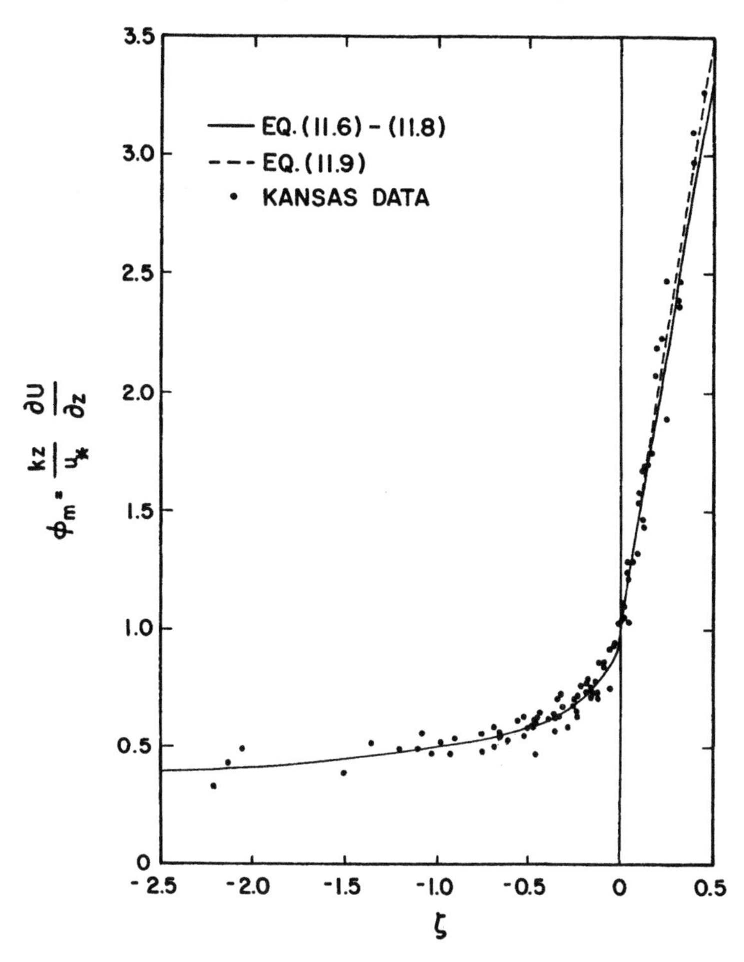

Figure 11.2 Dimensionless wind shear as a function of the M–O stability parameters. [Kansas data from Izumi (1971).]

In view of the above-mentioned uncertainties in measurements and in the determination of empirical similarity functions or constants, the following simpler flux-profile relations can be recommended for most practical applications in which great precision is not warranted:

$$\begin{aligned} \phi_h &= \phi_m^2 = (1 - 15\zeta)^{-1/2}, \quad \text{for } \zeta < 0 \\ \phi_h &= \phi_m = 1 + 5\zeta, \quad \text{for } \zeta \geq 0 \end{aligned} \tag{11.9}$$

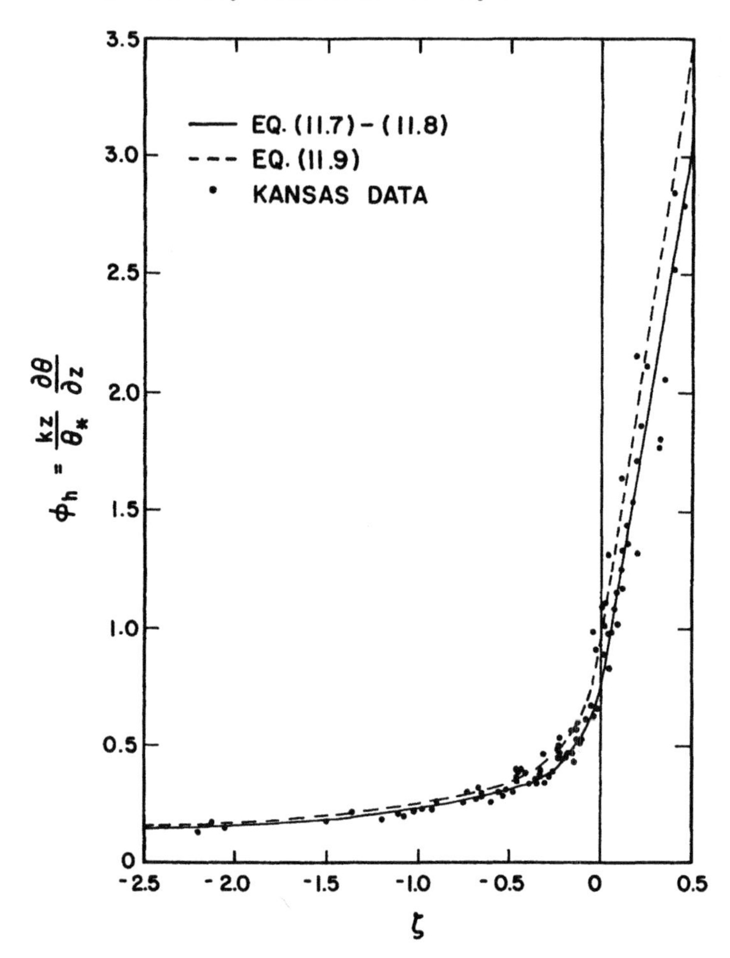

Figure 11.3 Dimensionless potential temperature gradient as a function of the M–O stability parameters. [Kansas data from Izumi (1971).]

These deviate from the Kansas relations [Equations (11.6)–(11.8)] only slightly, but have the advantage of relating ζ to Ri more simply and explicitly as

$$\begin{aligned} \zeta &= \mathrm{Ri}, && \text{for Ri} < 0 \\ \zeta &= \frac{\mathrm{Ri}}{1 - 5\mathrm{Ri}}, && \text{for } 0 \le \mathrm{Ri} \le 0.2 \end{aligned} \tag{11.10}$$

These are verified against the Kansas data in Figure 11.4.

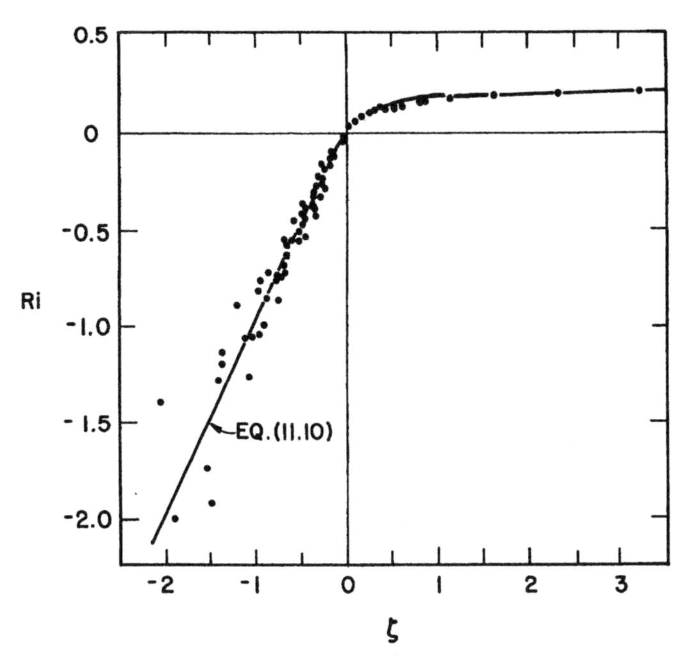

Figure 11.4 The Richardson number as a function of the M–O stability parameters. Equation (11.10) (—) is compared with Kansas data. [After Businger *et al.* (1971).]

After substituting from Equations (11.9) and (11.10) into Equation (11.5), the eddy diffusivities of heat and momentum can also be expressed as functions of the Richardson number

$$\frac{K_{\mathrm{m}}}{kzu_*} = \begin{cases} (1 - 15\mathrm{Ri})^{1/4}, & \text{for } \mathrm{Ri} < 0 \\ 1 - 5\mathrm{Ri}, & \text{for } 0 \leq \mathrm{Ri} \leq 0.2 \end{cases}$$
$$\frac{K_{\mathrm{h}}}{kzu_*} = \begin{cases} (1 - 15\mathrm{Ri})^{1/2} & \text{for } \mathrm{Ri} < 0 \\ 1 - 5\mathrm{Ri}, & \text{for } 0 \leq \mathrm{Ri} \leq 0.2 \end{cases} \tag{11.11}$$

These are represented in Figure 11.5. Note that the eddy diffusivities of heat and momentum are equal under stably stratified conditions; they decrease rapidly with increasing stability and vanish as the Richardson number approaches its critical value of $\mathrm{Ri_c} = 1/\beta \cong 0.2$.

There is a question about the validity of the above-mentioned empirically estimated similarity functions under extremely unstable (approaching free convection) and extremely stable (approaching critical Ri) conditions.

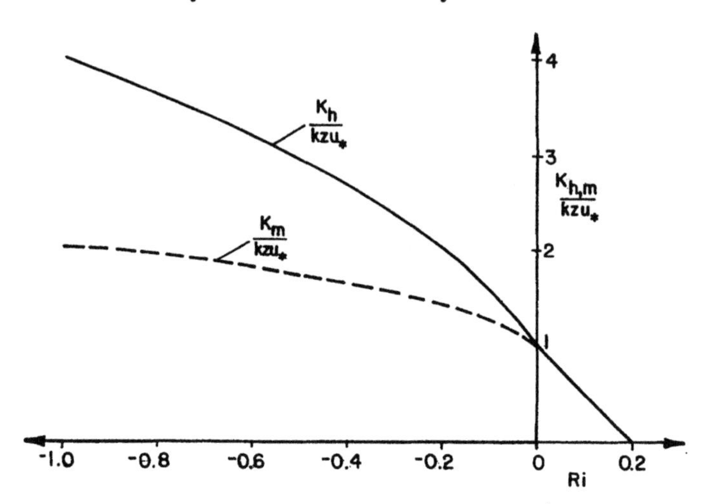

Figure 11.5 Variation of the dimensionless eddy diffusivities of heat and momentum with Richardson number according to Equations (11.11).

Micrometeorological data used in the determination of the above M–O similarity functions have generally been limited to the moderate stability range of $-5 < \zeta < 2$, and, as such, are strictly valid in this range. As a matter of fact, in conditions approaching free convection ($-\zeta \gg 1$, or $\zeta \to -\infty$), the above expressions for ϕ_h and K_h are not quite consistent with the local free convection similarity relations for $\partial\Theta/\partial z$ and K_h derived in Chapter 9. In order for them to be consistent one must have $\phi_h \sim (-\zeta)^{-1/3}$ for $-\zeta \gg 1$. Some of the proposed forms of ϕ_h satisfy this condition, but do not represent the data well, especially for $-\zeta < 2$.

On the other side of strong stability conditions, there is some evidence of a linear velocity profile in the lower part of the PBL, including the surface layer, which appears to be consistent with the M–O relation for $\phi_m(\zeta)$ for $\zeta \gg 1$. There is also other experimental evidence suggesting the limiting value of $\zeta \cong 1$ for the validity of the M–O relation $\phi_m = 1 + \beta\zeta$ (Webb, 1970; Hicks, 1976). In any case, strong stability conditions usually occur at nighttime under clear skies and weak winds. Because the temperature profile under such conditions is strongly influenced by longwave atmospheric radiation, which is ignored in the M–O similarity theory, the temperature field may not follow the M–O similarity. Experimental estimates of both $\phi_m(\zeta)$ and $\phi_h(\zeta)$ indicate that these functions become increasingly flatter as ζ increases above 0.5 (Hicks, 1976; Holtslag, 1984).

11.3 Wind and Temperature Profiles

Integration of Equations (11.2) with respect to height yields the velocity and potential temperature profiles in the form

$$\begin{aligned} U/u_* &= (1/k)[\ln(z/z_0) - \psi_m(z/L)] \\ (\Theta - \Theta_0)/\theta_* &= (1/k)[\ln(z/z_0) - \psi_h(z/L)] \end{aligned} \tag{11.12}$$

in which Θ_0 is the extrapolated temperature at $z = z_0$ and ψ_m and ψ_h are different similarity functions related to ϕ_m and ϕ_h, respectively, as

$$\begin{aligned} \psi_m\left(\frac{z}{L}\right) &= \int_{z_0/L}^{z/L} [1 - \phi_m(\zeta)]\frac{d\zeta}{\zeta} \\ \psi_h\left(\frac{z}{L}\right) &= \int_{z_0/L}^{z/L} [1 - \phi_h(\zeta)]\frac{d\zeta}{\zeta} \end{aligned} \tag{11.13}$$

For smooth and moderately rough surfaces, z_0/L is usually quite small and the integrands in Equation (11.13) are well behaved at small ζ, so that the lower limit of integration can essentially be replaced by zero. With this approximation, the ψ functions can be determined for any appropriate forms of ϕ functions. For example, corresponding to Equation (11.9), we obtain

$$\begin{aligned} \psi_m &= \psi_h = -5\frac{z}{L}, \quad \text{for } \frac{z}{L} \geq 0 \\ \psi_m &= \ln\left[\left(\frac{1+x^2}{2}\right)\left(\frac{1+x}{2}\right)^2\right] - 2\tan^{-1}x + \frac{\pi}{2}, \quad \text{for } \frac{z}{L} < 0 \\ \psi_h &= 2\ln\left(\frac{1+x^2}{2}\right), \quad \text{for } \frac{z}{L} < 0 \end{aligned} \tag{11.14}$$

where $x = (1 - 15z/L)^{1/4}$.

Note that the deviations in profiles from the log law increase with increasing magnitude of z/L. Under stable conditions the profiles are log linear and tend to become linear for large values of z/L.

Under unstable conditions ($\zeta < 0$), on the other hand, ψ_m and ψ_h are positive, so that the deviations from the log law are of opposite sign (negative). Consequently, the velocity and temperature profiles in the surface layer are expected to become more and more curvilinear as instability increases. The observed winds and temperatures are usually plotted against $\log z$ or $\ln z$, instead of a linear height scale, in order to reduce the profile curvature and to clearly show their deviations from the log law due to buoyancy effects in the surface layer.

If the wind profile in the stratified surface layer is represented by the power-law relation (10.2), as is often done in air pollution/dispersion models, the exponent m can be specified as a function of surface roughness and stability as

$$m = \phi_m(\zeta_r)/[\ln(z_r/z_0) - \psi_m(\zeta_r)] \tag{11.15}$$

where $\zeta_r = z_r/L$ and the reference height z_r is of the order of 10 m, so that it falls within the constant flux surface layer. Comparing the above expression for m with Equation (10.8) for neutral stability, it is obvious that m increases with the increase in both roughness and stability. For example, for a moderately rough surface with $m = 0.2$ under near-neutral stability conditions, it can be shown from Equation (11.15) that the exponent would be reduced to about 0.1 under very unstable conditions and increased to about 0.6 under very stable ($z_r/L = 1$) conditions.

11.4 Drag and Heat Transfer Relations

The surface stress and sensible heat flux are usually expressed or parameterized in terms of the bulk drag and heat transfer relations

$$\begin{aligned} \tau_0 &= \rho C_D U_r^2 \\ H_0 &= -\rho c_p C_H U_r(\Theta_r - \Theta_0) \end{aligned} \tag{11.16}$$

in which C_D and C_H are the drag and heat transfer coefficients, respectively, and U_r and Θ_r are the mean wind speed and potential temperature at the reference measurement height z_r. When z_r falls within the surface layer, it is easy to show from Equations (11.12) and (11.16) that

$$\begin{aligned} C_D &= k^2[\ln(z_r/z_0) - \psi_m(z_r/L)]^{-2} \\ C_H &= k^2[\ln(z_r/z_0) - \psi_m(z_r/z_0)]^{-1}[\ln(z_r/z_0) - \psi_h(z_r/L)]^{-1} \end{aligned} \tag{11.17}$$

Thus, according to the Monin–Obukhov similarity theory, the drag and heat transfer coefficients are some universal functions of z_r/z_0 and z_r/L. Because the latter parameter involves the difficult-to-determine Obukhov length L, it should be useful and desirable to relate z_r/L to a more easily estimated parameter from the known variables, U_r, Θ_r, Θ_0, and z_r. One such parameter is the bulk Richardson number

$$\mathrm{Ri_B} = (g/T_0)[(\Theta_r - \Theta_0)z_r/U_r^2] \tag{11.18}$$

Which can be related to z_r/L by substituting into Equation (11.18) from the M–O profile relations given by Equation (11.12). It is easy to show that

$$\mathrm{Ri_B} = \frac{z_r}{L}\left[\ln\frac{z_r}{z_0} - \psi_h\left(\frac{z_r}{L}\right)\right]\left[\ln\frac{z_r}{z_0} - \psi_m\left(\frac{z_r}{L}\right)\right]^{-2} \tag{11.19}$$

which is of the form $\mathrm{Ri_B} = F(z_r/z_0, z_r/L)$. The inverse of this function can be used to determine z_r/L for given values of z_r/z_0 and $\mathrm{Ri_B}$. For this a graphical representation of Equation (11.19) would be useful.

Knowing the general forms of ψ functions one can infer from Equation (11.17) that both the drag and heat transfer coefficients increase with increasing surface roughness (decreasing z_r/z_0) and decrease with increasing stability (see Figure 11.6). Their typical values over comparatively smooth water surfaces range from 1.0×10^{-3} to 2×10^{-3}. For land surfaces, however, the range of

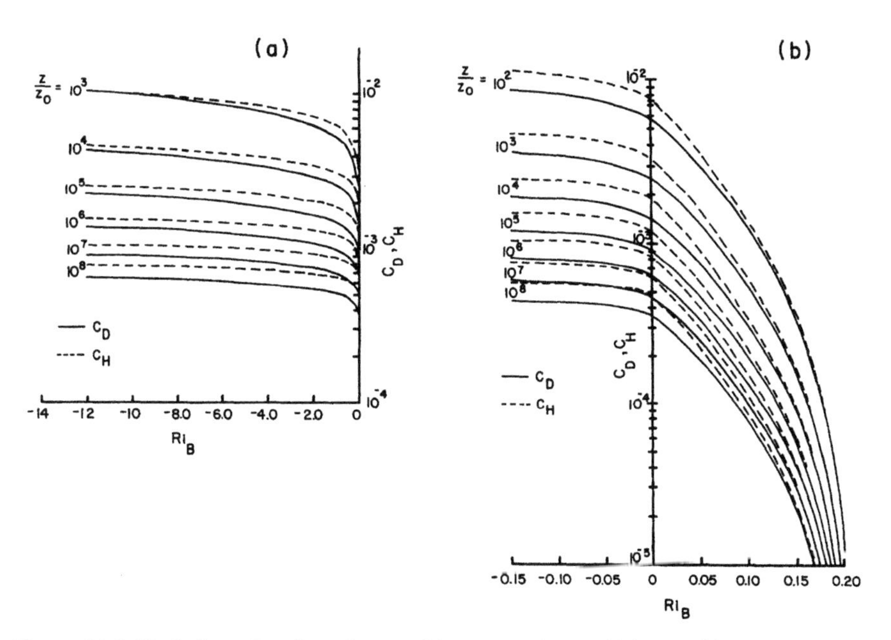

Figure 11.6 Variation of surface drag and heat transfer coefficients with surface roughness and the bulk Richardson number for (a) unstable conditions, and (b) near-neutral and stable conditions. [After Arya (1977).]

C_D and C_H values is considerably wider (say, 0 to 0.01). These are only weakly dependent on the choice of observation level in the surface layer; the most frequently recommended observation height for surface meteorological observations is 10 m.

11.5 Methods of Determining Momentum and Heat Fluxes

Determination of turbulent exchanges taking place between the earth and the atmosphere near their interface (surface) is of primary concern in micrometeorology. A number of methods have been devised with varying degrees of sophistication, some of which will be described here only briefly. The emphasis here is on principles and techniques, rather than on the details of instrumentation and measurements (Lenschow, 1986).

11.5.1 Surface-drag measurements

The only direct method of measuring shearing stress on a small sample of the surface is through the use of a carefully installed drag plate. The drag measured on the sample surface (plate) area must be representative of the whole area under consideration. This presumes a reasonably uniform ground cover with small roughness elements, which remain relatively undisturbed by the installation of the drag plate. The original roughness characteristics of the surface are retained or duplicated on the top of the drag plate. The drag plate has a small annular ring around it to isolate it from the surrounding area for drag force measurements using strain gauges and electromechanical transducers. The installation and successful operation of a drag plate requires considerable care, skill, and experience (Bradley, 1968). For this reason, such measurements have been made only in the context of a few micrometeorological experiments or research expeditions (e.g., the 1968 Kansas Field Program).

11.5.2 Energy balance method

No method exists for directly measuring the surface heat flux. A thin heat flux plate of known conductivity, with an embedded thermopile to measure temperature gradient across the plate, is often used to measure heat flux through the subsurface medium (e.g., soil, ice, and snow). But, to avoid radiative and convective effects the plate must be buried at least 10 mm below the surface. One can determine the ground heat flux H_G after applying an appropriate correction to the heat flux plate measurements. The sensible heat flux at the surface to or from air can be estimated indirectly, using the surface energy balance or other appropriate energy budget equation discussed in Chapter 2, provided other components of the energy balance are measured or can otherwise be estimated. The energy balance method is often used in conjunction with the estimated Bowen ratio from mean temperature and humidity measurements at two heights near the surface.

11.5.3 Eddy correlation method

The most reliable and direct measurements of turbulent exchanges of momentum and heat in the atmosphere are usually made with sophisticated fast-response turbulence instrumentation. If all fluctuations of velocity and temperature that contribute to the desired momentum and heat fluxes are faithfully sensed and recorded, one can determine their covariances simply by averaging the products of the appropriate fluctuations over any desired averaging time. In particular, the vertical fluxes of momentum and heat over a homogeneous surface are given by

$$\begin{aligned} \tau &= -\rho\overline{uw} \\ H &= \rho c_p \overline{\theta w} \end{aligned} \tag{11.20}$$

with the x axis oriented along the mean wind. Because the above fluxes are expected to remain constant, independent of height, in a horizontally homogeneous surface layer eddy correlation measurements also provide a means of determining the surface fluxes ($\tau_0 \cong -\rho\overline{uw}$ and $H_0 \cong \rho c_p \overline{\theta w}$).

Although the eddy correlation method of determining fluxes is simple, in practice, it requires expensive research-grade instrumentation, such as sonic, laser, or hot-wire anemometers and thin-wire resistance thermometers, as well as rapid (sampling rates of 10–100 s^{-1}) data acquisition systems. The closer to the surface the turbulence measurements are made, the more severe become the instrument response problems. At heights greater than 10 m or so, light cup, vane, and propeller anemometers may also be adequate for measuring variances and fluxes. Larger averaging times may be required, however, with increasing height of measurement, because the characteristic size of large eddies usually increases with height in the PBL. The requirements of instrument leveling, orientation, calibration, and maintenance are also quite severe for accurate eddy correlation measurements (Lenschow, 1986).

The above-mentioned instrumental requirements have kept the eddy correlation method from being widely used, except in special research expeditions. But, continued advances in instruments and data processing have already resulted in their increased use for many applications. The method has the advantage of measuring turbulent exchanges directly, without too many restrictive assumptions about the nature of the surface (such as uniform, flat, and homogeneous) or of the atmosphere. It is the only method available for measuring turbulent fluxes inside plant canopies, or in the wakes of hills and buildings.

11.5.4 Bulk transfer method

Indirect methods of estimating fluxes from more easily measured mean winds and temperatures in the surface layer or the whole PBL are based on the appropriate flux-profile relations. The simplest and the most widely used method is the bulk aerodynamic approach, which is based on the bulk transfer formulas [Equation (11.16)]. This method can be used when measurements or computations (e.g., in a numerical model) of mean velocity and temperature are available only at one level, in conjunction with the desired surface properties (e.g., the surface roughness and temperature). If the observation height is low enough for it to fall in the surface layer, the appropriate drag and heat transfer coefficients in Equation (11.16) can be parameterized on the basis of the Monin–Obukhov similarity relations [Equation (11.17)], as described in Section 11.4. This will indeed be the case for most routine surface meteorological observations over land and ocean areas. On the other hand, if the observation level is near the top of the PBL, or the reference wind is geostrophic, C_D and C_H may be parameterized on the basis of the PBL similarity theory (Deardorff 1972b; Arya, 1984). In practice C_D and C_H over ocean surfaces are usually prescribed some constant average values (e.g., $C_D \cong C_H \cong 1.5 \times 10^{-3}$) derived from major marine meteorological experiments. Over land areas, C_D and C_H are found to vary over a much greater range, due to the effects of surface roughness and stability, but in many applications constant values are still prescribed for the sake of simplicity. It is not difficult, however, to incorporate in such parameterizations the stability-dependent correction factors, such as C_D/C_{DN} and C_H/C_{HN} as functions Ri_B (Deardorff, 1968), where C_{DN} and C_{HN} are the prescribed coefficients for neutral stability.

11.5.5 Gradient method

In order to use the bulk transfer method described above, one needs to know the surface roughness and the surface temperature. Such information is not always easy to come by. In fact, for very rough and uneven surfaces, the surface itself is not well defined and its temperature cannot be measured directly. This difficulty can be avoided by making measurements at two or more heights in the surface layer. Here, we describe a simple gradient or aerodynamic method of determining fluxes from measurements of mean differences or gradients of velocity and temperature between any two heights z_1 and z_2 within the surface layer, but well above the tops of roughness elements.

The two most commonly used finite-difference approximations for the vertical gradient of a mean micrometeorological variable M are:

(1) Linear approximation

$$\left(\frac{\partial M}{\partial z}\right)_{z_a} \cong \frac{\Delta M}{\Delta z} = \frac{M_2 - M_1}{z_2 - z_1} \tag{11.21}$$

which is applicable at the arithmetic mean height $z_a = (z_1 + z_2)/2$.

(2) Logarithmic approximation

$$\left(\frac{\partial M}{\partial \ln z}\right)_{z_m} \cong \frac{\Delta M}{\Delta \ln z} = \frac{M_2 - M_1}{\ln(z_2/z_1)} \tag{11.22}$$

or

$$\left(\frac{\partial M}{\partial z}\right)_{z_m} \cong \frac{\Delta M}{z_m \ln(z_2/z_1)} \tag{11.23}$$

which is applicable at the geometric mean height $z_m = (z_1 z_2)^{1/2}$.

Note that the above two approximations may not be directly compared as they pertain to vertical gradients at different heights. However, each of them is likely to have some error associated with it, because the actual profile of the mean variable may not be either linear or logarithmic in height between z_1 and z_2. The error might be expected to increase with increasing deviation of the profile from that assumed in the finite-difference approximation and also with the increasing height interval $z_2 - z_1$, or the ratio z_2/z_1 (Arya, 1991). A comparative assessment of finite-difference errors in the estimation of velocity and potential temperature gradients in the atmospheric surface layer shows that the logarithmic approximation is far superior to the linear approximation in near-neutral, unstable, and convective conditions (Arya, 1991). Therefore, the former is more widely used and often recommended in the micrometeorological literature. In very stable conditions, however, the linear approximation might be superior, because mean velocity and temperature profiles become more nearly linear, especially in the upper part of the surface layer. Still, the errors associated with the use of the logarithmic approximation are estimated to remain small ($<8\%$) enough to be generally acceptable. In most applications of the gradient method, the same gradient approximation is used, irrespective of atmospheric stability.

Using the logarithmic finite difference approximation for both the velocity and potential temperature gradients, the gradient Richardson number at the geometric mean height z_m can be estimated as

$$\mathrm{Ri_m} = \frac{g}{T_0} \frac{\Delta\Theta z_m}{(\Delta U)^2} \ln\frac{z_2}{z_1} \tag{11.24}$$

Then, the corresponding value of the M–O stability parameter $\zeta_m = z_m/L$ can be determined from Equation (11.10)

$$\begin{aligned} \zeta_m &= \mathrm{Ri_m}, \quad \mathrm{Ri_m} < 0 \\ \zeta_m &= \mathrm{Ri_m}/(1 - 5\mathrm{Ri_m}), \quad \text{for } 0 \leq \mathrm{Ri_m} < 0.2 \end{aligned} \tag{11.25}$$

Knowing ζ_m, the M–O similarity functions ϕ_m and ϕ_h can be evaluated from Equation (11.9) and, then, the friction velocity and temperature scales can be calculated from Equation (11.2) as

$$\begin{aligned} u_* &= k\Delta U \Big/ \left(\phi_m \ln\frac{z_2}{z_1}\right) \\ \theta_* &= k\Delta\Theta \Big/ \left(\phi_h \ln\frac{z_2}{z_1}\right) \end{aligned} \tag{11.26}$$

The surface fluxes can be determined from the above estimated scales as $\tau_0 = \rho u_*^2$, and $H_0 = -\rho c_p u_* \theta_*$.

The finite-difference approximations [Equation (11.22)] may not be good enough when the ratio z_2/z_1 becomes very large. On the other hand, if the two height levels are too close to each other, the differences in velocities and temperatures at the two levels may not be well resolved. A good compromise is to specify the measurement heights such that $z_2/z_1 = 2$ to 4. Still, there must be expected some errors in the measurements of ΔU and $\Delta\Theta$, as well as in the estimation of $\partial U/\partial z$, $\partial\Theta/\partial z$, and $\mathrm{Ri_m}$. These, in conjunction with uncertainties in the M–O flux-profile relations, may lead to errors of more than 20% in the estimates of surface fluxes. Due to the above-mentioned errors and other reasons, the estimated value of $\mathrm{Ri_m}$ in stably stratified conditions may even turn out to be greater than the critical value of 0.2. The gradient method becomes invalid and should not be used in such cases, unless more appropriate forms of the M–O similarity functions for very stable conditions are used for this purpose.

Example Problem 1

The following measurements were made of the hourly-averaged wind speed and temperature at a homogeneous rural site:

z (m)	U (m s^{-1})	T (°C)
2	3.34	29.04
8	3.98	28.10

Using the gradient method, estimate the following:
(a) Velocity and potential temperature gradients at 4 m.
(b) The gradient Richardson number at 4 m and the Obukhov length.
(c) The friction velocity, temperature scale, and the surface heat flux.

Solution:

(a) Using the logarithmic finite-difference approximation, the velocity and temperature gradients at $z_m = \sqrt{z_1 z_2} = 4$ m are

$$\frac{\partial U}{\partial z} = \frac{1}{z_m} \frac{\Delta U}{\ln(z_2/z_1)} = \frac{0.64}{4 \ln 4} = 0.115 \text{ s}^{-1}$$

$$\frac{\partial T}{\partial z} = \frac{1}{z_m} \frac{\Delta T}{\ln(z_2/z_1)} = -\frac{0.94}{4 \ln 4} = -0.170 \text{ K m}^{-1}$$

$$\frac{\partial \Theta}{\partial z} \cong \frac{\partial T}{\partial z} + \Gamma = -0.170 + 0.01 = -0.16 \text{ K m}^{-1}$$

(b)

$$\text{Ri}_4 = \frac{g}{T_0} \frac{\partial \Theta/\partial z}{(\partial U/\partial z)^2} = -\frac{9.81 \times 0.16}{302.2 \times 0.0132} = -0.387$$

$$\frac{z_m}{L} = \text{Ri}_4 = -0.387$$

$$L = -\frac{4}{0.387} = -10.33 \text{ m}$$

(c) At $\zeta_m = -0.387$, the M–O similarity functions can be estimated as

$$\phi_m = (1 - 15\zeta_m)^{-1/4} \cong 0.619$$
$$\phi_h = (1 - 15\zeta_m)^{-1/2} \cong 0.383$$

Then, from Equation (11.26)

$$u_* = \frac{0.4 \times 0.64}{0.619 \times 1.386} = 0.298 \text{ m s}^{-1}$$

$$\theta_* = \frac{0.4 \times 0.92}{0.383 \times 1.386} = -0.693 \text{ K}$$

$$H_0 = -\rho c_p u_* \theta_* \cong 248 \text{ W m}^{-2}$$

11.5.6 Profile method

In order to minimize errors in the estimated fluxes, it is highly desirable to make measurements of mean velocity and temperature at several (more than two) levels within the surface layer. A good procedure for determining fluxes from such profile measurements is to fit the appropriate flux-profile relations to the observations, using the least-square technique. In this way, the effect of random experimental errors on flux estimates can be minimized. A simple graphical procedure can also be used for the same purpose, which is described here in the following.

First, calculate Ri from measurements of U and Θ at each pair of consecutive levels and obtain the best estimate of the Obukhov length L by fitting a straight line through the data points of z_m versus Ri plot for unstable conditions, or z_m versus Ri/(1 − 5Ri) plot for stable conditions. Note that in either case, according to Equation (11.25), the slope of the best-fitted line will be L. The second step is to plot U versus $\ln z - \psi_m(z/L)$ and Θ versus $\ln z - \psi_h(z/L)$ and draw best-fitted straight lines through these data points (see Figures 11.7 and 11.8). Note that the slopes of these lines, according to Equation (11.12), must be, k/u_* and k/θ_*, respectively, which readily determine u_*, θ_* and the surface fluxes. The additional information on intercepts can be used to determine z_0 and Θ_0 if desired. This follows from writing Equation (11.12) in the form

$$\begin{aligned} \ln z - \psi_m(z/L) &= \frac{k}{u_*} U + \ln z_0 \\ \ln z - \psi_h(z/L) &= \frac{k}{\theta_*} \Theta - \frac{k}{\theta_*} \Theta_0 + \ln z_0 \end{aligned} \qquad (11.27)$$

Note that for a fixed value of $\ln z - \psi_m$ and $\ln z - \psi_h$, the values of U and Θ can be determined from the best-fitted lines through the data points, from which $\ln z_0$ and Θ_0 can be estimated, knowing k/u_* and k/θ_* from the slopes of the best-fitted lines.

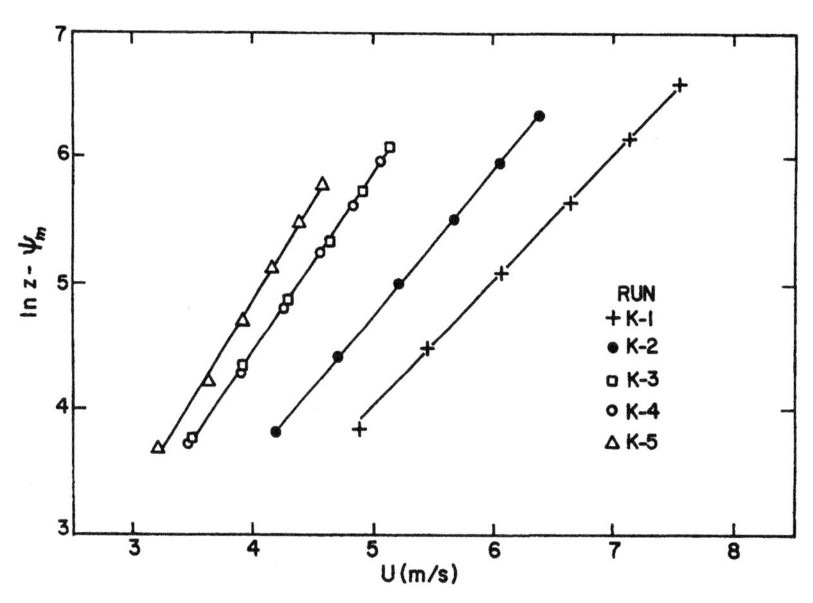

Figure 11.7 Least-square fitting of the M–O flux-profile relation (modified log law) to the observed mean velocity profiles at Kerang, Australia. [After Paulson (1967).]

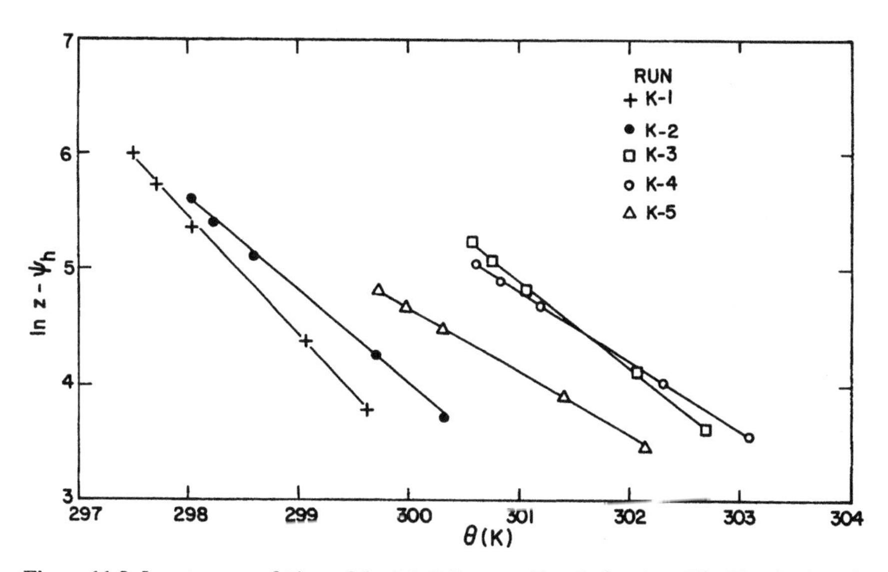

Figure 11.8 Least-square fitting of the M–O flux-profile relation (modified log law) to the observed mean potential temperature profiles at Kerang, Australia. [After Paulson (1967).]

The above procedure is an extension of the more familiar graphical method of determining the friction velocity or the surface stress from the observed wind profile in a neutral surface layer (see Chapter 10). The stability correction is made here to obtain a modified height coordinate in which the wind and potential temperature profiles are expected to be linear. If the surface roughness parameter is known or can be estimated, one can use a simpler version of the profile method that requires minimal measurements of wind speed at one level and temperatures at two height levels in the surface layer (Arya, 1999, Chapter 4).

11.5.7 Geostrophic departure method

The equations of mean motion (Chapter 9), with certain assumptions, can be used to determine the surface stress, as well as the profiles of turbulent momentum fluxes in the PBL. Here, we demonstrate this method for a horizontally homogeneous and quasistationary PBL for which the mean flow equations reduce to Equation (9.9). Their integration with respect to height from some level z to the top of the PBL, where the turbulent fluxes may be expected to vanish, gives momentum fluxes as functions of the height z

$$\begin{aligned} \overline{uw} &= -f \int_z^h (V - V_g)dz \\ \overline{vw} &= f \int_z^h (U - U_g)dz \end{aligned} \tag{11.28}$$

In particular, the surface stress is given by

$$\tau_0 = \rho f \int_0^h (V - V_g)dz \tag{11.29}$$

where the x axis is taken parallel to the surface wind or stress (this implies that $\overline{vw} = 0$ at $z = 0$, so that from Equation (11.28), $\int_0^h (U - U_g)dz = 0$ constitutes an integral constraint on the U profile]. The above relations are, of course, restricted to steady-state flow with no advection, but can easily be modified to include acceleration terms.

Note that Equations (11.28) and (11.29) express the local momentum fluxes and the surface stress in terms of the integrated departures of the actual wind components from the geostrophic wind components. Hence, the name 'geostrophic departure method' is given to this simple technique of determining turbulent fluxes from more easily measured wind profiles (e.g., from pibals, rawinsondes, or remote-sensing wind profilers). The presence of accelerations, and uncertainties in the determination of actual mean winds, geostrophic winds, and the PBL height, frequently lead to deviations from the ideal case and, hence, to large errors in flux determinations. Because the top of the PBL where turbulent momentum fluxes are expected to vanish may not be easy to identify from mean wind profiles alone, the upper limit of integrals in Equations (11.28) and (11.29) is sometimes replaced by the heights z_u and z_v, where the U and V profiles have a maximum or minimum, provided such maxima or minima exist. This is based on the assumption, implied by the gradient-transport hypothesis, that $\overline{uw}$ and $\overline{vw}$ must vanish at the levels where $\partial U/\partial z$ and $\partial V/\partial z$ respectively become zero.

11.5.8 Thermodynamic energy equation method

A similar method for determining the sensible heat flux at the surface and its vertical distribution (profile) in the PBL is based on the thermodynamic energy equation which, in the absence of temperature advection, reduces to

$$\frac{\partial T}{\partial t} = \frac{1}{\rho c_p} \frac{\partial R_N}{\partial z} - \frac{\partial \overline{w\theta}}{\partial z} \tag{11.30}$$

Note that the time-tendency (warming or cooling rate) term is retained here, because it is often found to be significant even when the flow field may be considered quasistationary. It is a manifestation of the diurnal heating and cooling cycle which is responsible for important stability and buoyancy effects in the PBL. According to Equation (11.30), the rate of warming or cooling essentially balances the convergence or divergence of radiative and sensible heat fluxes. The radiative flux divergence is usually ignored in the daytime unstable or convective PBL, especially in the absence of fog and clouds within the layer. It becomes more significant in the stably stratified nocturnal boundary layer.

In simpler situations, where the radiative flux divergence can be ignored, the integration of Equation (11.30) with respect to height yields

$$\begin{aligned} \overline{w\theta} &= \int_z^h \frac{\partial T}{\partial t} dz \\ H_0 &= \rho c_p \int_0^h \frac{\partial T}{\partial t} dz = \rho c_p h \left(\frac{\partial T}{\partial t}\right)_m \end{aligned} \tag{11.31}$$

where $(\partial T/\partial z)_m$ denotes the mean rate of warming in the PBL and we have assumed that the sensible heat flux vanishes at the top of the PBL, i.e., at $z = h$, $\overline{w\theta} = 0$. Thus, temperature soundings in the PBL at close intervals (say 1–3 h) may be used with Equation (11.31) to determine H_0 and the $\overline{w\theta}$ profile. Alternatively, the rate of warming can be estimated from continuous temperature measurements at a height above which the rate of warming becomes almost independent of height in the PBL. This is usually the case in the unstable or convective mixed layer at $z \geq 50$ m. However, the rate of warming at the standard reference height of 10 m may not be significantly different from that in the mixed layer, so that the former may provide a good approximation for $(\partial T/\partial t)_m$. One still needs an estimate of the PBL height h, either from a sounding or from an appropriate model or parameterization of h (Stull, 1988; Arya, 1999).

Both the geostrophic departure and thermodynamic energy methods are quite useful and reliable in that these are based on the fundamental conservation equations and measurements of mean wind and temperature profiles, without any restrictive assumptions, as implied in the empirical flux-profile relations. The simplifying assumptions of steady state and no advection can be relaxed whenever they cannot be justified and advection/acceleration terms can be evaluated from mean profile measurements.

11.5.9 Variance method

In a horizontally homogeneous and quasistationary PBL the turbulent fluxes or covariances are intimately related to the variances or standard deviations of velocity and temperature fluctuations through the various similarity relations for turbulence structure. Since variances can be measured more easily and accurately than covariances or turbulent fluxes, the latter may be determined indirectly by measuring the former and using the appropriate similarity relations. Fast-response instruments are required, which is a disadvantage in comparison with other indirect methods requiring measurements of mean winds and temperatures. However, the fact that observations at a single level may yield momentum and heat fluxes is an added attraction of the variance method.

The method is best suited for determining the surface fluxes from measurements of σ_w and σ_θ at an appropriate level (say, 10 m) in the fully developed surface layer, because the similarity relations are simpler and well established for a wide range of surface types and stability conditions. For example, in the unstable surface layer both σ_w and σ_θ follow the local free convection similarity relations [Equation (9.41)] (see Figure 11.9), either of which can be used to determine the surface heat flux. In the stably stratified surface layer, on the other hand, the local similarity scaling of turbulence structure implies that the ratios

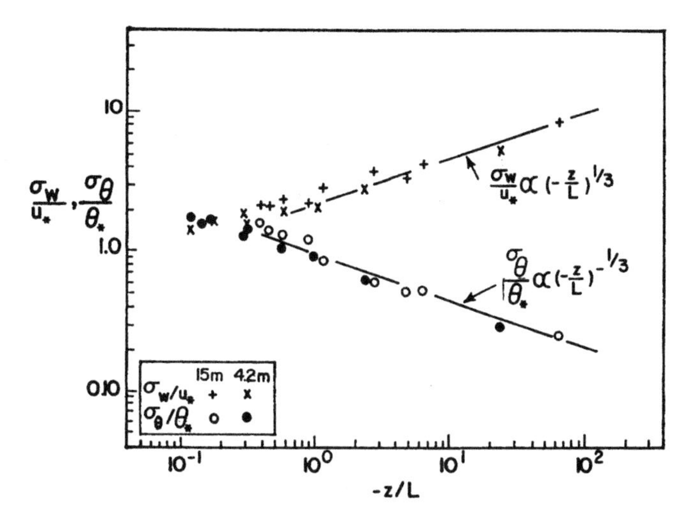

Figure 11.9 Normalized standard deviations of vertical velocity and temperature fluctuations as functions of $-z/L$, compared with local free convection similarity relations. [From Businger (1973); after Monji (1972).]

σ_w/u_* and σ_θ/θ_* must be constants, independent of height, where $u_* = (-\overline{uw})^{1/2}$ and $\theta_* = -\overline{w\theta}/u_*$. This can be used to estimate the local fluxes from measurements of σ_w and σ_θ. In slightly unstable to slightly stable conditions, the Monin–Obukhov similarity theory predicts σ_w/u_* and σ_θ/θ_* to be some unique functions of z/L, which have been evaluated on the basis of micrometeorological data from homogeneous sites. In particular, the empirical formula

$$\frac{\sigma_w}{u_*} = 1.25[1 - 3(z/L)]^{1/3} \qquad (11.32)$$

obtained from a compilation of many aircraft and tower data taken over land and sea surfaces can be recommended for unstable conditions (see also Panofsky and Dutton, 1984, Chapter 7). A plot of Equation (11.32) with the experimental data on σ_w/u_* as a function of $-z/L$ is shown in Figure 11.10.

The surface heat flux in a convective boundary layer (CBL) can also be estimated from measurements of σ_u or σ_v at any convenient height in the lower part of the CBL where the empirical similarity relations $\sigma_u \cong \sigma_v \cong 0.6W_*$ are found to be valid (Caughey and Palmer, 1979). Since, the convective velocity scale W_* is related to the mixed-layer height and the surface heat flux, the latter can be estimated if the mixed-layer height is measured independently (e.g., from temperature sounding).

In some applications, the above-mentioned similarity relations between the turbulent fluxes and variances are used to determine the latter from indirect

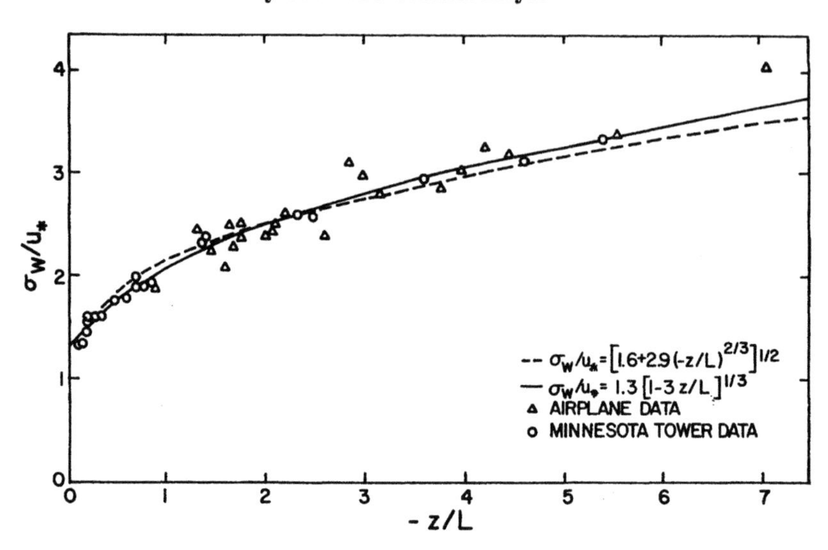

Figure 11.10 Normalized standard deviation of vertical velocity fluctuations in the surface layer as a function of z/L. [After Panofsky *et al.* (1977).]

estimates of the former. For example, the horizontal and vertical spread of a chimney plume or a pollutant cloud is intimately related to velocity variances, which need to be indirectly estimated in the absence of turbulence measurements.

Example Problem 2

The following measurements were made at a suburban site during the afternoon (say, at 2:00 PM) convective conditions:

Mean temperature at 10 m height = 17.0°C.

Average rate of warming at 10 m = 0.60 K h^{-1}.

Mixing height from the temperature sounding = 1500 m.

Using the appropriate thermodynamic and similarity relations, estimate the following at the time of above measurements:

(a) The surface heat flux.

(b) σ_u, σ_v, and σ_w at 10 m and 100 m heights.

Solution

(a) Assuming that the rate of warming at 10 m is representative of that in the whole convective boundary layer, we can use Equation (11.31) to estimate the surface heat flux as

$$H_0 = \rho c_p h \left(\frac{\partial T}{\partial t}\right)_m = \frac{1200 \times 1500 \times 0.6}{3600} = 300 \text{ W m}^{-2}$$

(b) $\sigma_u = \sigma_v \cong 0.6W_*$, which is independent of height, where

$$W_* = \left(\frac{g}{T_0}\frac{H}{\rho c_p}h\right)^{1/3} = \left(\frac{9.81}{290}\times\frac{300}{1200}\times 1500\right)^{1/3} = 2.31 \text{ m s}^{-1}$$

Thus, $\sigma_u = \sigma_v \cong 1.39$ m s^{-1} at both the 10 m and 100 m heights.
For estimating σ_w in the convective surface layer, the local free convection similarity relation (9.41) is more appropriate, so that

$$\sigma_w = 1.4\left(\frac{g}{T_0}\frac{H_0}{\rho c_p}z\right)^{1/3}$$

From this we can estimate $\sigma_w = 0.61$ and 1.31 m s^{-1} at the 10 m and 100 m heights, respectively.

11.6 Applications

Theory and observations of momentum and heat exchanges between the atmosphere and the earth's surface, presented in this chapter, may have the following practical applications:

- Systematic ordering of mean wind, temperature, and turbulence data with stability in the surface layer.
- Quantitative description of the surface layer wind and temperature profiles and their relationship to the surface fluxes of momentum and heat.
- Specifying eddy diffusivities and bulk transfer coefficients as functions of stability in the parameterizations of fluxes.
- Determining surface and PBL fluxes from different types of mean flow and turbulence measurements.
- Parameterizing turbulence and diffusion in the atmospheric surface layer.
- Determining the surface energy budget.

Problems and Exercises

1. What maximum error or uncertainty might be expected in the determination of the Obukhov length, if the fluxes of the momentum and heat have uncertainties of $\pm 20\%$ associated with them?

2. Using the definitions of the various scales and parameters and the basic

Monin–Obukhov similarity relations [Equation (11.2)], verify or derive Equations (11.4) and (11.5).

3. Using the similarity functions obtained by Businger *et al.* (1971) from the 1968 Kansas Field Program, calculate and plot the Richardson number as a function of z/L in the range $-2 \leq z/L \leq 2$, and compare it with the graph based on the simpler relations [Equation (11.10)]. What conclusions can you draw from this?

4. Substituting from the M–O similarity profile relations in the bulk transfer formulas [Equation (11.16)],

(a) derive the expressions given by Equation (11.17) for C_D and C_H as functions of z/L and z/z_0;

(b) write the same as functions of the bulk Richardson number $\mathrm{Ri_B}$ for stable conditions ($\mathrm{Ri_B} > 0$), and plot C_D/C_{DN} as a function of $\mathrm{Ri_B}$ for $z_r/z_0 = 10^3$ and 10^6.

What conclusions can you draw from this?

5. The following measurements of mean wind and potential temperature were taken around noon during the 1968 Kansas Field Program:

z (m)	2	4	8	16	32
U (m s^{-1})	5.81	6.70	7.49	8.14	8.66
Θ (K)	307.20	306.65	306.28	305.88	305.62

Using the M–O similarity relations $\phi_h = \phi_m^2 = [1 - 15(z/L)]^{-1/2}$ for the unstable surface layer.

(a) Calculate and plot Ri as a function of height.

(b) Determine the Obukhov length from the above plot.

(c) Calculate the surface shear stress τ_0 and the heat flux H_0, using the gradient method with the data from the lowest two (2 and 4 m) heights; take $\rho = 1.2$ kg m^{-3} and $c_p = 10^3$ J kg^{-1} K^{-1}.

6. The following observations were taken on a summer evening during the Kansas Experiment (with a surface pressure of 1000 mbar):

z (m)	2	4	8	16
U (m s^{-1})	2.84	3.39	4.01	4.85
T (°C)	33.09	33.31	33.57	33.82

(a) Use the graphical version of the profile method discussed in the text to determine u_* and θ_*.
(b) Use the gradient method for the same purpose, using information from the lowest two (2 and 4 m) heights only.
(c) Compare the results of the above two methods with the eddy correlation measurements of $\overline{uw} = -0.044\ \mathrm{m^2\ s^{-2}}$ and $\overline{w\theta} = -0.023\ \mathrm{m\ s^{-1}\ K}$.

7. During the Minnesota 1973 Experiment, the following mean temperature measurements were made from a tethered balloon at 75 min intervals:

	T (°C)	
z (m)	1217–1332 (CDT)	1332–1447 (CDT)
2	21.75	22.46
61	19.54	20.25
305	17.16	17.86
610	14.43	15.16
914	11.66	12.34
1219	8.99	9.61

(a) Calculate the warming rate at various levels and the average value for the whole PBL. What conclusion can you draw on the variation of the rate of warming with height?
(b) Calculate the surface heat flux in the middle of the observation period when the PBL height was 1400 m, assuming a constant (average) warming rate, independent of height. Compare this with the direct (eddy correlation) measurement of $\overline{w\theta} = 0.20\ \mathrm{m\ s^{-1}\ K}$.
(c) Suggest a simpler method of estimating the surface heat flux in daytime, using temperature measurements at only one level and remote sensing of the PBL height from the surface.

Chapter 12 Evaporation from Homogeneous Surfaces

12.1 The Process of Evaporation

The evaporation in the atmosphere and consequent phenomena of condensation, cloud formation, and precipitation have fascinated scientists, philosophers, and common people since time immemorial. Early philosophers and scientists proposed many theories and explanations of the evaporation process; for a brief review of the history the reader may refer to Brutsaert (1982). As early as the fourth century BC, Aristotle recognized that 'wind is more influential in evaporation than the sun.' However, it was not until the nineteenth century that the foundations of the modern quantitative theories of evaporation and other transport phenomena were laid by J. Dalton, A. Fick, and O. Reynolds.

Evaporation is the phenomenon by which a substance is converted from the liquid state into vapor state. In the atmospheric surface layer evaporation may take place from a free water surface or a moist soil surface, as well as from the leaves of living plants and trees (here, we are ignoring the distinction between evaporation and transpiration). The details of the physical process of conversion from the liquid to vapor state may differ in different cases, but the basic mechanisms of transport of vapor away from any interface are the same. Very close to the interface, vapor is transferred through molecular exchanges in the same manner as heat and momentum are transferred. Most of the transfer takes place within a few molecular free path lengths of the surface. The vigorous molecular activity in this region is indicated by the observation that an estimated 2 to 3 kg of water moves across a free water interface in each direction every second, for each square meter of the interface (Munn, 1966, Chapter 10). However, the net transport of water vapor is only a tiny fraction of the total transport in each direction. According to Fick's law, the net flux of a material (e.g., water vapor) in a given direction is proportional to its concentration (specific humidity) gradient in that direction, the coefficient of proportionality being defined as the molecular diffusivity. Thus, the evaporation rate at a horizontal surface is given by

$$E_0 = -\rho\alpha_w(\partial Q/\partial z) \quad (12.1)$$

in which α_w is the molecular diffusivity of water vapor. The above relationship is expected to be valid only in the laminar or viscous sublayer in which the molecular exchange remains the primary (perhaps the only) transport mechanism. The same mechanism is operating at the air–soil and air–leaf interfaces. However, natural surfaces, not being very smooth and uniform, may not have well-defined molecular sublayers which could be amenable to direct measurements. Therefore, the practical utility of Equation (12.1) for estimating evaporation in the atmosphere remains questionable. Some attempts have been made by physical chemists and fluid dynamists to study evaporation and the other transport processes on the molecular scale. Micrometeorologists are more interested, however, in turbulent exchange processes over the much larger scales that are encountered in the atmospheric surface layer and the PBL.

In the lower part (surface layer) of the atmospheric boundary layer the air motion is almost always turbulent. Here, the water vapor is efficiently transported away from the interfacial molecular sublayer by turbulent eddies. Because eddy transport and the intensity of turbulent mixing depend on the surface roughness, wind shear or friction velocity, and thermal stratification, the evaporation rate also depends on the above factors, as well as on the average specific humidity gradient. The frequently used gradient-transport relation for the same is

$$E = -\rho K_w(\partial Q/\partial z) \tag{12.2}$$

in which the eddy diffusivity K_w replaces the molecular diffusivity in Equation (12.1).

While Fick's law or Equation (12.1) has a firm theoretical (e.g., the kinetic theory of gases) and experimental basis, the analogous relation [Equation (12.2)] for a turbulent flow is based on a phenomenological gradient-transport theory whose limitations have already been pointed out in Chapter 9. Better flux-gradient relations obtained on the basis of the Monin–Obukhov similarity theory and carefully conducted micrometeorological experiments will be discussed in a later section.

12.2 Potential Evaporation and Evapotranspiration

Over a bare land surface with no standing water, the soil moisture is the only source of water for evaporation. Therefore, the rate of evaporation must depend on the moisture content of the topmost layer of the soil. When the soil surface is fully saturated and the soil moisture content is not a limiting factor in evaporation, this maximum rate of evaporation for the given surface weather conditions is called potential evaporation (E_p). When the soil becomes drier, the rate of evaporation for the given atmospheric conditions (particularly the surface temperature, near-surface wind speed, specific humidity, and stability) also depends on the moisture content of the topmost soil layer which, in turn, depends on the soil moisture flow through the soil. The relationship between the rate of evaporation (E_0) and the soil moisture flow rate (M) is given by the instantaneous water balance equation at the surface

$$E_0 = M_0 \tag{12.3}$$

or the water balance equation for the subsurface layer,

$$E_0 = M_b - \Delta M \tag{12.4}$$

in which ΔM is the rate of storage of soil moisture in the layer per unit area of the surface, and M_b is the moisture flow rate at the bottom of the layer. The above simplified water budget equations would be valid only during the periods of no precipitation, irrigation, or runoff at the surface.

The vertical moisture flow rate, or the flux of soil moisture, is related to the vertical gradient of the soil moisture content (S) through Darcy's law

$$M = -k_m(\partial S/\partial z) \tag{12.5}$$

where k_m is the hydraulic conductivity through the soil medium. Note that Equation (12.5) (Darcy's law) is analogous to Equation (4.1) (Fourier's law) for heat conduction. Similarly, following the derivation of Fourier's heat conduction equation [Equation (4.2) or (4.3)], one can obtain

$$(\partial/\partial t)\,(\rho S) = -\partial M/\partial z \tag{12.6}$$

or, after substituting from Equation (12.5),

$$\partial S/\partial t = (\partial/\partial z)[\alpha_m(\partial S/\partial z)] \tag{12.7}$$

where $\alpha_m = k_m/\rho$ is the diffusivity of soil moisture.

For surfaces with live vegetation, a significant portion of water vapor transfer to the atmosphere is by transpiration from the vegetation, most of which takes place through the stomata of leaves. Transpiration may be considered an important by-product of the process of photosynthesis and respiration, in which plants take CO_2 during daytime and give out the same at nighttime. Transpiration may also be a physiological necessity for plants to draw up dissolved nutrients from the soil through their roots. Moisture transfer through roots, stands, and leaves is considerably more efficient than that through the soil pores, because the former is forced by a strong osmotic pressure gradient. Still, evaporation from the bare soil surface in between the plants is by no means negligible. Because evaporation and transpiration occur simultaneously and it is not easy to distinguish between the vapor transferred by the two processes, the term evapotranspiration is sometimes used to describe the total water vapor transfer to the atmosphere. More often, though, meteorologists use evaporation as a substitute or synonym for evapotranspiration. The distinction between the two disappears, anyway, when one considers the vegetative surface as a composite of soil and leaf surfaces, which contribute to the total water vapor transfer (evaporation) to the atmosphere per unit horizontal area of the composite surface, per unit time. In this gross sense, details of the vegetation canopy (e.g., leaves and branches) and small-scale transfer processes around the individual surface elements are essentially glossed over and the canopy is considered as an idealized, homogeneous material layer which has a finite thickness and mass, and which can store or release heat and water vapor. In particular, the detailed water budget for a vegetative surface is often too complicated to be used as a practical tool for estimating evaporation.

The concept of potential evaporation (E_p) can also be extended to any vegetative surface for which E_p represents the maximum evapotranspiration likely from the vegetative surface for a given set of surface weather conditions. The potential evapotranspiration is generally less than the free water surface evaporation under the same weather conditions, especially in humid regions. The former can exceed the latter, however, in certain arid regions and under certain weather conditions (Rosenberg *et al.*, 1983, Chapter 7).

Actual evaporation usually differs from the potential evaporation, because the surface may not be saturated and the plants may not be drawing water from the soil and transpiring at their maximum rate. Still, the concept of potential evaporation is quite useful in agricultural and hydrological applications.

12.3 Modified Monin–Obukhov Similarity Relations

The original Monin–Obukhov similarity hypothesis and subsequent relations based on the same, as discussed in the preceding chapter, are strictly valid when buoyancy effects of water vapor can be ignored. When substantial evaporation occurs and it affects the density stratification in the surface layer, modified M–O similarity relations incorporating the buoyancy effects of water vapor would be more appropriate.

Some buoyancy effects of water vapor have already been discussed in Chapter 5, where the concepts of virtual temperature and virtual potential temperature were introduced. A similar concept is that of the virtual heat flux H_v, which may be interpreted as the flux of virtual temperature ($H_v = \rho c_p \overline{w\theta_v}$). The modified Monin–Obukhov similarity hypothesis states that, in a homogeneous and stationary atmospheric surface layer, the mean gradients and turbulence structure depend only on four independent variables: z, u_*, g/T_{v0}, and $H_{v0}/\rho c_p$. As mentioned in Chapter 5, the difference between the virtual temperature and actual temperature is no more than 7 K and frequently less than 2 K, so that the buoyancy parameter g/T_{v0} does not differ much from g/T_0. However, the virtual heat flux may differ significantly from the actual heat flux, and the two are approximately related as

$$H_v \cong H + 0.61 c_p \Theta E \tag{12.8}$$

This follows from the approximate relationship between the fluctuations of virtual temperature, actual temperature, and specific humidity

$$\theta_v \cong \theta + 0.61 \Theta q \tag{12.9}$$

which, in turn, can be derived from the relationship between $\tilde{\Theta}_v$ and $\tilde{\Theta}$, noting that $\tilde{\Theta}_v = \Theta_v + \theta_v$ and $\tilde{\Theta} = \Theta + \theta$. These derivations are left as an exercise for the reader.

The relation given by Equation (12.8) can also be expressed in terms of the latent heat flux H_L, or the Bowen ratio B, as

$$H_v = \mathrm{H} + \alpha_\theta H_L = H(1 + \alpha_\theta B^{-1}) \tag{12.10}$$

in which $\alpha_\theta = 0.61 c_p \Theta / L_e$ is a dimensionless coefficient. Since α_θ is only weakly dependent on the absolute temperature, a constant value of $\alpha_\theta = 0.07$, corresponding to $\Theta = 280$ K, is frequently used in the above relations. From Equation (12.10) it is obvious that the suggested modification to the M–O hypothesis is necessary only when $|B| < 1$, i.e., when the latent heat flux exceeds the sensible heat flux. This is usually the case over the oceans, but not so common over the land areas.

Note that in the modified similarity hypothesis both the sensible heat flux and water vapor flux are considered together in an appropriate combination (the virtual heat flux) and not separately. Therefore, a modified Obukhov length is defined as

$$L = -u_*^3 \Big/ \left(k \frac{g}{T_{v0}} \frac{H_{v0}}{\rho c_p} \right) \tag{12.11}$$

However, the scales of temperature, virtual temperature, and specific humidity have to be defined from their respective fluxes, i.e.,

$$\begin{aligned} \theta_* &= -H_0/(\rho c_p u_*) \\ \theta_{v*} &= -H_{v0}/(\rho c_p u_*) \\ q_* &= -E_0/(\rho u_*) \end{aligned} \tag{12.12}$$

The corresponding flux-profile relations for the transfer of water vapor in the atmospheric surface layer are

$$\begin{aligned} \frac{kz}{q_*}\frac{\partial Q}{\partial z} &= \phi_w\left(\frac{z}{L}\right) \\ \frac{Q - Q_0}{q_*} &= \frac{1}{k}\left[\ln\frac{z}{z_0} - \psi_w\left(\frac{z}{L}\right)\right] \\ E_0 &= -\rho C_W U(Q - Q_0) \end{aligned} \tag{12.13}$$

which are analogous to the heat transfer relations discussed in Chapter 11. Thus, the transfer of water vapor and other gaseous substances from the surface to the atmosphere or vice versa can be treated in the same way as the transfer of heat. The similarity between water vapor and heat transfer implies that

$$\begin{aligned} \phi_w(z/L) = \phi_h(z/L); \quad \psi_w(z/L) = \psi_h(z/L) \\ K_w = K_h; \quad C_W = C_H \end{aligned} \tag{12.14}$$

Experimental verification of Equations (12.14) has been done in a few micrometeorological studies. For example, simultaneous determinations of $\phi_h(z/L)$ and $\phi_w(z/L)$ from micrometeorological observations at Kerang, Australia, are compared in Figure 12.1. Note that there is no systematic and significant difference between these two functions, although there is more scatter in the estimates of $\phi_w(z/L)$, probably due to larger uncertainties in the eddy correlation measurements of $\overline{wq}$ as compared to $\overline{w\theta}$. Both of the empirical functions are well represented by the simpler Businger–Dyer relations [Equation (11.9)]. There is also some observational evidence for the equality of K_h and K_w under stable conditions (Oke, 1970; Webb, 1970). A number of air–sea interaction studies in the marine atmospheric surface layer have also confirmed the approximate equality of eddy diffusivities, as well as bulk transfer coefficients of heat and water vapor, especially in the absence of such complicating factors as breaking waves and water spray.

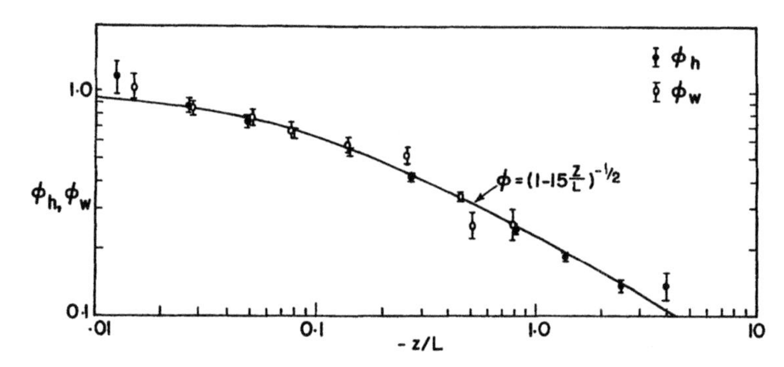

Figure 12.1 Comparison of the observed dimensionless potential temperature and specific humidity similarity functions for unstable conditions. [After Dyer (1967).]

There may be some exceptions to the above-mentioned similarity between heat and water vapor transfers and the implied equality of their eddy exchange coefficients and M–O similarity functions. For example, micrometeorological measurements made over irrigated fields under conditions of warm air advection, when heat and water vapor are transferred in opposite directions (e.g., downward sensible heat flux and upward water vapor flux), have shown K_h values to be generally larger than K_w (Verma *et al.*, 1978). Under such strong advection conditions, which are generally accompanied by strong winds and are representative of near-neutral or mild stability (Ri < 0.01), the ratio K_h/K_w is found to range between 1 and 3.

12.4 Micrometeorological Methods of Determining Evaporation

Depending on the application for which the rate of evaporation is to be estimated and the associated temporal and spatial scales, a large number of hydrological, climatological, and micrometeorological methods have been proposed in the literature for determining evaporation (Brutsaert, 1982, Chapters 8–11; Rosenberg *et al.*, 1983, Chapter 7; Garratt, 1992, Chapter 5). Here, we will restrict ourselves mainly to the micrometeorological methods which can be used to determine or measure the rate of evaporation at small time (order of an hour or less) and space (order of 1000 m or smaller) scales.

12.4.1 Direct measurements by lysimeters and evaporation pans

Instruments used for direct measurement of evaporation or potential evaporation from land surfaces are called lysimeters. A lysimeter is a cylindrical container (1–2 m deep and 1–6 m in diameter) in which an undisturbed block of soil and vegetation is isolated and its water budget is carefully monitored and controlled. Changes in the mass of the lysimeter are monitored either by a sensitive balance (mechanical or electronic) installed underneath, or a manometer measuring differences of hydrostatic pressures. There are two types of lysimeters, namely, weighing and floating lysimeters. In the latter, representative samples are floated on liquids such as water, oil, or solutions of zinc chloride, and their weight changes are determined from the fluid displacements resulting from the changes in buoyancy of the floating sample. Precision lysimeters with accuracies of 0.01–0.02 mm equivalent of water evaporated are found to be most suitable for measuring evaporation on short time scales (less than an hour). The lysimeter dimensions depend on the type of vegetation on the surface and the depth of the root zone. An extreme example is that of a giant floating lysimeter containing a mature Douglas fir tree in a forest at Cedar River, Washington, which is used for measuring evapotranspiration from the tree (Oke, 1987).

Evaporation pans are used for measuring free water evaporation from small lakes and ponds. Many types of pans have been used over the years and some have been standardized. The diameters of cylindrical evaporation pans used in practice range from 1 to 5 m and their depths range from 0.25 to 1.0 m. Due to limitations of size and spurious edge effects, the evaporation rate from a pan is usually 5–30% larger than that from a small lake or a pond. Large, sunken pans are most representative of free water evaporation in the absence of breaking waves and spray.

Evaporation pans have also been used for estimating potential evapotranspiration E_p from vegetative surfaces. Since, evaporation from a pan depends on its size, location, and exposure to sun and winds, calibration of a pan against measured potential evapotranspiration may be necessary for this method to be reliable. The ratio of potential evapotranspiration to pan evaporation on time scales of 1–30 days has been found to vary between 0.5 and 1.5 (this range may be expected to be even larger for small time scales of interest in micrometeorology). It depends on the pan size and exposure, geographical location, weather conditions, season, and the type of vegetation and its growth stage.

Other inexpensive instruments for estimating potential evaporation are atmometers, which are essentially porous ceramic or paper evaporating surfaces. The evaporating surface is continuously supplied with water; the measured rate at which water must be supplied to keep the porous material saturated is a measure of potential evaporation for given weather conditions. Again, calibration against a standard instrument or technique is necessary for an atmometer to be useful.

12.4.2 Eddy correlation method

A direct method of measuring the local water vapor flux over a homogeneous or nonhomogeneous surface is to measure simultaneously turbulent velocity and specific humidity fluctuations and determine their covariance over the desired sampling or averaging time. It is based on the relationship

$$E = \rho \overline{wq} \tag{12.15}$$

where E is the vertical flux of water vapor. Over a homogeneous surface, if eddy correlation measurements are made in the constant flux surface layer, the rate of evaporation from the surface is also given by Equation (12.15), as $E \cong E_0$.

This method is simple in theory, but very difficult to use in practice, because the fast-response instruments required for measuring vertical velocity and specific humidity fluctuations need great care to install, maintain, calibrate, and operate. Reliable fast-response humidity instruments, such as the Lyman–Alpha humidiometer, microwave refractometer/hygrometer, and infrared hygrometer have been developed (Hay, 1980; Lenschow, 1986). Such sophisticated instruments and the eddy correlation method have so far been used only during certain research expeditions and not for routine measurements of evaporation.

12.4.3 Energy balance/Bowen ratio method

The energy budgets of various types of surfaces, as well as of interfacial layers, have been discussed in Chapter 1. As mentioned in Chapter 11, the appropriate energy budget equation can be used to determine the sum of sensible and latent heat fluxes from measurements or estimates of the rest of the budget terms. Further partitioning of this into sensible and latent heat fluxes can be done if their ratio $B = H/H_L$ is measured or can otherwise be estimated. Then, Equation (2.3) can be used to determine the latent heat flux H_L from which

$$E_0 = H_L/L_e \tag{12.16}$$

The Bowen ratio can be estimated from the gradient transport relations for H and E, with the assumption of the equality of the eddy exchange coefficients K_h and K_w. It is easy to show that

$$B = \frac{c_p}{L_e} \frac{\partial \Theta / \partial z}{\partial Q / \partial z} \cong \frac{c_p}{L_e} \frac{\Delta \Theta}{\Delta Q} \tag{12.17}$$

where $\Delta\Theta = \Theta_2 - \Theta_1$ and $\Delta Q = Q_2 - Q_1$. Thus, in order to determine B one needs to measure the differences in temperature and specific humidities at two levels in the surface layer. This can be easily done through the use of dry- and wet-bulb thermometers or thermocouples, preferably in a difference circuit.

Another method of estimating the Bowen ratio is discussed later in the context of the Penman approach.

12.4.4 Bulk transfer approach

This is similar to the bulk transfer method for determining the surface heat flux. The appropriate bulk transfer formula for the water vapor flux is

$$E_0 = \rho C_w U_r (Q_0 - Q_r) \tag{12.18}$$

in which Q_0 is the mean specific humidity very close to the surface (more appropriately, at $z = z_0$) and C_w is the bulk transfer coefficient for water vapor, which can be specified or parameterized in the same manner as C_H (Reynolds' analogy between water vapor and heat transfers implies $C_W = C_H$). This approach requires measurements of wind speed, temperature, and specific humidity at a reference height, a good estimate of the surface roughness z_0, and the temperature and specific humidity at z_0. The main difficulty is that Θ_0 and Q_0 are not easy to measure or estimate, especially for vegetative surfaces. For such surfaces the zero-plane displacement height d_0 must also be estimated, as Θ_0 and Q_0 refer to the height $d_0 + z_0$.

Alternative forms of the bulk-transfer relations that are frequently used to parameterize the surface fluxes of sensible heat and water vapor are

$$\begin{aligned} H_0 &= \rho c_p C_H U_r (\Theta_s - \Theta_r) \\ E_0 &= \rho C_W U_r (Q_s - Q_r) \end{aligned} \tag{12.19}$$

in which Θ_s and Q_s denote the potential temperature and specific humidity right at the surface. The use of Θ_s and Q_s is more appropriate if their values are readily available from observations or can be obtained from the prognostic equations for Θ_s and Q_s.

Over large lake, sea and ocean surfaces, the water surface temperature T_s is routinely available from satellite measurements. It has been estimated by bucket water temperature in most field experiments on air–sea interactions. Due to very small values of z_0 over water surfaces, no distinction is usually made between Θ_0 and Θ_s, or between Q_0 and Q_s. The specific humidity near the water surface is determined simply as the saturation value at T_s. An experimental verification of Equation (12.19) is given in Figures 12.2 and 12.3, in which vertical fluxes of

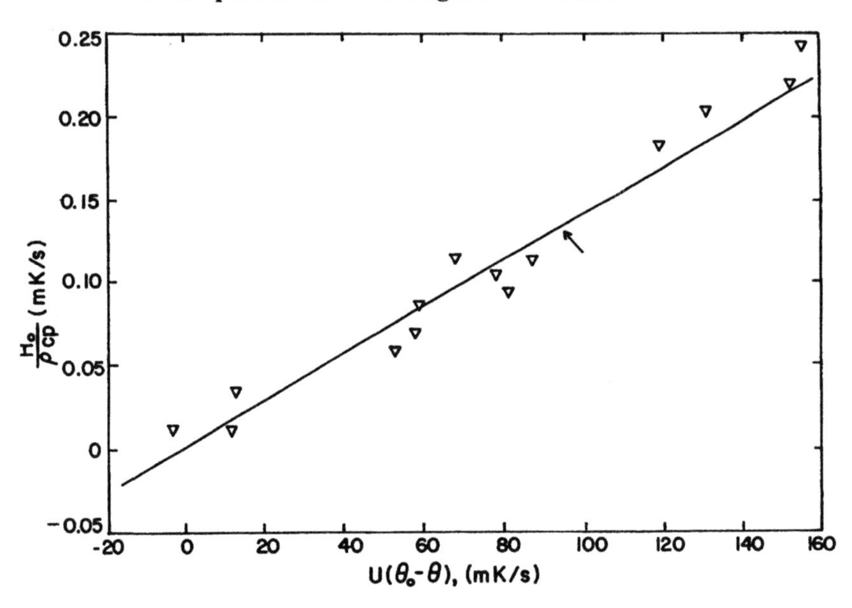

Figure 12.2 Sensible heat flux at the sea surface as a function of $U(\Theta_0 - \Theta)$ in strong winds compared with Equation (12.19) with $C_H = 1.14 \times 10^{-3}$. [After Friehe and Schmitt (1976); data from Smith and Banke (1975).]

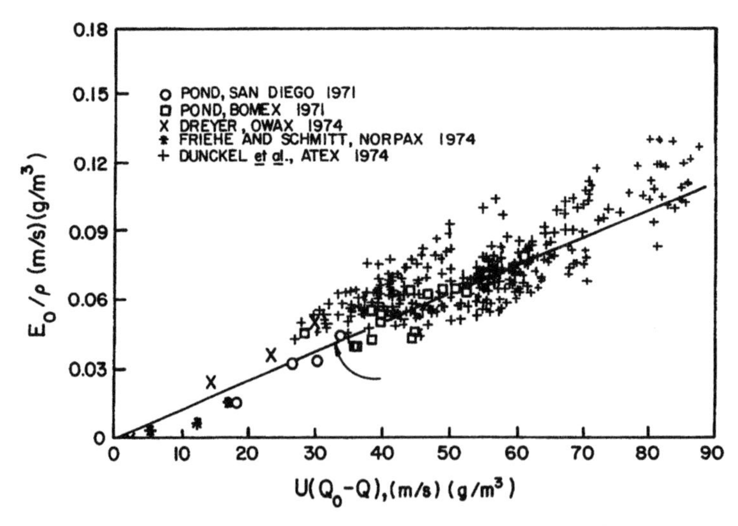

Figure 12.3 Observed moisture flux at the sea surface as a function of $U(Q_0 - Q)$ compared with Equation (12.18) with $C_W = 1.32 \times 10^{-3}$. [After Friehe and Schmitt (1976).]

heat and moisture are shown to be well correlated (nearly proportional) with the products $U_r(\Theta_s - \Theta_r)$ and $U_r(Q_s - Q_r)$, respectively, at a reference height $z_r = 10$ m. Note that the bulk transfer coefficients C_H and C_W are given by the slopes of the regression lines passing through the origin in Figures 12.2 and 12.3, respectively.

A central objective of many air–sea interaction experiments and expeditions has been the empirical determination of the drag, heat, and water vapor transfer coefficients (C_D, C_H, and C_W), covering a wide range of surface roughness, sea state, and stability conditions. Under moderate winds ($6 < U_{10} < 12 < \text{m s}^{-1}$) and near-neutral stability, there is an overwhelming experimental evidence indicating that $C_{DN} \cong C_{HN} \cong C_{WN} \cong 1.2 \times 10^{-3}$. Under more typical, slightly unstable conditions prevailing over the oceans, however, transfer coefficients are somewhat larger, but still approximately equal to each other ($C_D \cong C_H \cong C_W \cong 1.5 \times 10^{-3}$), at least within the estimated margin of error ($\pm 20\%$) of their determination (Pond, 1975). These results point to a rather simple parameterization of air–sea fluxes in terms of wind speed, temperature, and specific humidity at the 10 m level and the sea-surface temperature.

Atypical or exceptional stability conditions may be encountered, of course, over certain oceanic areas and during certain weather conditions when C_D, C_H, and C_W may differ considerably from their above typical values. For example, in the areas and periods of intense cold-air advection over the warmer ocean, free convection may occur in which C_H and C_W are expected to become considerably larger than $C_D > 1.5 \times 10^{-3}$. On the other hand, under inversion conditions and light winds, all the transfer coefficients are expected to become much smaller than their neutral value.

The stability effects on the drag and other transfer coefficients can be considered in the framework of the Monin–Obukhov similarity theory. For example, Equation (11.17) can be expressed in the form

$$\begin{aligned} C_D/C_{DN} &= [1 - k^{-1}C_{DN}^{1/2}\psi_m(z/L)]^{-2} \\ C_H/C_{DN} &= [1 - k^{-1}C_{DN}^{1/2}\psi_m(z/L)]^{-1}[1 - k^{-1}C_{DN}^{1/2}\psi_h(z/L)]^{-1} \end{aligned} \tag{12.20}$$

expressing the ratios C_D/C_{DN} and C_H/C_{DN} as functions of C_{DN} and z/L. These can also be expressed in terms of the more convenient bulk Richardson number $\text{Ri}_B = gz(\Theta_v - \Theta_{v0})/T_{v0}U^2$, using the following relations between z/L and Ri_B:

$$\begin{aligned} z/L &\cong kC_{DN}^{-1/2}\text{Ri}_B, & &\text{for } \text{Ri}_B < 0 \\ z/L &\cong kC_{DN}^{-1/2}\text{Ri}_B(1 - 5\text{Ri}_B)^{-1} & &\text{for } \text{Ri}_B \geq 0 \end{aligned} \tag{12.21}$$

The first of these relations is only an approximation to the more exact but implicit relationship between z/L and $\mathrm{Ri_B}$. The ψ functions in Equation (12.20) are given by Equation (11.14), so that the ratios C_D/C_{DN} and C_H/C_{DN} can be evaluated as functions of $\mathrm{Ri_B}$. Figure 12.4 shows that these are only weakly dependent on the neutral value of the drag coefficient (C_{DN}), but have rather strong dependence on $\mathrm{Ri_B}$, particularly in stable conditions.

Over land surfaces, Θ_s can be determined from the prognostic equation [Equation (4.14)] for the ground surface temperature T_s, in which the sensible and latent heat fluxes are expressed using the bulk-transfer relations (12.19). The surface specific humidity, Q_s, is usually estimated from T_s and wetness or moisture content of the surface, using a variety of simple, noninteractive, or more complicated, interactive schemes that have been proposed (Garratt, 1992, Chapter 8).

For a bare saturated soil surface, Q_s is simply equal to the saturated specific humidity Q_s^* at T_s, and can be determined from Equations (5.11)–(5.13) or a corresponding thermodynamic (e.g., e_s versus T) diagram. For unsaturated or drying soil surfaces and for vegetated canopies in particular, Q_s is not so easily determined. Approximate diagnostic relations for Q_s, based on an assumed Bowen ratio, the surface relative humidity, or the ratio of actual to potential evaporation are often used. Sometimes, interactive schemes utilizing the so-called 'bucket' and 'force restore' methods are employed (Garratt, 1992, Chapter 8).

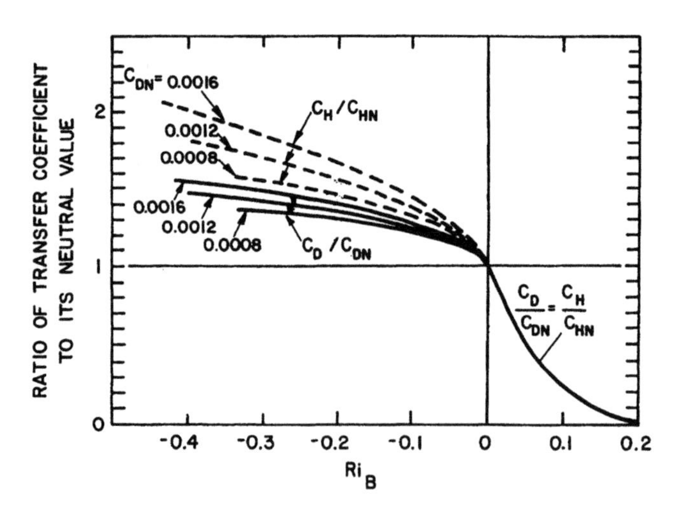

Figure 12.4 Calculated ratios of the drag and heat transfer coefficients to their neutral values as functions of the bulk Richardson number and C_{DN}. [After Deardorff (1968).]

The presence of a vegetation canopy makes the above bulk-transfer approach to flux parameterization very crude and somewhat dubious. The difficulty comes from our inability to define and determine the effective surface temperature and specific humidity without a detailed canopy model or parameterization. An introduction to canopy modeling and parameterization will be given later in Chapter 15.

Example Problem 1

The following meassurements were made on the research vessel FLIP during the 1969 Barbados Oceanographic and Meteorological Experiment (BOMEX):

Mean wind speed at 10.9 m above the sea surface = 7.96 m s^{-1}.
Mean temperature at 10.9 m = 27.34°C.
Mean specific humidity at 10.9 m = 16.04 g kg^{-1}.
Sea-surface temperature = 28.35°C.
Surface pressure = 1000 mbar.

Using the bulk-transfer relations with $C_D = C_H = C_W = 1.5 \times 10^{-3}$, calculate the surface stress, the sensible heat flux, and the rate of evaporation at the time of these observations. Also estimate the Bowen ratio and the virtual heat flux at the surface.

Solution

First, we calculate the saturation specific humidity Q_s corresponding to the observed sea-surface temperature $T_s = 301.50$ K. For this, we can use Equation (5.13) for determining the saturation water vapor pressure e_s as

$$\ln \frac{e_s}{6.11} = \frac{m_w L_e}{R_*}\left(\frac{1}{273.2} - \frac{1}{301.5}\right)$$
$$= \frac{18.02 \times 10^{-3} \times 2.43 \times 10^6 (0.00366 - 0.00332)}{8.314}$$
$$= 1.791$$

so that $e_s = 36.62$ mbar

Then, from Equation (5.11),

$$Q_s = 0.622 \frac{e_s}{P} = 0.02278$$

Next, the density of moist air near the surface can be determined from the equation of state as

$$\rho = \frac{P}{RT_v}$$

where,

$$T_v = T(1 + 0.61Q) = 301.5(1 + 0.61 \times 0.0228) = 305.7 \text{ K}$$

Thus,

$$\rho = \frac{1000 \times 100}{287.04 \times 305.7} \cong 1.14 \text{ kg m}^{-3}$$

Finally, using the bulk-transfer relations (11.16) and (12.19), the surface stress and heat and water vapor fluxes can be estimated as follows:

$$\begin{aligned}
\tau_0 &= \rho C_D U_r^2 = 1.14 \times 1.5 \times 10^{-3} \times (7.96)^2 \cong 0.108 \text{ N m}^{-2} \\
H_0 &= \rho c_p C_H U_r(\Theta_s - \Theta_r) \\
&= 1.14 \times 1004 \times 1.5 \times 10^{-3} \times 7.96(301.50 - 300.49 - 0.0098 \times 10.9) \\
&= 12.3 \text{ W m}^{-2} \\
E_0 &= \rho C_E U_r(Q_s - Q_r) \\
&= 1.14 \times 1.5 \times 10^{-3} \times 7.96(0.02278 - 0.01604) \\
&\cong 9.17 \times 10^{-5} \text{ kg m}^{-2} \text{ s}^{-1}
\end{aligned}$$

The Bowen ratio can be determined as

$$B = \frac{H_0}{L_e E_0} = \frac{12.3}{2.45 \times 10^6 \times 9.17 \times 10^{-5}} \cong 0.055$$

which may be compared with an indirect estimate from Equation (12.17)

$$B = \frac{c_p}{L_e} \frac{\Delta\Theta}{\Delta Q} = \frac{1004(301.50 - 300.60)}{2.45 \times 10^6(0.02278 - 0.01604)} \cong 0.055$$

The virtual heat flux at the surface can be estimated from Equation (12.10) with $\alpha_\theta = 0.07$ as

$$H_{v0} = H_0(1 + 0.07B^{-1}) = 12.3(1 + 1.28) \cong 28.0 \text{ W m}^{-2}$$

12.4.5 Penman approach

Using a combination of the energy balance and bulk transfer formulas, Penman (1948) derived a formula for evaporation from open water and saturated land surfaces, which considerably simplifies the observational procedure. Penman's formula, either in the original or slightly modified form, is widely used for estimating potential evaporation or evapotranspiration, especially in hydrological and agricultural applications. A modern derivation of the formula is given in the following discussion.

Equation (12.18) can be written in the form

$$
\begin{aligned}
E_0 &= \rho C_W U_r (Q_0 - Q_r^*) + E_a \\
E_a &= \rho C_W U_r (Q_r^* - Q_r)
\end{aligned}
\tag{12.22}
$$

where Q_r^* represents the specific humidity at the reference height if air were saturated there, and E_a can be interpreted as the drying power of air, because it is proportional to the difference between the saturation humidity and the actual specific humidity of air at the reference height z_r.

If the measurement height is low (say, a few meters), the bulk transfer relation for surface heat flux can be approximated as

$$
H_0 = \rho c_p C_H U_r (T_0 - T_r) \tag{12.23}
$$

From Equations (12.22) and (12.23), assuming $C_W = C_H$, we obtain

$$
\frac{L_e E_0}{H_0} = \frac{L_e}{c_p}\left(\frac{Q_0 - Q_r^*}{T_0 - T_r}\right) + \frac{L_e E_a}{H_0}
$$

or, after converting specific humidities into water vapor pressures through Equation (5.11),

$$
B^{-1} = \frac{0.622 L_e}{P c_p}\left(\frac{e_0 - e_r^*}{T_0 - T_r}\right) + B^{-1}\frac{E_a}{E_0}
$$

From this one obtains Penman's expression for the Bowen ratio

$$
B = \frac{\gamma}{\Delta}\left(\frac{E_0 - E_a}{E_0}\right) \tag{12.24}
$$

where

$$\gamma = \frac{c_p P}{0.622 L_e} \tag{12.25}$$

is the psychrometer constant whose value is about 0.66 mbar K^{-1} at $T = 273$ K and $P = 1000$ mbar and

$$\Delta = (e_0 - e_r^*)/(T_0 - T_r) \cong de_s/dT \tag{12.26}$$

is the slope of the saturated vapor pressure versus temperature curve at $(T_r + T_0)/2$. The crucial approximation in Equation (12.26), originally suggested by Penman (1948), could be justified only when the surface is wet, so that the water vapor pressure e_0 is close to its saturation value at the surface temperature T_0, and the measurement height above the surface is low.

After substituting from Equation (12.24) into Equation (2.3), we obtain the well-known Penman (1948) formula

$$E_0 = \frac{\Delta}{\Delta + \gamma}\left(\frac{R_N - H_G}{L_e}\right) + \frac{\gamma}{\Delta + \gamma} E_a \tag{12.27}$$

It is clear from the above derivation that Penman's formula has a sound theoretical and physical basis, although some investigators have considered it only empirical. In order to determine the evaporation rate from Equation (12.27), one needs measurements or estimates of net radiation and soil heat flux at ground level, together with the measurements of wind speed, temperature, and humidity at a low level (usually 2 m) above the surface. The observational procedure is simplified by the ingenuous way in which the surface temperature has been eliminated from the final result.

Several further simplifications of Penman's approach to determining potential evaporation have been suggested in the literature. For example, over a water surface or a wet land surface the air might be close to saturation in water vapor, so that E_a would be expected to be negligibly small in comparison to E_0. Then, the simplified relation

$$E_0 = \frac{\Delta}{\Delta + \gamma}\left(\frac{R_N - H_G}{L_e}\right) \tag{12.28}$$

should be adequate for determining the potential evaporation or evapotranspiration. The above formula was first suggested by Slatyer and McIlroy (1961), who called the above estimate 'equilibrium evapotranspiration.' It has been observed that the simpler formula [Equation (12.28)] is valid over a wide range of meteorological conditions and even for less than fully saturated surfaces with vegetation. Note that the only meteorological observation required is that of mean temperature at a low height of about 2 m, in addition to the measurement or estimate of $R_N - H_G$.

When Equation (12.28) is valid, the Bowen ratio is simply estimated as

$$B \cong \gamma/\Delta \tag{12.29}$$

which also follows from Equation (12.24) when $E_a \ll E_0$. Equation (12.29) can be used in conjunction with Equation (12.28) to obtain the sensible heat flux as

$$H_0 \cong [\gamma/(\Delta + \gamma)](R_N - H_G) \tag{12.30}$$

Equations (12.28) and (12.30) are most suitable for determining the surface fluxes of water vapor and heat from routine micrometeorological measurements. Their applicability is, however, limited to wet, bare surfaces and vegetative surfaces with wet foliage over which evapotranspiration is near its potential rate and both H_0 and E_0 are positive (upward) fluxes. For water surfaces and those with tall vegetation, the use of the surface energy balance equation [Equation (2.1)] may not be appropriate. For these, the energy budget for a layer can be used instead, and Equations (12.16), (12.27), (12.28), and (12.30) can be modified accordingly (replacing $R_N - H_G$ with $R_N - H_G - \Delta H_s$).

12.4.6 Gradient or aerodynamic method

This is similar to the gradient method of determining surface heat flux described in Chapter 11. It requires measurements of specific humidity, in addition to those of wind speed and temperature, at two levels in the surface layer. Under neutral stability conditions, both the wind speed and specific humidity follow similar logarithmic profile laws. It is easy to show that the evaporation rate is given by the Thornthwaite–Holtzman formula

$$E_0 = -\rho k^2 \Delta U \Delta Q / [\ln(z_2/z_1)]^2 \tag{12.31}$$

For the more general stratified conditions, following the same procedure as in the determination of momentum and heat fluxes in Chapter 11, the water vapor flux is given by

$$E_0 = -\rho k^2 \Delta U \Delta Q / \phi_m(\zeta_m)\phi_w(\zeta_m)[\ln(z_2/z_1)]^2 \tag{12.32}$$

which may be considered as a generalization of Equation (12.31). Alternatively, one can use the gradient-transport relation [Equation (12.2)] with $K_w = K_H$. The eddy diffusivity may be specified on the basis of Monin–Obukhov similarity relations, as described in Chapter 11.

12.4.7 Profile method

Again, this is similar to the profile method of determining momentum and heat fluxes. The additional profile relationship to be fitted through the specific humidity data is

$$Q = (q_*/k)[\ln z - \psi_w(z/L)] + [Q_0 - (q_*/k)\ln z_0] \tag{12.33}$$

which suggests a plot of Q versus $\ln z - \psi_w(z/L)$ for determining q_* and, if desired, Q_0. The friction velocity u_* is determined from a similar plot of wind profile and, then, the rate of evaporation $E_0 = -\rho u_* q_*$.

Wind, temperature, and specific humidity profiles over bare land and vegetated surface can easily be measured using small masts and portable towers. But, similar profile measurements over the open ocean are made extremely difficult by undesirable motions of any floating platforms and distortion of the air flow around them. Even more difficult to take and interpret are observations in the lowest layer of direct wave influence. Accurate profile measurements in the fully developed region of the atmospheric surface layer have been made from shallow-water towers and platforms, as well as from specially designed stable buoys and research vessels (Paulson *et al.*, 1972). These are found to be consistent and in good agreement with the Monin–Obukhov similarity relations [Equations (11.9)–(11.14)], provided the buoyancy effects of water vapor are included in the definitions of the M–O stability parameter, as described in Section 12.2. For example, Figures 12.5 and 12.6 show average profiles for different stability classes ranging from slightly stable (I) to moderately unstable (V), taken during the 1964 International Indian Ocean Expedition (IIOE). Note that according to the M–O similarity relations [Equation (11.12)], plots of U versus $\ln z - \psi_m$ and Θ versus $\ln z - \psi_h$ are expected to be linear and are shown to be so. Also note the much smaller gradients of wind and temperature over the ocean, compared to those over land surfaces, which limit the accuracy of flux determinations from mean profile measurements.

Example Problem 2

In addition to the wind, temperature, and humidity observations at 10.9 m, given in Example Problem 1, the following measurements were made at a lower height of 2.42 m:

Mean wind speed = 7.33 m s^{-1}.
Mean temperature = 27.55°C.
Mean specific humidity = 16.72 g kg^{-1}.

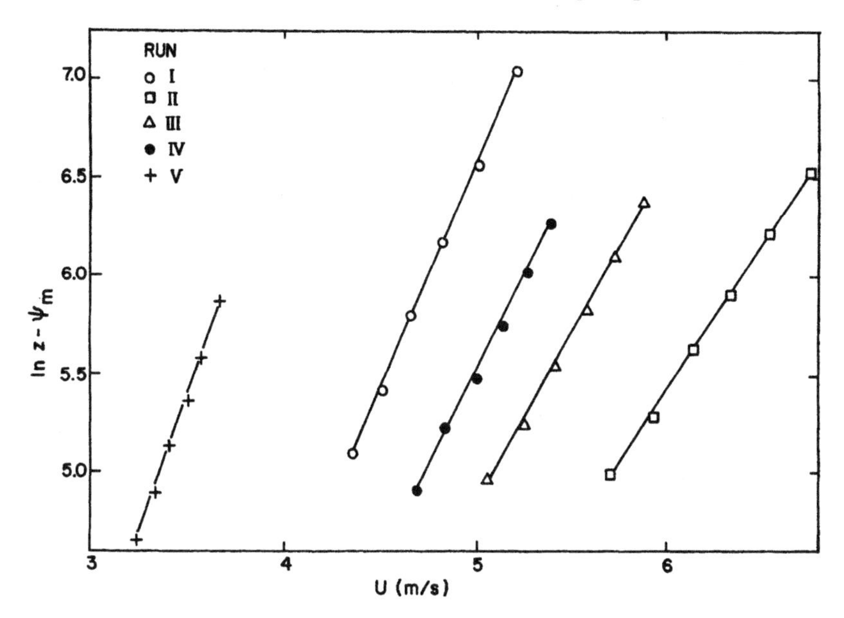

Figure 12.5 Composite wind speeds as functions of $\ln z - \psi_m$ for different stability classes (I–V) during IIOE. [After Paulson (1967).]

Use the measurements at the two height levels with the gradient method for estimating the near-surface fluxes of momentum, heat, and water vapor and compare them with those from the bulk-transfer approach used in Example Problem 1.

Solution

Here, $z_1 = 2.42$ m, $z_2 = 10.90$ m, so that

$$z_m = (2.42 \times 10.9)^{1/2} \cong 5.136 \text{ m}$$
$$z_m \ln(z_2/z_1) \cong 7.73 \text{ m}$$

In order to determine the Richardson number at z_m, we calculate the virtual temperatures at the two heights:

$$T_{v1} = 300.70(1 + 0.61 \times 0.01672) \cong 303.77 \text{ K}$$
$$T_{v2} = 300.49(1 + 0.61 \times 0.01604) \cong 304.43 \text{ K}$$

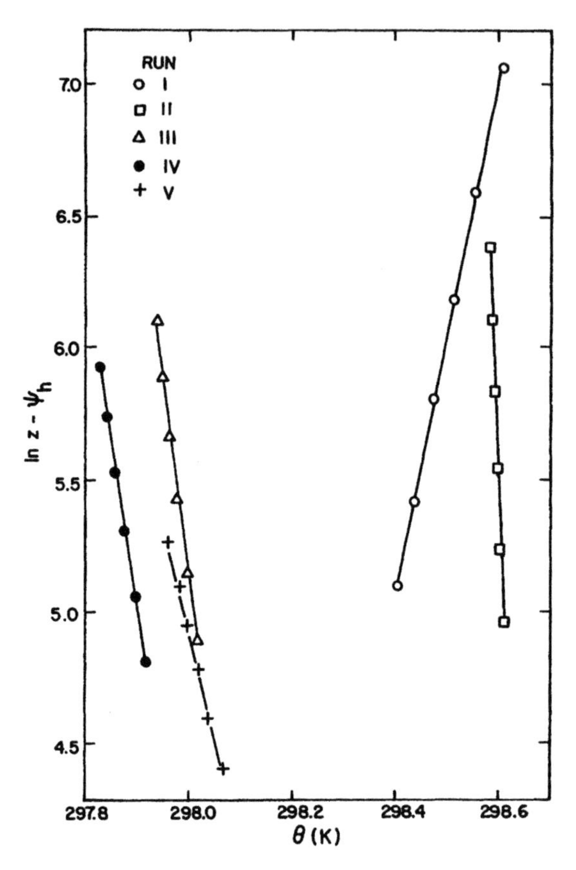

Figure 12.6 Composite potential temperatures as functions of $\ln z - \psi_h$ for different stability classes (I–V) during IIOE. [After Paulson (1967).]

Then, using Equation (11.23),

$$\frac{\partial T_v}{\partial z} = \frac{T_{v2} - T_{v1}}{z_m \ln(z_2/z_1)} = -\frac{0.66}{7.73} \cong -0.0854 \text{ K m}^{-1}$$

$$\frac{\partial \Theta_v}{\partial z} = \frac{\partial T_v}{\partial z} + \Gamma = -0.0854 + 0.0098 = -0.0756 \text{ K m}^{-1}$$

$$\frac{\partial U}{\partial z} = \frac{7.96 - 7.33}{7.73} \cong 0.0815 \text{ s}^{-1}$$

$$\text{Ri}_m = -\frac{9.81 \times 0.0756}{304.1 \times (0.0815)^2} \cong -0.367$$

Using Equation (11.25), the M–O stability parameter can be estimated as

$$\zeta_m = \mathrm{Ri}_m \cong -0.367$$

and the corresponding M–O similarity functions are estimated as

$$\phi_m = (1 - 15\zeta_m)^{-1/4} \cong 0.626$$
$$\phi_h = \phi_w = (1 - 15\zeta_m)^{-1/2} \cong 0.392$$

Then, from Equation (11.26), we obtain

$$u_* = \frac{0.4(7.96 - 7.33)}{0.626 \times 1.505} \cong 0.268 \text{ m s}^{-1}$$

$$\theta_* = \frac{0.4(300.60 - 300.72)}{0.392 \times 1.505} \cong -0.0814 \text{ K}$$

$$q_* = \frac{0.4(0.01604 - 0.01672)}{0.392 \times 1.505} \cong -4.61 \times 10^{-4}$$

Using the previously estimated value of $\rho = 1.14$ kg m^{-3}, the surface stress and other fluxes can be determined as

$$\tau_0 = \rho u_*^2 = 1.14 \times (0.268)^2 \cong 0.082 \text{ N m}^{-2}$$
$$H_0 = \rho c_p u_* \theta_* = -1.14 \times 1004 \times 0.268 \times 0.0814 \cong 24.97 \text{ W m}^{-2}$$
$$E_0 = -\rho u_* q_* = 1.14 \times 0.268 \times 4.61 \times 10^{-4} \cong 1.41 \times 10^{-4} \text{ kg m}^{-2} \text{ s}^{-1}$$

These may be compared with the earlier estimated values from the bulk-transfer approach:

$$\tau_0 \cong 0.108 \text{ N m}^{-2};\ H_0 \cong 12.3 \text{ W m}^{-2};\ E_0 \cong 0.92 \times 10^{-4} \text{ kg m}^{-2} \text{ s}^{-1}$$

The largest discrepancy between the two methods is found in the determination of the sensible heat flux, presumably due to the large uncertainty in the measurement of the rather small temperature gradient. The given sea-surface temperature may also be in error, because it actually represents the bucket water temperature, rather than the skin surface temperature. The two estimates of the rate of evaporation also differ by 40%, while the discrepancy in the surface stress is about 27%. These differences are consistent with the generally estimated uncertainties of $\pm 20\%$ in estimates of surface fluxes using the various micrometeorological methods.

12.5 Applications

The material presented in this chapter may have the following practical applications:

- Estimating evaporation or evapotranspiration over various types of surfaces.
- Parameterizing the transfer of water vapor to the atmosphere in large-scale atmospheric models.
- Systematic ordering of specific humidity observations in the atmospheric surface layer.
- Determining the water budget of the earth's surface.
- Incorporating the buoyancy effects of water vapor in flux-profile relations in the lower atmosphere.

Problems and Exercises

1.

(a) Considering the flow of moisture only in the vertical direction through a uniform and homogeneous subsurface stratum, derive Equation (12.7) for the soil moisture content.

(b) Describe a plausible method of estimating the rate of evaporation from a bare soil surface, using measurements of the soil moisture content in the topmost layer.

2.

(a) Derive an expression for the virtual heat flux in terms of the sensible and latent heat fluxes.

(b) If evaporation is taking place from a sea surface of temperature 27°C, which is slightly cooler than the air at 10 m height, what value of the Bowen ratio would correspond to the neutral stability? What would be the corresponding heat flux if the rate of evaporation is 2 mm day^{-1}?

3.

(a) If the normalized potential temperature and specific humidity gradients in the surface layer are equal [i.e., $(kz/\theta_*)(\partial\Theta/\partial z) = (kz/q_*)(\partial Q/\partial z)$], show that the eddy diffusivities of heat and water vapor must also be equal (i.e., $K_h = K_w$).

(b) How would you expect the ratio K_w/K_m to vary with stability? Express this ratio as a function of Ri.

(c) Calculate and compare the values of eddy diffusivity K_h or K_w during the typical daytime unstable conditions ($u_* = 0.5$ m s^{-1}, $H_{v0} = 500$ W m^{-2})

and nighttime stable conditions ($u_* = 0.1$ m s^{-1}, $H_{v0} = -50$ W m^{-2}) at a height of 10 m over a bare soil surface.

4. Starting from the instantaneous vertical flux of water vapor, derive the eddy correlation formula for the rate of evaporation $E_0 = \rho\overline{wq}$, and state all the assumptions you have to make.

5. Compare and contrast the energy balance/Bowen ratio method with the bulk transfer method for estimating heat and water vapor fluxes over (a) a short grass surface and (b) an ocean surface. Which method would you recommend in each case?

6. The following micrometeorological measurements of mean wind speed, temperature, specific humidity, and the surface energy budget were made on October 7, 1967, 1200 PST, on a large grass field near Davis, California:

z (m)	U (m s^{-1})	T (°C)	Q (g kg^{-1})
0.25	1.11	23.52	8.22
0.50	1.34	23.35	7.73
1.00	1.51	23.30	6.88
2.00	1.63	23.14	6.65
6.00	1.80	22.95	6.05

Net radiation flux = 450 W m^{-2}
Ground heat flux = 43 W m^{-2}
Average grass height = 0.07 m

(a) Determine the average value of the Bowen ratio.
(b) Calculate the sensible heat flux and the evaporation rate, using the surface energy balance/Bowen ratio method.
(c) Calculate the potential evapotranspiration from the grass surface using the simplified Penman formula [Equation (12.28)] with the observed air temperature at 2 m.
(d) Do the same, but using the original Penman formula [Equation (12.27)] and taking the neutral value of $C_W = C_D = k^2/(\ln z/z_0)^2$.

7. Given the profile data of Problem 6, calculate the sensible heat flux and the rate of evaporation using the following methods:
(a) The gradient or aerodynamic method with observations at heights of 0.5 m and 2 m.
(b) The profile method.

Chapter 13 | Stratified Atmospheric Boundary Layers

13.1 Types of Atmospheric Boundary Layers

Atmospheric boundary layers are usually classified into different types, depending on the atmospheric stability and the dominating mechanism of turbulence generation in the boundary layer. Turbulence is moderately strong and more or less continuous throughout the PBL under very windy and near-neutral stability conditions. It is essentially generated by surface friction and wind shear and, hence, is referred to as shear-generated or mechanical turbulence. The resulting near-neutral PBL has already been described in Chapter 10.

Vigorous turbulence can also be generated by convection over a heated surface and is generally referred to as convective turbulence. More often, over a heated surface, both the shear and convective mechanisms of turbulence generation operate simultaneously, especially in the surface layer. Wind shear is strongest near the surface and its influence is largely confined to the surface layer. Convective turbulence dominates in the overlying, deep mixed layer, while much weaker shear-generated turbulence occurs in the transition layer near the top of the PBL. Such a PBL is generally referred to as the convective boundary layer (CBL) whose three-layer structure is schematically shown in Figure 13.1.

The surface layer constitutes the lowest 5–10% of the CBL in which significant variations (gradients) of wind speed, potential temperature and specific humidity with height are observed to occur, but mean wind direction remains nearly constant with height. Immediately above the shallow surface layer lies the much deeper mixed layer, which extends almost up to the base of the capping inversion. It is characterized by a uniform potential or virtual potential temperature and nearly uniform winds and specific humidity. The mixed-layer depth or the height of inversion base z_i constitutes the most important height or length scale in the mixed-layer similarity theory. The mixed layer is capped by a stabley stratified transition layer in which turbulence is suppressed with increasing height and completely vanishes at the top of the PBL. The transition layer is characterized by intermittently penetrating thermals from below and entrainment of the nonturbulent free-atmospheric

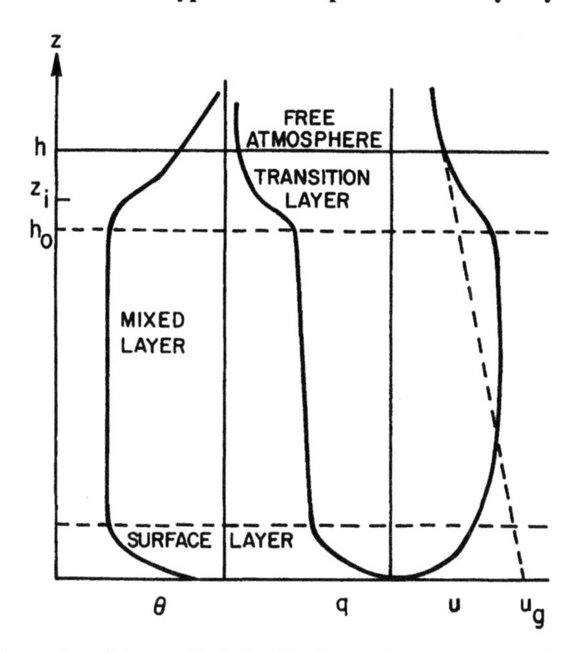

Figure 13.1 Schematic of the vertical distributions of mean potential temperature (Θ), specific humidity (Q), wind speed (U), and geostrophic wind (U_g), showing the three-layer structure of the convective boundary layer.

warmer and drier air from above, particularly during the late morning period of rapidly growing PBL. The transition layer typically extends from $0.9z_i$ to $1.2z_i$, although both the lower and upper boundaries can vary over a wider range.

In contrast to the CBL, the atmospheric boundary layer over a cold surface (e.g., any land surface during a clear night) is called the stable boundary layer (SBL), because it is stably stratified due to increasing temperature or potential temperature with height (also referred to as an inversion condition). Turbulence in the SBL is essentially shear generated, but is much weakened and sometimes entirely suppressed in the upper part by the action of buoyancy. Above the shallow surface layer, turbulence in the SBL is highly intermittent (episodic), patchy, and frequently mixed with internal gravity waves. In terms of the boundary layer height or thickness (h), the SBL is much thinner (typically, $h \cong 100$ m) than the CBL (typically, $h \cong 1000$ m).

The atmospheric boundary layer during the transitional near-neutral stability regime between unstable and stable conditions is also called the near-neutral boundary layer (NNBL). The surface is either slightly warmer or cooler than the air above. The morning and evening transition periods, when the sensible heat exchange between the surface and the atmosphere vanishes momentarily, as it changes sign from negative (downward flux) to positive (upward flux) and vice versa, are often marked by rapid changes in the PBL height, mean winds, and turbulence. Thus, the transitional NNBL is rapidly evolving under changing surface energy fluxes and is essentially nonstationary. It is the least studied and the least understood of all the atmospheric boundary layers.

The above classification of the PBL according to three broad stability regimes is somewhat artificial and for convenience only. More specific and quantitative PBL stability parameters, such as the bulk Richardson number and the ratio h/L of the PBL height to the Obukhov length, are found to vary continuously over wide ranges. One of these can be used to represent more systematically the important effects of stability on the mean flow and turbulence structure of the PBL as it varies from extremely unstable or convective to extremely stable conditions in the atmosphere. Still, in describing the typical observed structure of the PBL under convective, near-neutral and stable conditions, we often use the broader, qualitative stability characterizations and also the corresponding designations of the atmospheric boundary layer as CBL, NNBL and SBL. Since most of the PBL theories and models are more generic in their formulation and concepts, however, these will be reviewed without regard to the PBL stability. But, their lack of validity and other limitations under certain stability conditions will be discussed. In describing the PBL theories and models, it is also important to specify the presence or lack of certain simplifying conditions, such as stationarity and horizontal homogeneity of the PBL.

13.1.1 Horizontally homogeneous PBL

The atmospheric boundary layer formed under and driven by large-scale flows (e.g., geostrophic or gradient winds) over a reasonably flat and uniform surface can essentially be considered horizontally homogeneous, especially over horizontal dimensions of up to 100 km, i.e., smaller than the typical resolution of the synoptic observing network. Such an idealized PBL may evolve gradually with time in response to the similar evolution of the large-scale flow with time. Even when the large-scale flow is steady over several days, the PBL evolves diurnally as a result of the diurnal heating and cooling of the surface. Diurnal changes in the PBL are observed to become small and, sometimes, even insignificant over open ocean and sea surfaces and also extensive snow- and ice-covered surfaces over the polar regions during winter. Over a short period of the order of an hour, however, even a diurnally evolving PBL over a bare and dry land surface can be considered to be quasi-stationary when one considers its short-term averaged properties. This idealized (horizontally homogeneous and quasi-stationary) PBL has been the focus of many experiments as well as simpler theories and models that will be discussed in more detail in later sections.

The PBL depth or mixing height is the most important length scale which determines the size of the largest energy-containing eddies, average wind speed and direction shears in the PBL, as well as the overall stability of the PBL. The PBL depth shows strong diurnal variations in response to the diurnal heating and cooling of the surface. It increases throughout the day, becomes maximum (typically, 1 km) in the late afternoon, and then rapidly collapses to its typical nighttime value of the order of 100 m. At night, mixing height rarely stays constant, but usually oscillates around its mean value in response to periodic turbulence generation episodes. The maximum mixing height during the afternoon also shows large seasonal and spatial variations as shown in Figure 13.2 for the continental United States (Holzworth, 1964).

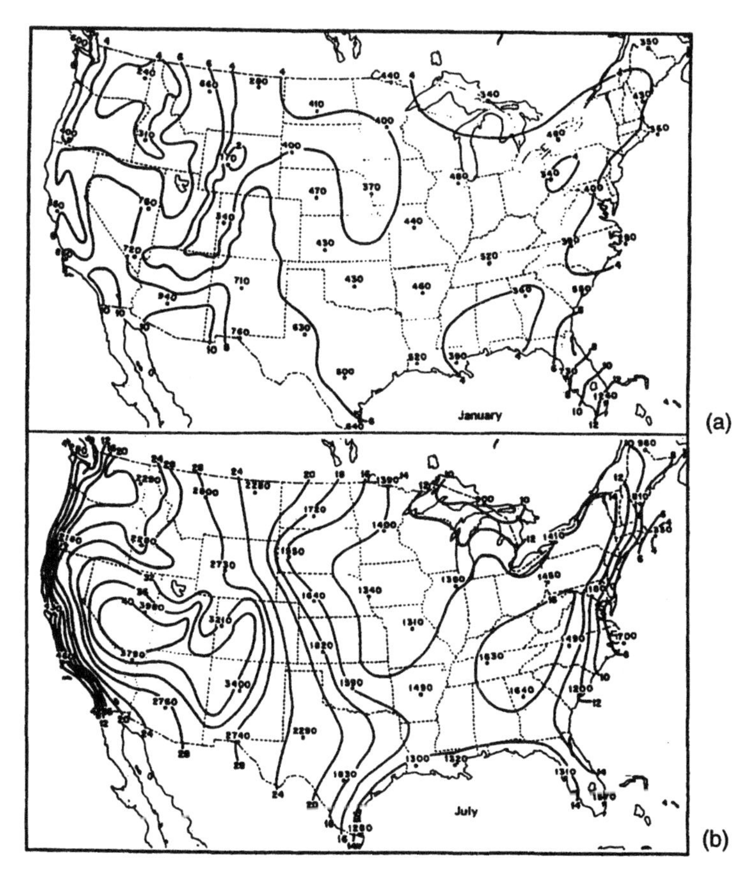

Figure 13.2 Spatial variation of mean maximum mixing height in meters over the continental United States for the months of (a) January and (b) July. [From Holzworth (1964).]

13.1.2 Nonhomogeneous boundary layers

The idealized horizontally homogeneous PBL may be expected to develop over extensive flat and uniform water surfaces, such as open seas, oceans and large lakes, as well as over a variety of land surfaces, such as deserts, grasslands, and forests, covering more than 80% of the earth's surface. However, over most of the developed land areas, which are marked by residential, industrial, commercial, agricultural, and recreational use, the atmospheric boundary layer is nonhomogeneous and frequently in a state of transition from one land use type to another. Surface inhomogeneities affecting the PBL flow include boundaries between land and water surfaces (coastlines); urban and rural areas (urban heat and roughness islands); different types of vegetation; and hills and valleys. In going over these land use and terrain inhomogeneities, the PBL flow encounters sudden or gradual changes in surface roughness, temperature, wetness, or elevation. Quite often, changes in several surface properties occur simultaneously. The mesoscale atmospheric circulation systems, such as land-sea breezes, urban heat island, mountain and valley winds, and others associated with complex topography, are likely to have nonhomogeneous boundary layers in them. Some of the cases of nonhomogeneous atmospheric boundary layers will be covered in more detail in Chapter 14.

13.2 Similarity Theories of the PBL

The observed mean wind, temperature and specific humidity profiles in the homogeneous atmospheric boundary layer, such as those presented earlier in Chapters 5 and 6, as well as turbulence structure (e.g., variances and covariances) are best represented in an appropriate similarity theory framework. A similarity theory, based on dimensional analysis, provides the most appropriate scales for normalizing variables into dimensionless groups or similarity parameters, organizing and presenting experimental data in the most efficient manner, and empirically deriving the various similarity relations or predictions. Most generalized parameterizations of the PBL including drag and heat transfer relations are also based on some form of similarity scaling. The surface layer similarity theories and scaling have already been reviewed and discussed in Chapters 9–12. Chapter 10 also covers the simpler case of the neutral PBL whose depth is not restricted by any low-level inversion. In this section, we will review the similarity theories and scaling for the more frequently encountered stratified atmospheric boundary layers including the CBL and the SBL.

13.2.1 Mixed-layer similarity theory/scaling

In unstable and convective PBLs, there is a mixed layer above the surface layer in which the mean potential temperature is observed to be nearly uniform (Stull, 1988; Garratt, 1992; Kaimal and Finnigan, 1994). Variations in specific humidity and mean velocity are also observed to be small over the depth of the mixed layer. Figure 13.3 shows such a convective mixed layer within the marine boundary layer whose depth (h) can be identified from mean potential temperature and specific humidity profiles.

Deardorff (1970b) proposed a similarity hypothesis that turbulence structure in the convective mixed layer depends only on the virtual heat flux $H_{v0}/\rho c_p$, the buoyancy variable g/T_{v0}, the height above the surface z, and the height z_i of the base of inversion which caps the mixed layer. The appropriate scaling parameters for the mixed-layer similarity are z_i and the convective velocity

$$W_* = \left(\frac{g}{T_{v0}} \frac{H_{v0}}{\rho c_p} z_i\right)^{1/3} \tag{13.1}$$

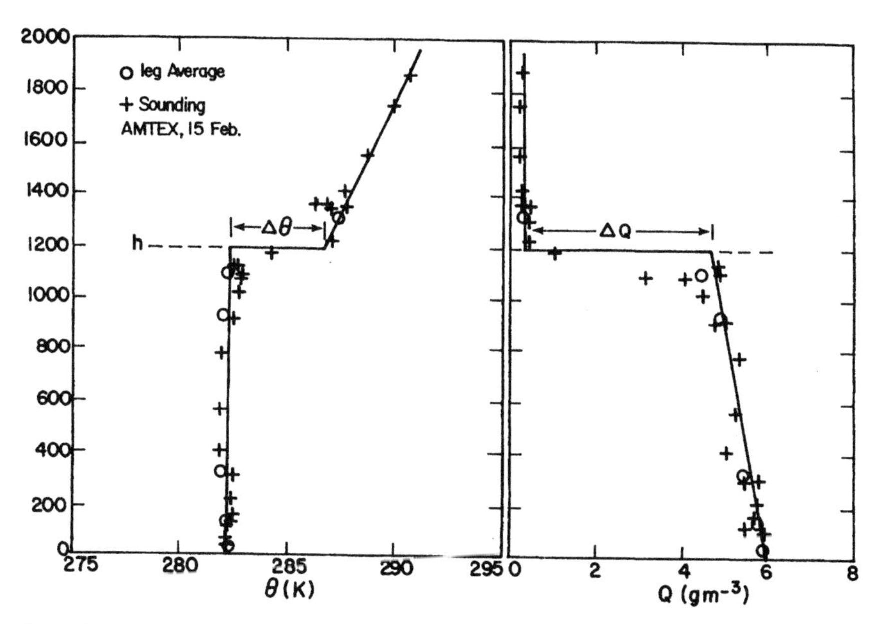

Figure 13.3 Observed mean potential temperature (θ) and specific humidity (Q) profiles under convective conditions on February 15, 1975, during AMTEX. [After Wyngaard *et al.* (1978).]

The mixed-layer similarity theory predicts that σ_u/W_*, σ_v/W_*, etc., and other appropriately scaled turbulence statistics must be unique functions of z/z_i only. This is strictly valid only within the mixed layer ($0.1z_i < z \leq z_i$) of the CBL with $-z_i/L > 10$. Actual observations of turbulence in the CBL (Figure 13.4), however, indicate that there is hardly any dependence of σ_u/W_* and σ_v/W_* on z/z_i and, for all practical purposes,

$$\sigma_u/W_* \cong \sigma_v/W_* \cong 0.60 \tag{13.2}$$

The standard deviation of vertical velocity fluctuations in the CBL has been observed to show a much stronger dependence on z/z_i (see Figure 13.4). It increases with height, attains a broad maximum in the middle of the mixed layer, and decreases again in the upper part of the CBL (Caughey and Palmer, 1979; Sorbjan, 1989). For most practical applications, $\sigma_w/W_* \cong 0.60$ represents an average value for the whole mixed layer; more detailed empirical expressions of the same in different layers are given by Hanna (1982).

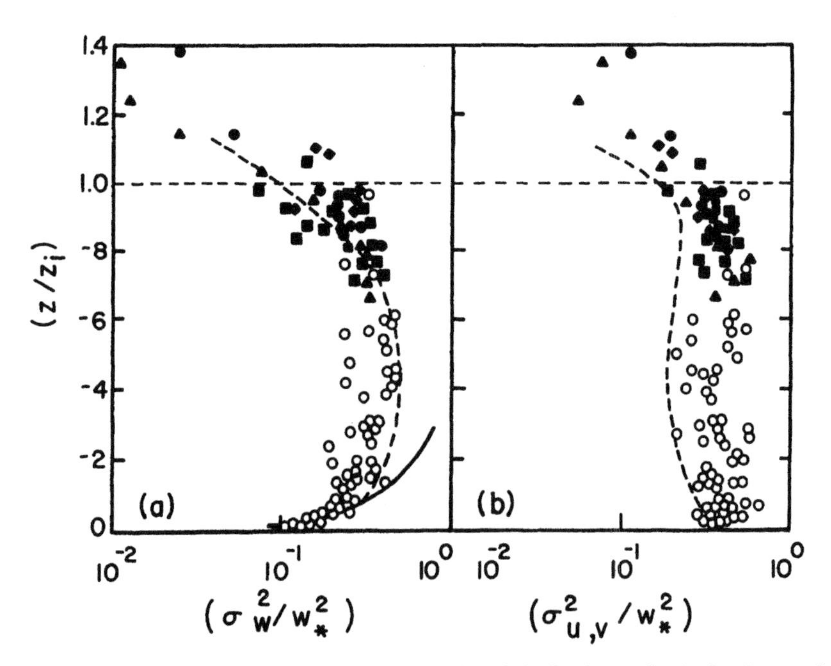

Figure 13.4 Normalized variances of (a) vertical and (b) horizontal velocity fluctuations as functions of z/z_i. Solid line represents the local free convection prediction and dashed lines represent the convection tank results of Willis and Deardorff (1974). [From Caughey and Palmer (1979).]

Combining the mixed-layer similarity relations for σ_u, σ_v and σ_w into the definition of TKE, one obtains

$$E = 0.36W_*^2 + 0.9\left(\frac{z}{z_i}\right)^{2/3}\left(1 - 0.8\frac{z}{z_i}\right)^2 W_*^2 \tag{13.3}$$

Alternatively, for most practical purposes,

$$E \cong 0.54W_*^2 \tag{13.4}$$

may represent the layer-averaged value of the TKE in the convective mixed layer. An empirical expression for rate of energy dissipation in the same is:

$$\varepsilon = \frac{W_*^3}{z_i}\left(0.8 - 0.3\frac{z}{z_i}\right) \tag{13.5}$$

which yields the layer-averaged value of $0.5W_*^3/z_i$.

Higher moments of turbulence in the mixed layer are also found to be scaled by W_*. In particular, the normalized third moment of vertical velocity $\overline{w^3}/W_*^3$, or skewness $S = \overline{w^3}/\sigma_w^3$, is observed to be significantly positive, indicating much larger positive velocities in convective updrafts compared to the average negative velocity in downdrafts. Since the average vertical velocity is essentially zero in a horizontally homogeneous CBL, the horizontal area occupied by downdrafts is much (about 50%) larger than that occupied by updrafts. This is also confirmed by estimated probability density functions of vertical velocity in the CBL (Arya, 1999, Chapter 5).

The mixed-layer similarity scaling may not be valid in the transition layer above the inversion base. Turbulence and diffusion processes in this layer are also influenced by entrainment of free atmospheric air, stronger wind speed and direction shears, and potential temperature gradients observed in the transition layer (Stull, 1988).

13.2.2 Rossby-number similarity theory

The Rossby-number similarity theory for the stratified PBL is based on a similarity hypothesis proposed by Kazanski and Monin (1961) that the PBL structure is determined by all the variables that were included in the M–O similarity hypothesis plus the Coriolis parameter f. The latter yields an addition length scale $u_*/|f|$ which is a measure of the height of the neutral PBL

($h \cong 0.25u_*/|f|$). The normalized height of the stratified PBL is predicted to be a unique function of the stability parameter, i.e.,

$$h|f|/u_* = F_h(\mu_*) \tag{13.6}$$

where $\mu_* = u_*|f|/L$ is the relevant stability parameter. The similarity theory also predicts that in a barotropic PBL, the dimensionless ageostrophic wind components $(U - U_g)/u_*$ and $(V - V_g)/u_*$, as well as turbulence statistics, must be unique functions of the dimensionless height, $z|f|/u_*$, and μ_* only. An appropriate matching of the PBL similarity profiles with those of the surface layer also yields geostrophic drag or resistance laws (Arya, 1975; Zilitinkevich *et al.*, 1976) which can be used for parameterizing the PBL in large-scale atmospheric circulation models. These resistance laws contain the surface Rossby number, $G|f|/z_0$, as an important parameter, which explains the name given to this similarity theory and scaling.

Tennekes (1982) has given a formal and elegant derivation of the velocity-defect profile relations (10.11) and the geostrophic drag relations (10.20) for the barotropic, neutral PBL. The proposed extensions of the same relations consider the dependence of the various similarity functions, including A, B, etc., on the stability parameter μ_* (Clarke and Hess, 1974; Arya, 1975). This also implies that the dimensionless PBL height must also be a unique function of μ_* as in Equation (13.6). However, Deardorff (1972a, b, 1974) and others have pointed out that it was not true under unstable and convective conditions in which the PBL height is not at all related to u_*/f or μ_*, but is determined by the height of the lowest inversion base. For this reason, the Rossby-number similarity scaling cannot be valid under unstable and convective conditions. Its validity in the stable boundary layer (SBL) is also somewhat limited to slightly and moderately stable conditions during which there is continuous exchange between adjacent layers in the vertical, i.e., turbulence is continuous throughout the SBL. Under very stable conditions, turbulence becomes highly intermittent and sporadic, so that upper layers may become decoupled from the surface layer. Any similarity scaling based on the surface fluxes will not be appropriate.

For the SBL with more or less continuous turbulence, Zilitinkevich (1972) derived the following diagnostic relationship for steady state or stationary conditions:

$$h = d(u_* L/|f|)^{1/2} \tag{13.7}$$

in which d is an empirical constant. The above relationship implies a simple form of the similarity function in Equation (13.6) as

$$h|f|/u_* = d\mu_*^{-1/2} \tag{13.8}$$

These SBL height relations have been tested against experimental data obtained at several flat and approximately homogeneous sites (Garratt, 1982). The estimated values of d range from 0.2 to 0.7 with an average value of about 0.4 which is also the value suggested by numerical modeling studies (Brost and Wyngaard, 1978). It has been shown that d may depend on the time after transition, surface cooling rate, terrain slope and possibly other factors.

The above diagnostic similarity relation may not be valid for an evolving SBL for several hours after the transition time. Prognostic equations for the SBL height have been proposed for such an evolving boundary layer (Stull, 1988, Chapter 12). The SBL evolves rather slowly in time and rarely reaches a steady or equilibrium state. This may explain why correlations between the observed and theoretically predicted SBL heights are not usually very strong (Arya, 1981; Koracin and Berkowicz, 1988).

The mean velocity profiles in the similarity form are usually expressed as

$$\begin{aligned} (U - U_g)/u_* &= F_u(fz/u_*, \mu_*) \\ (V - V_g)/u_* &= F_v(fz/u_*, \mu_*) \end{aligned} \tag{13.9}$$

The corresponding geostrophic drag relations are

$$\begin{aligned} U_g/u_* &= \frac{1}{k}[\ln(u_*/fz_0) - A(\mu_*)] \\ V_g/u_* &= -B(\mu_*)/k \end{aligned} \tag{13.10}$$

Geostrophic drag and heat transfer relations in the SBL have been extensively studied and the similarity functions $A(\mu_*)$, $B(\mu_*)$, etc., have been evaluated from experimental data as well as numerical models of the SBL (Arya, 1975, 1977, 1984). Experimental estimates of the same show large scatter, reflecting possible effects of baroclinicity and other factors (Sorbjan, 1989; Garratt, 1992).

13.2.3 Generalized similarity theory

Recognizing the importance of the PBL height (h) as an independent height scale and the lack of general validity of Equation (13.6), Deardorff (1972a) utilized z/h and h/L as the dimensionless height and stability parameters. Zilitinkevich and Deardorff (1974) also proposed that $u_*/|f|$ be replaced by h in the PBL similarity formulation for the entire range of stability conditions. It was later recognized, however, that $u_*/|f|$ cannot be ignored entirely, and $h|f|/u_*$ should also be considered as an additional parameter in any generalized similarity theory. Further consideration of baroclinicity added two more similarity parameters to the expanding list of the same (Arya and Wyngaard, 1975). Subsequently, the similarity functions A, B, etc., have been evaluated as functions of h/L (Yamada, 1976). These are also characterized by large scatter, possibly due to baroclinicity and other parameters. The relations involving layer-averaged (over the PBL depth) winds, temperature, etc., are found to be less sensitive to the effects of baroclinicity (Arya, 1978; Garratt *et al.*, 1982). For example, the layer-averaged wind speed in an unstable PBL is given by

$$V_m/u_* = \frac{1}{k}\left[\ln\frac{h}{z_0} - 0.5\ln|h/L| - 2.3\right] \quad (13.11)$$

For representing the turbulence structure of unstable and convective boundary layers, it is much simpler to use the mixed-layer similarity scaling than the generalized similarity scaling discussed above. Therefore, most turbulence and diffusion data in the CBL are presented using the mixed-layer scaling.

Turbulence data in slightly and moderately stable boundary layers with more or less continuous turbulence are best represented in the generalized similarity framework, i.e., as functions of z/h and h/L. The parameter $h|f|/u_*$ appears to be uniquely related to h/L and may not be an independent parameter affecting turbulence structure. Even the influence of the stability parameter does not appear to be overly significant, when normalized turbulence statistics are represented as functions of z/h. For example, in a steady PBL over a flat terrain, the vertical profiles of TKE and fluxes can be represented as (Lenschow *et al.*, 1988; Smedman, 1991; Rao and Nappo, 1998):

$$E \cong 5.75u_*^2\left(1 - \frac{z}{h}\right)^{\alpha_1} \quad (13.12)$$

$$\tau = \rho u_*^2\left(1 - \frac{z}{h}\right)^{\alpha_1} \quad (13.13)$$

$$H = H_0\left(1 - \frac{z}{h}\right)^{\alpha_2} \quad (13.14)$$

in which the values of exponents α_1 and α_2 may depend on the temporal development or evolution of the SBL, terrain slope, and other factors. For a well developed or steady SBL, $\alpha_1 \cong 1.75$ and $\alpha_2 \cong 1.5$ (Lenschow *et al.*, 1988; Smedman, 1991). Equations (13.12)–(13.14) may be considered as extensions of the surface-layer similarity relations to the whole SBL.

13.2.4 Local similarity theory/scaling

For the stable boundary layer, Nieuwstadt (1984) proposed a local similarity hypothesis that, above the surface layer, mean gradients and turbulence statistics in the SBL follow a local similarity scaling based on the local (not surface) fluxes. Thus, analogous to the Monin–Obukhov similarity scales, the local scaling parameters are defined as

$$u_{*\ell} = (\tau/\rho)^{1/2};\ \theta_{*\ell} = -H/\rho c_p u_{*\ell}$$
$$\Lambda = T_v u_{*\ell}^2 / k g \theta_{*\ell} \tag{13.15}$$

where $u_{*\ell}$ and $\theta_{*\ell}$ are the local friction velocity and temperature scale respectively, and Λ is a local buoyancy length scale which is analogous to the Obukhov length in the surface layer.

The local similarity theory predicts that appropriately scaled (normalized) mean velocity and temperature gradients, as well as turbulence statistics, at any height in the SBL must be some universal functions of the local stability parameter z/Λ. Furthermore, under very stable conditions when upper layers become decoupled from the surface layer and even from adjacent layers, and turbulence is sporadic and becomes independent of z, i.e., under the so-called z-less stratification (Monin and Yaglom, 1971; Stull, 1988), the locally scaled gradients and turbulence statistics must become constants, independent of z or z/Λ. Nieuwstadt (1984) has shown that this is indeed the case for $z/\Lambda \gg 1$.

The practical utility of the local similarity theory predictions is somewhat questionable, because in order to predict velocity variances and the TKE one needs to know the local fluxes of momentum and heat, which are even more difficult to measure or estimate. Nevertheless, it provides a convenient framework of representing and comparing mean gradients and turbulence data in the SBL.

13.3 Mathematical Models of the PBL

Most of the quantitative PBL models are based on the fundamental laws of conservation of mass, energy, and momentum, which are expressed in the form of partial differential equations containing velocity, pressure, temperature or density, and concentrations of water vapor and trace gases, if needed. With the usual simplifying assumptions or approximations, these conservation equations have already been reviewed in Chapter 9. The difficulties in obtaining their particular or general solution using analytical and numerical methods have also been pointed out. Here, we further elaborate on some of the simpler first-order closure, integral, and TKE closure models of the PBL, which are frequently used in practical applications of micrometeorology, as well as more sophisticated models employed in fundamental research on the PBL turbulence.

13.3.1 First-order closure models

For the idealized horizontally homogeneous PBL, in the absence of any active condensation and precipitation processes, the equations of mean motion and scalars reduce to the following:

$$\begin{aligned} \frac{\partial U}{\partial t} - f(V - V_g) &= -\frac{\partial \overline{uw}}{\partial z} \\ \frac{\partial V}{\partial t} + f(U - U_g) &= -\frac{\partial \overline{vw}}{\partial z} \end{aligned} \tag{13.16}$$

$$\begin{aligned} \frac{\partial \Theta}{\partial t} &= -\frac{\partial \overline{\theta w}}{\partial z} \\ \frac{\partial Q}{\partial t} &= -\frac{\partial \overline{qw}}{\partial z} \end{aligned} \tag{13.17}$$

in which horizontal pressure gradients have been expressed in terms of the specified geostrophic wind components U_g and V_g. The above equations for the horizontally homogeneous PBL still contain twice as many unknowns as the number of equations. Equations (13.16) of motion can further be simplified for the stationary flow by dropping the time-dependent terms; these terms are usually retained in scalar equations (13.17).

Local closure models In order to close the above set of equations of mean motion and other scalars, turbulent covariances or fluxes are expressed in terms of the gradients of mean velocity and scalars, using the eddy diffusivity and/or mixing length hypotheses introduced in Chapter 9. This type of closure is known as the local closure, because local fluxes are assumed to be related to only local gradients. The implied assumption and local closure models based on local gradient-transport hypotheses may be justified in the presence of large gradients as in the SBL, but not so in the presence of vanishingly small gradients that are usually encountered in the CBL. The basic premise of down-gradient transport becomes highly questionable and even invalid when large-scale convective motions dominate turbulent transports. It is a well-known fact that, in the convective mixed layer, the vertical momentum, heat, and water vapor fluxes are not related to the vanishingly small gradients of mean velocity, potential temperature and specific humidity in that layer. In the neutral and stably stratified boundary layers with strong gradients of mean variables, however, there are no such conceptual difficulties, but eddy diffusivities and mixing lengths are specified largely on an ad hoc basis.

One can specify mixing length more rationally by identifying it with the characteristic large-eddy length scale which can be determined from measurements of turbulence. Both the mixing length and the large-eddy length scale are found to be proportional to the height above the surface, especially closer to the surface, and approach some constant values which are either proportional to the PBL depth or limited by stability in the upper part of the PBL. Simple interpolation formulas of the form

$$\frac{1}{l_m} = \frac{1}{kz} + \frac{1}{l_b} + \frac{1}{l_o} \tag{13.18}$$

have been proposed, in which $k \cong 0.4$ is the von Karman constant, l_b is a buoyancy-limited mixing length, and l_o is another limiting value which is assumed proportional to h, $u_*/|f|$, or $G_0/|f|$. The buoyancy-limited mixing length term is included only for modeling the SBL, in which l_b is taken proportional to either the Obukhov length L, or σ_w/N, where N is the Brunt–Vaisala frequency.

A comprehensive review of the many first-order local closure models that have been reported in the literature is given by Holt and Raman (1988). Some are based on explicit or implicit specification of eddy diffusivity profiles in the PBL. In most of the models eddy diffusivities are expressed in terms of a mixing length or large-eddy length scale whose profile (variation with height) in the PBL is then specified, as in Equation (13.18) or through other expression. The most commonly used relation between eddy viscosity and mixing length is of the form

$$K_m = l_m^2 \left| \frac{\partial \mathbf{V}}{\partial z} \right| f(\text{Ri}) \tag{13.19}$$

where $f(\text{Ri})$ is a decreasing function of Richardson number in the SBL, such that $f(0) = 1$.

As an example of a first-order closure model of the SBL, used in conjunction with the Rossby-number similarity scaling, Businger and Arya (1974) proposed the following expression for the normalized eddy viscosity $K_* = K_m f/u_*^2$:

$$K_* = k\xi(1 + \beta\xi\mu_*)^{-1} \exp(-|V_g/u_*|\xi) \tag{13.20}$$

where $\xi = fz/u_*$, $\mu_* = u_*/fL$, and $\beta \cong 5$ is an empirical constant. For the stationary (steady state) SBL, they expressed Equation (13.16) in the normalized form as

$$K_* \frac{d^2 T_x}{d\xi^2} + T_y = 0$$
$$K_* \frac{d^2 T_y}{d\xi^2} - T_x = 0 \qquad (13.21)$$

in which $T_x = -\overline{uw}/u_*^2$ and $T_y = -\overline{vw}/u_*^2$ represent the normalized stresses in x and y directions, respectively.

The numerical solution of Equations (13.21) with K_* specified by Equation (13.20) yielded the vertical profiles of K_*, $(U - U_g)/u_*$, $(V - V_g)/u_*$, $\overline{uw}/u_*^2$, and $\overline{vw}/u_*^2$ for the various specified values of μ_* (Businger and Arya, 1974). The model results for the normalized eddy viscosity, shown in Figure 13.5, give some idea about the wide range of K_m values that may occur in the SBL, as well as about the strong dependence of K_m on height and stability. The model results of the PBL height (see Figure 13.6) also confirm the validity of Equations (13.7) and (13.8), although the implied value of $d \cong 0.7$ falls on the higher side of the empirical estimates. However, it also depends on the definition of the PBL height. Figure 13.6 shows that different heights represent different salient features of mean velocity and momentum flux profiles. The most satisfactory and frequently used definition is the height where the magnitude of the normalized stress reduces to an arbitrarily specified small fraction, e.g., 0.01 used by Businger and Arya (1974). The model estimated value of d will be smaller if a higher fraction (say, 0.05) is used for the normalized stress in defining the PBL height.

Figure 13.7 shows that ξ_τ, representing the normalized PBL height where shear stress falls to 1% of the surface stress, is uniquely related to the maximum dimensionless eddy viscosity ($K_{*\max}$). The best fitted line, also shown in the same figure, represents a simple power-law relationship, $\xi_\tau \cong 5.0 K_{*\max}^{0.55}$, which is not much different from that obtained from an analytical solution for a constant K_* (see Example Problem 1). The normalized stress profiles are found to be strongly dependent on the stability parameter μ_*, if they are represented as functions of fz/u_* (Businger and Arya, 1974). However, Figure 13.8 shows that the normalized stress profiles become essentially independent of μ_*, if they are represented as functions of z/h instead of ξ. This points to the merits of using the generalized similarity scaling, even for the SBL.

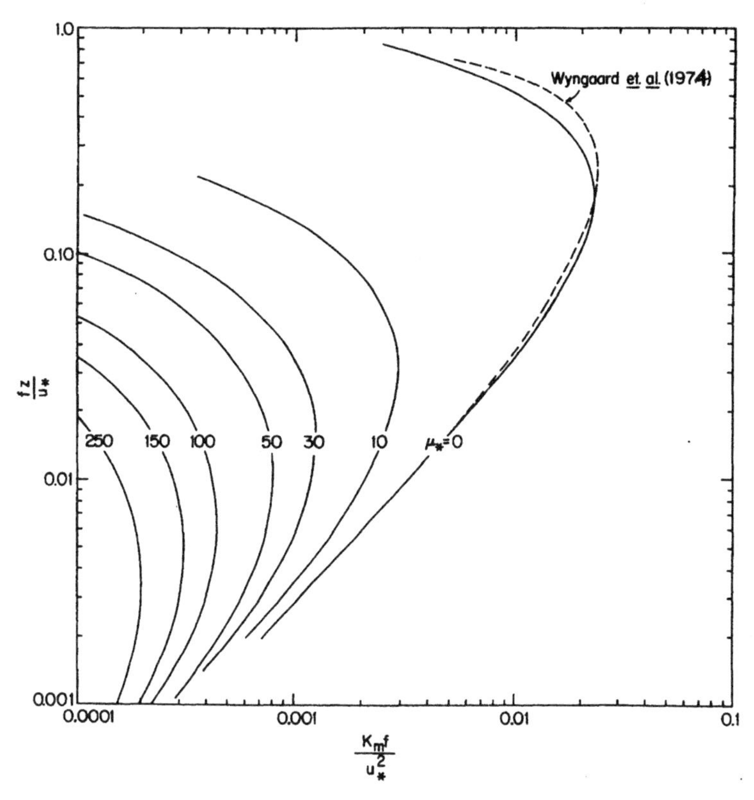

Figure 13.5 Model-predicted normalized eddy viscosity as a function of normalized height for different values of the stability parameter μ_*. The dashed curve is the result of the second-order closure model of Wyngaard *et al.* (1974) for the neutral PBL. [From Businger and Arya (1974).]

Example Problem 1

(a) Using the equations of mean motion in a stationary, horizontally homogeneous, and barotropic SBL, derive the corresponding equations (13.21) for the normalized stresses T_x and T_y.

(b) Obtain an analytical solution of Equations (13.21) for constant eddy diffusivity (K_*) and appropriate boundary conditions.

(c) Utilizing the above solution, calculate and compare the normalized stress profiles for $K_* = 0.002$ and 0.0002 representing slightly stable and moderately stable conditions, respectively. Also compare the dimensionless height of the SBL for the above two cases.

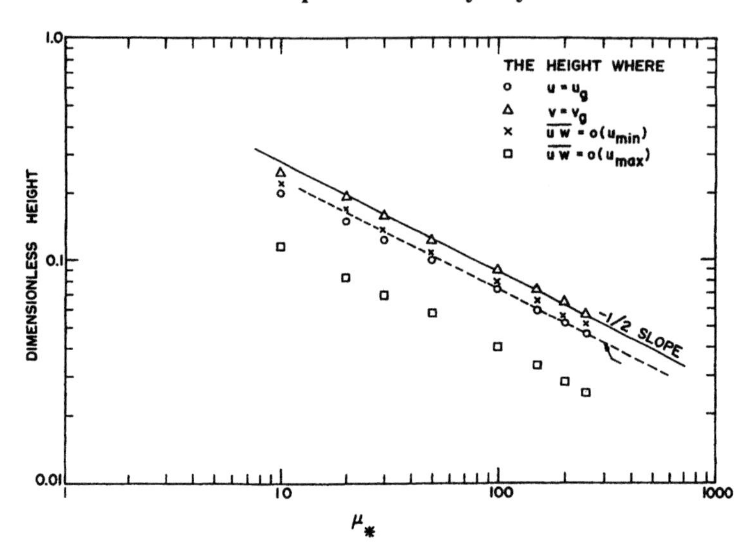

Figure 13.6 Dimensionless heights, representing salient features of mean velocity and momentum flux profiles in the SBL, as functions of stability parameter. The dashed line (indicated by arrow) represents Equation (13.8) with $d = 0.72$. [From Businger and Arya (1974).]

Solution

(a) For the stationary, horizontally homogeneous PBL, Equations (13.16) reduce to

$$-f(V - V_g) = -\frac{\partial \overline{uw}}{\partial z}$$

$$f(U - U_g) = -\frac{\partial \overline{vw}}{\partial z}$$

Differentiating these with respect to z and recognizing that in a barotropic PBL $\partial U_g/\partial z = 0$ and $\partial V_g/\partial z = 0$, we can write

$$-f\frac{\partial V}{\partial z} = -\frac{\partial^2 \overline{uw}}{\partial z^2}$$

$$f\frac{\partial U}{\partial z} = -\frac{\partial^2 \overline{vw}}{\partial z^2}$$

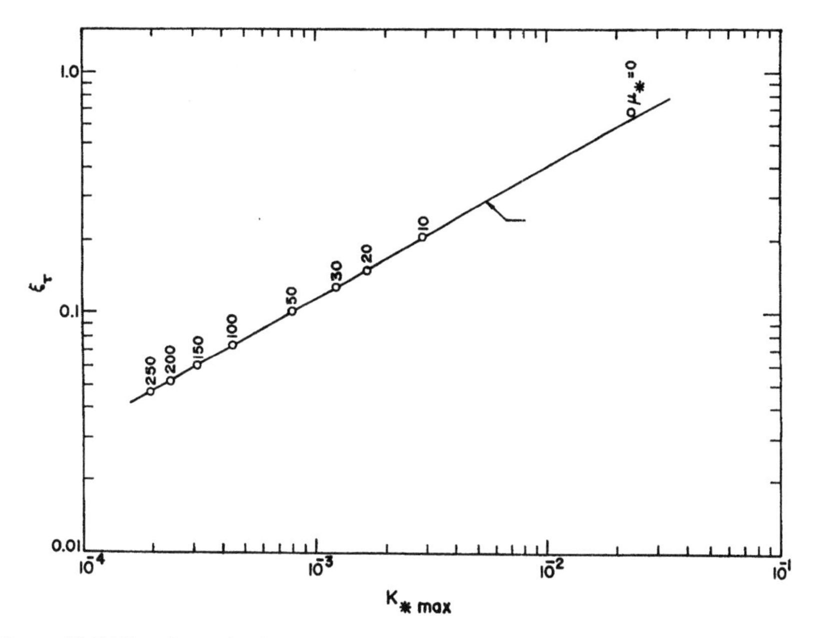

Figure 13.7 The dimensionless SBL height, where stress reduces to 1% of the surface stress, as a function of the maximum dimensionless eddy viscosity. The best-fitted line (indicated by arrow) represents a power-law relationship between the two parameters. [From Businger and Arya (1974).]

Multiplying both sides of equations by K_m and utilizing the eddy-viscosity hypothesis, the above equations can be expressed in terms of the momentum fluxes only as

$$f\,\overline{vw} = -\frac{\partial^2 \overline{uw}}{\partial z^2} K_m$$

$$-f\,\overline{uw} = -\frac{\partial^2 \overline{vw}}{\partial z^2} K_m$$

Now, using the normalized stresses $T_x = -\overline{uw}/u_*^2$ and $T_y = -\overline{vw}/u_*^2$ and the normalized height coordinate $\xi = fz/u_*$, the above equations are easily expressed in the form of Equations (13.21). Note that, in the above equations for the stationary and homogeneous PBL, partial derivatives can be written as ordinary derivatives.

(b) The solution of Equations (13.21) can more easily be obtained by combining the two coupled ordinary differential equations into one for the complex stress variable

$$T_c = T_x + iT_y$$

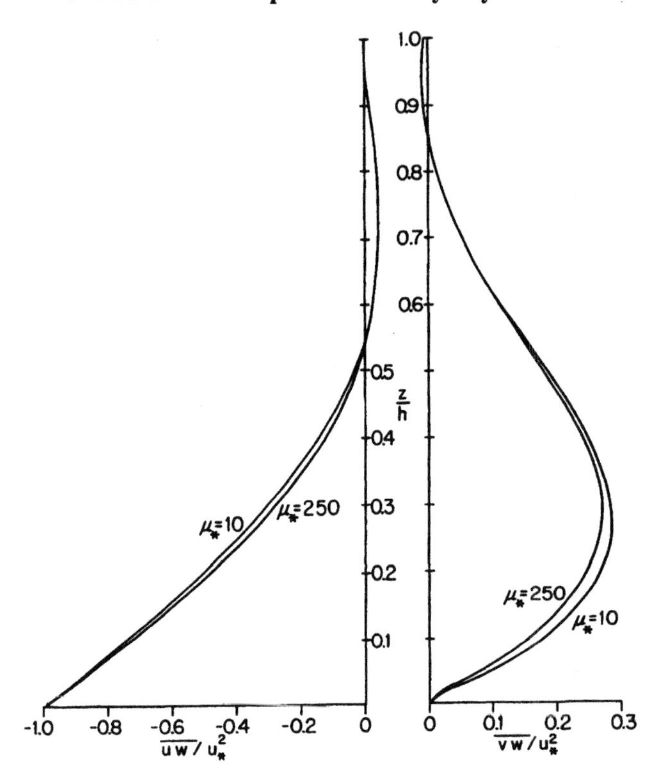

Figure 13.8 Normalized stresses (T_x and T_y) as functions of the normalized height (z/h) for $\mu_* = 10$ and 250, based on the numerical model results of Businger and Arya (1974).

We can follow the procedure used in Section 7.5 to express Equations (13.21) as

$$\frac{d^2 T_c}{d\xi^2} - \frac{i}{K_*} T_c = 0$$

The particular solution satisfying the upper boundary condition that $T_c \to 0$ as $\xi \to \infty$, is given by

$$T_c = A \exp[-(1+i)a\xi]$$

where $a = (2K_*)^{-1/2}$, and A is a complex constant to be evaluated from the lower boundary condition. Here, we use a surface-layer coordinate system, so that $T_c = 1$ at $\xi = 0$, which yields $A = 1$. Then, the above solution can be expressed as

$$T_x + iT_y = \exp(-a\xi)[\cos(a\xi) - i\sin(a\xi)]$$

so that,

$$T_x = \exp(-a\xi)\cos(a\xi)$$
$$T_y = -\exp(-a\xi)\sin(a\xi)$$

and the total stress

$$T = (T_x^2 + T_y^2)^{1/2} = \exp(-a\xi)$$

(c) According to the above solution, the total stress decreases exponentially with the normalized height $a\xi$, while the two components (T_x and T_y) are modulated by $\cos(a\xi)$ and $-\sin(a\xi)$ terms, respectively. The stress profiles are self-similar in terms of the modified similarity parameter $a\xi$ which is proportional to the ratio ξ/ξ_τ, where $\xi_\tau = fh/u_*$ is the dimensionless PBL height where the stress has decreased to about 1% of the surface value (i.e., $T = 0.01$). From the exponential decrease of stress, it is easy to see that

$$\xi_\tau \cong 4.6\sqrt{2K_*} \cong 6.51K_*^{1/2}$$

which gives the values of $\xi_\tau \cong 0.29$ and 0.09 for the slightly stable ($K_* = 0.002$) and moderately stable ($K_* = 0.0002$) cases, respectively. Note that the heights where the stress has decreased to only 5% of the surface stress are 35% smaller than the above values of ξ_τ.

Nonlocal closure models Recognizing the limitations of local closure hypotheses, especially under unstable and convective conditions, several nonlocal closure formulations have been proposed for modeling mixing processes in the CBL (Stull, 1988, 1993; Pleim and Chang, 1992). Nonlocal closures are based on the basic premise that turbulent eddies of varying sizes can transport flow properties, such as momentum, heat and mass, across finite vertical distances, comparable to their size. Thus, the largest eddies in the CBL with the vertical dimension equal to the PBL depth h, can transfer properties all the way from the surface layer to the top of the CBL and vice versa, irrespective of the vanishingly small gradients in the mixed layer. Consequently, mean velocity, potential temperature, and scalar concentrations at any location (height level) are a result of eddy transport and mixing of fluid parcels coming from many cross-stream locations (height levels).

Stull (1993) has proposed a transilient turbulence model of nonlocal mixing across finite distances. In this, a vertical column of air is assumed to be divided between a finite number (N) of equal size grid boxes. Turbulent mixing of fluid coming from any box (denoted by index j) into the reference grid box (denoted by index i) can change any conserved passive scalar property $\tilde{s}_i$ (e.g., potential temperature, specific humidity, etc.) of the reference box. If c_{ij} represents the fraction of air in box i that came from box j during a time interval Δt, one can express the mean value of $\tilde{s}_i$ at a future time as

$$S_i(t + \Delta t) = \sum_{j=1}^{N} c_{ij}(t, \Delta t) S_j(t) \tag{13.22}$$

in which the matrix of mixing coefficients c_{ij} is called the transilient matrix. By definition, the conservation of the transferred property requires that

$$\sum_{j=1}^{N} c_{ij} = \sum_{i=1}^{N} c_{ij} = 1 \tag{13.23}$$

The vertical turbulent flux of scalar at any height k (defined as the top of the grid box k) is given by

$$\overline{ws}(k) = \frac{\Delta z}{\Delta t} \sum_{i=1}^{k} \sum_{j=1}^{N} c_{ij}(S_i - S_j) \tag{13.24}$$

where Δz is the vertical grid spacing and Δt is time step for the $c_{ij}(t, \Delta t)$ matrix. The above discrete-grid expression can also be written in a continuous form (Stull, 1988).

The main closure problem in the above model is the specification or estimation of the transilient matrix c_{ij}. These mixing coefficients are expected to be different for different flow situations and cannot be generalized. Different approaches have been suggested for estimating the mixing coefficients, but none of them is very rigorous and completely satisfactory (Stull, 1988, 1993). For example, one approach utilizes the nonlocal (integral) form of the turbulent kinetic energy equation. After parameterizing the unknown terms in terms of the mean wind and potential temperature parameters, mixing potentials are obtained and c_{ij} are, then, expressed in terms of mixing potentials. The more non-zero elements of matrix in the model, the more complex and expensive would be the numerical solution.

Another approach aimed at reducing the computational cost is to specify the transilient matrix according to some simple conceptual model of mixing in the

CBL (Pleim and Chang, 1992). In this way, only a fraction of the elements in the transilient matrix need to be nonzero. Both symmetric and asymmetric convective mixing models have been proposed. For example, in the asymmetric convective mixing (ACM) model proposed by Pleim and Chang (1992) the upward transport originates in the bottom-most layer, which constitutes the surface layer, from which air is allowed to mix with each of the upper layers in the CBL directly. The downward transport and mixing, on the other hand, proceed from each layer to the next lower layer only in a cascading manner. This simulates rapid upward transport from the surface layer by buoyant plumes and more gradual compensatory subsidence in downdrafts. The ACM model has been shown to perform much better than some of the local closure schemes including the one based on the turbulence kinetic energy equation (Alapaty *et al.*, 1997).

Integral models Integral or slab models use the vertically integrated or layer-averaged forms of equations of motion and thermodynamic variables (Driedonks, 1982; Arya and Byun, 1987; Stull, 1988):

$$\frac{\partial U_{\mathrm{m}}}{\partial t} = f(V_{\mathrm{m}} - V_{\mathrm{gm}}) + \frac{1}{h}[(\overline{uw})_0 - (\overline{uw})_h]$$
$$\frac{\partial V_{\mathrm{m}}}{\partial t} = -f(U_{\mathrm{m}} - U_{\mathrm{gm}}) + \frac{1}{h}[(\overline{vw})_0 - (\overline{vw})_h] \tag{13.25}$$

$$\frac{\partial \Theta_{\mathrm{m}}}{\partial t} = \frac{1}{h}[(\overline{w\theta})_0 - (\overline{w\theta})_h] \tag{13.26}$$

in which the subscript m denotes the layer-averaged value for the mixed layer, e.g.,

$$\Theta_{\mathrm{m}} = \frac{1}{h}\int_0^h \Theta \, dz \tag{13.27}$$

The above equations for the layer-averaged wind components (U_{m}, V_{m}) and potential temperature (Θ_{m}), also contain the surface fluxes of momentum and heat, and those at the top of the mixed layer or the PBL. The surface fluxes can be parameterized through the usual bulk transfer relations. The fluxes at the top of the mixed layer are parameterized through the following entrainment relations:

$$
\begin{aligned}
(\overline{uw})_h &= -w_e \, \Delta U \\
(\overline{vw})_h &= -w_e \, \Delta V \\
(\overline{\theta w})_h &= -w_e \, \Delta \Theta
\end{aligned}
\tag{13.28}
$$

in which ΔU, ΔV, and $\Delta\Theta$ denote the jumps in U, V, and Θ across the interface or the transition layer, and w_e is the entrainment velocity which is intimately related to the rate of growth of the mixed layer as

$$
\frac{\partial h}{\partial t} = w_e + W_h \tag{13.29}
$$

Note that $w_e = \partial h/\partial t$ in the absence of any mean vertical velocity (W_h) at the top of the mixed layer; it represents the volume of air entrained into the top of the mixed layer per unit horizontal area per unit time.

The entrainment relations (13.28)–(13.29), in conjunction with the usual parameterization of the surface fluxes provide the necessary closure to the set of equations which can be numerically solved. In addition, the rate equations for the mixed layer depth or the PBL height, transition-layer thickness, and changes or jumps in the mean variables across the transition layer are also carried in some integral models. Depending on the sophistication with which the transition or entrainment layer at the top of the mixed layer is represented, the various zeroth-order and first-order slab models have been proposed for the horizontally homogeneous CBL (Driedonks, 1982; Arya and Byun, 1987; Stull, 1988). Simpler zeroth-order jump models may be adequate for predicting the mixed-layer depth and potential temperature, but more sophisticated integral models are required for better simulation of the thickness of and entrainment processes occurring in the transition layer.

The most widely used application of integral modeling has been in the development of mixed-layer height models of varying degrees of sophistication. The evolution of the mixed-layer depth over heated land surfaces during the day occurs in several stages or phases (Stull, 1988, Chapter 11): (1) the erosion or burning off of the nocturnal inversion and formation of a shallow mixed layer which slowly deepens in early morning hours after sunrise; (2) rapid growth of the mixed layer during the mid-morning period; (3) slow growth of the deep mixed layer in the afternoon period; and (4) decay of turbulence in the mixed layer of nearly constant depth during the late afternoon and evening transition period.

No simple integral model can faithfully simulate all the above stages of mixed-layer depth evolution. However, even some crude models have been shown to simulate, approximately, the mixed-layer growth during most of the daytime period, especially under fair-weather, clear sky conditions. For example, the thermodynamic method or model of mixed-layer growth is based on thermodynamic considerations only, while the dynamics of turbulent entrainment is neglected. The rate of growth of mixed-layer depth is simply related (proportional) to the rate of warming of the mixed layer as

$$\frac{\partial h}{\partial t} = \frac{1}{\gamma}\frac{\partial \Theta_m}{\partial t} \tag{13.30}$$

where $\gamma = \partial\Theta/\partial z$ is the potential temperature gradient above the top of the mixed layer.

Substituting from Equation (13.26) into Equation (13.30), one obtains

$$\frac{\partial h}{\partial t} = \frac{1}{\gamma h}[(\overline{w\theta})_0 - (\overline{w\theta})_h]$$

or

$$\frac{\partial h^2}{\partial t} = \frac{2}{\gamma}[(\overline{w\theta})_0 - (\overline{w\theta})_h] \tag{13.31}$$

Parameterizing the entrainment heat flux simply as a fraction of the surface heat flux, i.e.,

$$(\overline{w\theta})_h = -C(\overline{w\theta})_0 \tag{13.32}$$

and integrating Equation (13.31), one can obtain the following expression for the evolution of the mixed-layer depth:

$$h(t) = \left[h_0^2 + 2\int_{t_0}^{t} \frac{(1+C)}{\gamma}(\overline{w\theta})_0\, dt\right]^{1/2} \tag{13.33}$$

in which h_0 is the initial depth at time t_0. The integral can be evaluated numerically if C, γ, and $(\overline{w\theta})_0$ can be specified as functions of time. In the simplest case, C and γ are assumed constants (e.g., $C \cong 0.2$ and γ can be estimated from the initial temperature sounding at $t = t_0$ which can also be used to estimate h_0), so that Equation (13.33) can be simplified as

$$h(t) = \left[h_0^2 + \frac{2(1+C)}{\gamma}\int_{t_0}^{t} (\overline{w\theta})_0\, dt\right]^{1/2} \tag{13.34}$$

Thus, according to this thermodynamic model, the mixed-layer depth at any time essentially depends on the strength of the inversion above and the accumulation of surface heat flux up to that time. The most important parameter in this model is $(\overline{w\theta})_0/\gamma$ and its variation with time; it may vary from 0 to 10^5 m^2 h^{-1} with the maximum occurring near midday.

The integral of the surface heat flux in Equation (13.34) can easily be evaluated for the specified $(\overline{w\theta})_0$ as a function of time. For the special case of constant heat flux, which may be typical of overcast conditions, Equation (13.34) yields

$$h(t) = \left[h_0^2 + \frac{2}{\gamma}(1 + C)(\overline{w\theta})_0(t - t_0)\right]^{1/2} \tag{13.35}$$

For the more common fair-weather conditions, the surface heat flux may be expressed as a half-cosine wave function whose integral is also readily obtained.

Even though the simple thermodynamic model described above neglects the dynamics of turbulent entrainment, it is found to explain roughly 80–90% of the observed variation of the mixed-layer depth. Other more sophisticated mixed-layer height or depth models have also been proposed in the literature (Driedonks, 1982; Arya and Byun, 1987; Stull, 1988, Chapter 11). They mostly differ in the representation of the entrainment or transition layer and the parameterization of entrainment velocity.

Example Problem 2

Using the thermodynamic model of mixed-layer growth during the daytime unstable and convective conditions, calculate and compare the evolutions of the mixed-layer height with time after the breakup of the surface inversion at 10 AM, when the mixed layer was 200 m deep, for the following conditions:

(a) overcast skies with a constant surface heat flux of 150 W m^{-2} throughout the 12 h daylight period and $\gamma = 0.02$ K m^{-1};

(b) clear skies with a sinusoidally varying heat flux with its maximum value of 300 W m^{-2} at 2 PM (14 h), and a capping inversion with a constant potential temperature gradient of 0.02 K m^{-1}.

Solution

(a) For this case we can use Equation (13.35) in which $h_0 = 200$ m, $(\overline{w\theta})_0 = 150/1200 = 0.125$ K m s^{-1}, $\gamma = 0.02$ K m^{-1}, $(\overline{w\theta})_0/\gamma = 22\,500$ m^2 h^{-1}, and $t_0 = 10$ h. Assuming $C = 0.2$, the mixed-layer height is given by

$$h(t) = [40\,000 + 54\,000(t - t_0)]^{1/2}$$

which is used to calculate the following results:

t (h):	10	12	14	16	18	20
h (m):	200	385	506	603	687	762

(b) The surface heat flux can be expressed as

$$(\overline{w\theta})_0 = (\overline{w\theta})_{\max} \sin[\pi(t-8)/12]$$

in which $(\overline{w\theta})_{\max} = 300/1200 = 0.25$ K m s^{-1}.
Substituting the above expression in Equation (13.34), we get

$$\begin{aligned} h(t) &= [40\,000 + 108\,000 \int_{10}^{t} \sin[\pi(t-8)/12]dt]^{1/2} \\ &= [40\,000 + 108\,000 I(t)]^{1/2} \end{aligned}$$

in which the integral function $I(t)$ can be evaluated as

$$I(t) = \frac{12}{\pi}\left[\cos\frac{\pi}{6} - \cos\frac{\pi(t-8)}{12}\right]$$

The results of calculations are as follows:

t (h):	10	12	14	16	18	20
$I(t)$:	0	1.398	3.308	5.186	6.616	7.128
$h(t)$:	200	437	630	775	869	900

Comparing the model-predicted results for the two cases, it is clear that more rapid development of the mixed layer occurs during the clear sky conditions with sinusoidally varying surface heat flux.

13.3.2 Turbulent kinetic energy models

In the first-order local closure models, one has to specify eddy diffusivities or mixing lengths as functions of the characteristic length and velocity scales of turbulence. The most appropriate dynamical equation for the turbulence velocity scale, which can be taken as the square-root of TKE, is Equation (9.10) which can further be simplified for the horizontally homogeneous PBL as

$$\frac{\partial E}{\partial t} = -\left[\overline{uw}\frac{\partial U}{\partial z} + \overline{vw}\frac{\partial V}{\partial z}\right] + \frac{g}{T_{vo}}\overline{w\theta_v} - \frac{\partial}{\partial z}\left(\overline{we} + \frac{\overline{wp}}{\rho_o}\right) - \varepsilon \tag{13.36}$$

All the TKE models contain Equation (13.36), in addition to the equations for mean variables, in which turbulent fluxes or covariances are parameterized using the usual first-order closure relations. Thus, the turbulent transport term is parameterized as

$$\overline{we} + \frac{\overline{wp}}{\rho_o} = -K_E\frac{\partial E}{\partial z} \tag{13.37}$$

where K_E is the TKE diffusivity which is assumed equal or proportional to K_m. In the parameterized form Equation (13.36) can be written as

$$\frac{\partial E}{\partial t} = K_m\left[\left(\frac{\partial U}{\partial z}\right) + \left(\frac{\partial V}{\partial z}\right)^2\right] - \frac{g}{T_{vo}}K_h\frac{\partial \theta_v}{\partial z} + \frac{\partial}{\partial z}\left(K_E\frac{\partial E}{\partial z}\right) - \varepsilon \tag{13.38}$$

in which the ratios of eddy diffusivities, viz., $\sigma_h = K_m/K_h$ and $\sigma_E = K_m/K_E$, are generally specified as constants (e.g., $\sigma_h = 0.9$ and $\sigma_E = 1$ are the widely used standard values). The closure is still not complete without the specification of K_m and ε. Different approaches or models have been used for this purpose.

The parameterized length-scale model The simplest parameterizations of eddy viscosity and the rate of energy dissipation are based on the TKE and appropriate length scales; the most commonly used parametric relations are:

$$K_m = C\ell E^{1/2} \tag{13.39}$$

$$\varepsilon = C_\varepsilon E^{3/2}/\ell_\varepsilon \tag{13.40}$$

Here, ℓ is the large-eddy length scale, ℓ_ε is the dissipation length scale, and C and C_ε are empirical constants whose values may depend on how ℓ and ℓ_ε are specified. In most of the TKE models it is assumed that $\ell_\varepsilon = \ell = l_m$, where the mixing length is usually specified as in Equation (13.18), while a few investigators have used different expressions for ℓ and ℓ_ε (Therry and Lacarrere, 1983).

Substituting from the well-known similarity relations for the various parameters in the neutral surface layer (e.g., $K_m = kzu_*$, $\varepsilon = u_*^3/kz$, $\ell = \ell_\varepsilon = kz$, and $E \cong 5.5u_*^2$), one can estimate $C \cong 0.43$ and $C_\varepsilon = C^3 \cong 0.08$. Somewhat different values of these constants might be more appropriate when modeling the entire PBL under different stability conditions (Detering and Etling, 1985; Lacser and Arya, 1986).

A wide variety of length-scale parameterizations have been proposed in the literature (Holt and Raman, 1988). Sensitivity tests, including a comparative study of several different parameterizations, have shown that, while mean wind and temperature profiles may not be very sensitive to different formulations, the TKE and the rate of energy dissipation are quite sensitive to the same (Lacser and Arya, 1986).

The E–ε *model* In most computational fluid dynamics (CFD) models of turbulent flows, the dynamical equation for the rate of energy dissipation is carried in addition to the TKE equation. In these so-called k–ε or E–ε models, the eddy viscosity is usually expressed (parameterized) in terms of E and ε as

$$K_m = C_k E^2/\varepsilon \tag{13.41}$$

Where C_k is an empirical constant that may actually depend on the type of flow, e.g., for the neutral surface layer, $C_k \cong 0.033$. The derivation of the dynamical equation for ε is quite involved (Lumley, 1980); it contains many new unknown terms that are considerably simplified and parameterized. The highly parameterized form used in the standard E–ε model is given as

$$\frac{\partial \varepsilon}{\partial t} = C_1 \frac{\varepsilon}{E}\left[-\overline{uw}\frac{\partial U}{\partial z} - \overline{vw}\frac{\partial V}{\partial z} + \frac{g}{T_{vo}}\overline{\theta_v w}\right] - C_2 \frac{\varepsilon^2}{E} + \frac{\partial}{\partial z}\left(K_\varepsilon \frac{\partial \varepsilon}{\partial z}\right) \tag{13.42}$$

where C_1 and C_2 are empirical constants, and K_ε is the diffusivity of dissipation rate which is usually linked to that of momentum as $K_\varepsilon = K_m/\sigma_\varepsilon$, where σ_ε is another diffusivity ratio.

The various terms on the right-hand side of Equation (13.42) represent production, destruction and turbulent transport of the dissipation rate respectively. In their parameterized forms, represented in Equation (13.42), the production and destruction terms have been simply assumed to be proportional to the production and dissipation of the TKE. The various constants (C_k, C_1, C_2, etc.) in the E–ε model, often used in commercial CFD codes, have been estimated from the laboratory experimental data. Their 'standard' values ($C_k = 0.09$, $C_1 = 1.44$, $C_2 = 1.92$, etc.) may not be appropriate for the stratified, rotating atmospheric PBL (Detering and Etling, 1985). But sufficiently accurate measurements of the various terms in E and ε equations are not available for a direct empirical determination of these constants for the PBL. Different values have been suggested by different investigators on the basis of comparisons of model-predicted and observed winds and turbulence (Holt and Raman, 1988).

The ε–ℓ model In this model the eddy viscosity is determined by Equation (13.39), with both E and ℓ computed from their dynamical equations. Thus, instead of the ε equation, a prognostic equation for the length scale ℓ (more appropriately, the product $E\ell$) is included in the model. Mellor and Yamada (1974) proposed such a model as a compromise between their level 2 and level 3 models. The parameterized form of the prognostic equation for $E\ell$ looks similar to that for the dissipation rate ε (Mellor and Yamada, 1982); both have the same limitations and large uncertainties in the estimates of the various closure constants.

A comparison of different TKE closure schemes with limited field experimental data by Holt and Raman (1988) reveals that mean profiles of wind, potential temperature, and specific humidity show little sensitivity to the type of closure scheme. Thus, in the determination of K_m, using a diagnostic formulation of ℓ or a prognostic determination of ε makes very little difference. However, the E–ε model performs best in the simulation of turbulence structure, because the energy dissipation rate is calculated from its prognostic equation rather than simply parameterized by Equation (13.40) with a specified length scale.

13.3.3 Higher-order closure models

The gradient transport theory with local closure relations has some serious flaws and becomes invalid in convective conditions when turbulent transports are not necessarily down the mean gradients. Most of the TKE closure models also utilize the local gradient-transport concept and relations and, hence, fail to account for significant counter-gradient transports by large convective eddies. For this reason, second- and higher-order closure models of turbulence and diffusion in the PBL have also been proposed (Mellor and Yamada, 1974; Wyngaard *et al.*, 1974; Andre *et al.*, 1978; Lumley, 1980). Second-order closure models are based on the dynamical equations for turbulent transports or fluxes (Stull, 1988; Sorbjan, 1989). These equations contain the unknown third moments, correlations involving pressure and velocity or other scalar fluctuations, and the molecular dissipation terms, all of which have to be parameterized to close the set of equations. Thus, the closure problem shifts to the higher-order second-moment equations.

The second-moment equations contain much more physics of turbulent transports than the TKE equation. Therefore, it has been argued that the second-order closure approach should be more general and accurate than the first-order or one-and-a-half-order (TKE) closure approach. Similarly, a third-order closure model utilizing the dynamical equations for all the first, second and third moments might be considered even more general, because it contains more physics of turbulence. Unfortunately, the governing equations also become more complex and numerous, and their closure becomes much more difficult and ad hoc in the absence of experimental data on higher moments of turbulence (Andre *et al.*, 1978). Here we mention only some of the commonly used closure approaches to parameterize the turbulent transport (third moments), molecular dissipation, and pressure terms in the second-moment equations.

The third-moment terms, representing turbulent transports of second moments, are very important to the dynamics of buoyancy-driven unstable and convective boundary layers. They are found to be rather insignificant in stably stratified flows, such as the SBL. The simplest parameterization of turbulent transport terms is based on the gradient-transport concept, whereby third moments are expressed as products of the spatial gradients of second moments and some form of turbulent diffusivity. The diffusivity is expressed in terms of large-eddy length and velocity scales (e.g., $K \sim E^{1/2}\ell$). Alternatively, the turbulent diffusion time scale defined by the ratio E/ε can be used (e.g., $K \sim E^2/\varepsilon$). However, this type of turbulent-transport modeling has been found to be deficient and inconsistent with the observed turbulence structure in the CBL. A more complex, but general, functional-expansion approach (Lumley, 1980) can remove some of the limitations of the simpler approach, but it also introduces many more terms and empirical constants in the parametric relations.

An accurate representation or parameterization of molecular dissipation terms in second-moment equations, especially the variance equations, is important, because these terms determine the rate at which turbulence variances and the TKE are destroyed. In some models, ε also determines the turbulence length and time scales that are used in the parameterization of turbulent transport and pressure terms. In the simplest approach, dissipations are expressed in terms of the appropriate length and velocity scales as in Equation (13.40). Alternatively, dynamical equations are carried for the rates of dissipation of TKE and variances of scalar fluctuations. Such equations are very complex and only the highly simplified and parameterized forms are used in second-order closure models (Wyngaard *et al.*, 1974; Lumley, 1980). The molecular dissipation terms in the dynamical equations for the covariances or turbulent fluxes are usually neglected, using Kolmogorov's local isotropy hypothesis for large-Reynolds-number flows.

The satisfactory parameterization of the covariances between pressure fluctuations and spatial gradients of velocity, temperature, and other scalar fluctuations, which are also called return-to-isotropy terms, is probably the most difficult and challenging problem in higher-order closure modeling. These are the primary destruction terms in the covariance equations; they are also responsible for the distribution of turbulence kinetic energy among three components. In the simplest parameterizations, pressure terms are assumed to be proportional to the second moments (variances and covariances) they tend to destroy. But this approach does not properly account for the interactions of turbulence with mean flow shear and buoyancy. More refined parameterizations can account for the above interactions approximately, but contain many more terms and empirical constants.

Higher-order closure models provide a great deal of additional information on turbulence variances and covariances that cannot be obtained from any first-order closure model. This information is particularly useful in applications of micrometeorology to turbulent exchange and transport processes in the PBL. Higher-order closure models also permit counter-gradient transports under convective conditions, although only after introducing additional terms representing buoyancy-induced motions and mixing.

The major limitations of the second- and higher-order closure approach are the increasingly large number of dynamical equations for the second and higher moments that have to be solved and the lack of general validity and uniqueness of closure relations that have been developed and used. Some of the closure relations have been validated and closure constants evaluated using laboratory experimental data from near-neutral, nonrotating, turbulent shear flows. The same relations and values of empirical constants may not be applicable to the atmospheric PBL in which buoyancy and rotational effects are particularly important. There are too many closure constants of which only a few can be estimated from the available experimental data. This limitation becomes more and more severe as the order of closure is increased. Only limited measurements of third and fourth moments of turbulence in the atmosphere are available and large uncertainties in them do not permit their use in validating closure relations and empirically evaluating the various constants.

A major limitation of all the ensemble-averaged turbulence closure models is that ensemble averaging forces one to parameterize the effects of the entire spectrum of eddy motions, including the large energy-containing eddies. These large eddies in a turbulent flow are very sensitive to the geometry of flow, as well as to the dominant mechanisms of production and destruction of turbulence in the flow. Therefore, some of the turbulence modelers have expressed doubt whether any higher-order closure model could be applied to the wide range of conditions encountered in the atmosphere. With increasing computer power and widespread use of computer networks, large-eddy simulation of turbulence is likely to become a more practical and economically feasible modeling approach.

13.3.4 Large-eddy simulations

The large-eddy simulation (LES) approach to turbulence modeling has been briefly introduced in Chapter 9. Here we describe some aspects of grid-volume averaging of the equations of motion, the subgrid-scale (SGS) parameterizations or models that are invariably used in LES, and some results of large-eddy simulations of neutral, unstable, and stable atmosphere boundary layers. Merits and limitations of the LES approach will also be discussed.

Grid-volume-averaged equations An appropriate set of equations of motion, potential temperature and other scalar variables to be used in LES can be obtained by grid-volume averaging of the instantaneous equations. This operation is similar to ensemble averaging and utilizes the same Reynolds averaging rules or conditions as discussed in Chapter 9. But there are also some subtle and conceptual differences as discussed by Mason (1994). Grid-volume averaging amounts to applying a spatial filter which, for all practical purposes, is determined by the grid resolution used in the model. The optimum filter is not necessarily the one with sharp cutoffs at grid boundaries; such a filter was used by Deardorff (1972a, 1973). Gaussian and other smooth spatial filters have been found to be more efficient for mathematical operations.

If we define the grid-volume averaging or the spatial filter operation so that the average variables are denoted by an overbar and the SGS fluctuations around averages are denoted by a prime, instantaneous variables in Equations (9.1) can be expressed as

$$\begin{aligned} \tilde{u} &= \bar{u} + u', \quad \tilde{v} = \bar{v} + v' \\ \tilde{w} &= \bar{w} + w', \quad \tilde{\theta} = \bar{\theta} + \theta' \end{aligned} \tag{13.43}$$

and so on. Then, substituting from Equations (13.43) into (9.1) and averaging over each grid volume yields the grid-volume-averaged equations. For spatial filters preserving the average values of variables upon second-filter operation (e.g., $\bar{\bar{u}} = \bar{u}$, etc.), the filtered equations have exactly the same forms as ensemble-averaged equations (9.8), except for the difference in our notation of ensemble-averaged and grid-volume-averaged (filtered) variables. Thus, the latter can be written as

$$\frac{\partial \bar{u}}{\partial x} + \frac{\partial \bar{v}}{\partial y} + \frac{\partial \bar{w}}{\partial z} = 0 \tag{13.44}$$

$$\frac{\partial \bar{u}}{\partial t} + \frac{\partial(\bar{u}\bar{u})}{\partial x} + \frac{\partial(\bar{u}\bar{v})}{\partial y} + \frac{\partial(\bar{u}\bar{w})}{\partial z} - f\bar{v} = -\frac{1}{\rho_o}\frac{\partial \bar{p}_1}{\partial x} + \nu\nabla^2\bar{u} + \left(\frac{\partial \tau_{xx}}{\partial x} + \frac{\partial \tau_{xy}}{\partial y} + \frac{\partial \tau_{xz}}{\partial z}\right) \quad (13.45)$$

and so on for $\bar{v}$, $\bar{w}$, and $\bar{\theta}$. The Reynolds stresses on the right-hand side can, in general, be expressed as

$$\tau_{ij} = -\left(\overline{\tilde{u}_i \tilde{u}_j} - \bar{u}_i \bar{u}_j\right) \quad (13.46)$$

in which the indices i and j can take on values of 1, 2 and 3 to represent the velocity components ($u_1 = u$, $u_2 = v$, and $u_3 = w$) in the three directions ($x_1 = x$, $x_2 = y$, $x_3 = z$), using the Cartesian frame of reference.

The Reynolds stresses in the above LES equations essentially represent the contributions of subgrid-scale motions. This can easily be seen from the Reynolds decomposition of instantaneous variables and by expressing the grid-volume-averaged momentum flux as

$$\begin{aligned} \overline{\tilde{u}_i \tilde{u}_j} &= \overline{(\bar{u}_i + u_i')(\bar{u}_j + u_j')} \\ &= \overline{\bar{u}_i \bar{u}_j} + \overline{\bar{u}_i u_j'} + \overline{u_i' \bar{u}_j} + \overline{u_i' u_j'} \end{aligned} \quad (13.47)$$

Note that the first term on the right-hand side of Equation (13.47) may not be much different from $\bar{u}_i \bar{u}_j$, so that τ_{ij} largely consists of SGS contributions. It should be recognized, however, that the resolved LES variables $\bar{u}$. $\bar{v}$, etc., also include the random turbulent fluctuations in both space and time that are due to large-eddy motions with scales larger than the filter or grid size. The SGS stresses or fluxes also have this random variability, but on average, their magnitudes are much smaller than the ensemble-averaged Reynolds fluxes, because the former contain only minor contributions of SGS motions.

For the more general filter operations that may not preserve previously filtered average variables on a second filter operation or averaging, i.e., for which $\bar{\bar{u}} \neq \bar{u}$, etc., additional terms are generated by the averaging or filter operation on the instantaneous equations. We will not consider such filters here, but simply state that the grid-volume-averaged or filtered equations of conservation of mass, momentum, etc., used in LES are still expressed in the same forms as Equations (13.44) and (13.45), but each of the SGS flux terms contain different types of SGS contributions as indicated by Equations (13.46) and (13.47).

Subgrid-scale models In order to close the set of grid-volume-averaged or filtered equations, the unknown SGS Reynolds stresses and fluxes must be parameterized in terms of the resolved variables for which the set of LES equations is numerically solved. The various SGS closure models have been proposed in the literature; these have been reviewed by, among others, Deardorff (1973), Ferziger (1993), and Mason (1994). Here, we briefly describe only a few widely used, simple SGS models.

The simplest and the oldest SGS model is that proposed by Smagorinski (1963) in the context of general circulation modeling. The Smagorinski model is essentially a first-order closure model based on the classical Boussinesq concept of eddy viscosity, but applied to SGS fluxes. Using the generalized gradient-transport relationship with a scalar SGS eddy viscosity K, the Reynolds stresses can be expressed as

$$\tau_{ij} = K\left(\frac{\partial \bar{u}_i}{\partial x_j} + \frac{\partial \bar{u}_j}{\partial x_i}\right) = 2KS_{ij} \tag{13.48}$$

in which S_{ij} denotes the rate of SGS deformation.

The SGS eddy viscosity in Equation (13.48) should be distinguished from the overall eddy viscosity introduced in Chapter 9. Even when the ensemble-averaged fluxes may not be down gradient, as often occurs in the convective mixed layer, due to the presence of large-eddy convective motions, it is reasonable to expect that the gradient-transport hypothesis might be more generally valid for small-scale (SGS) motions. Indeed, Equation (13.48) can be derived in several different ways and not just heuristically as Smagorinski did in 1963 (Ferziger, 1993). These derivations suggest that the SGS eddy viscosity can be expressed as

$$K = \ell_o^2 S \tag{13.49}$$

$$S^2 = \frac{1}{2}\left(\frac{\partial \bar{u}_i}{\partial x_j} + \frac{\partial \bar{u}_j}{\partial x_i}\right)^2 \tag{13.50}$$

where ℓ_o is the SGS mixing-length scale which is related to filter scale ℓ_f, or the geometric mean grid size

$$\Delta = (\Delta x \Delta y \Delta z)^{1/3} \tag{13.51}$$

Deardorff (1972a) used the simplest proportionality relation for the mixing length

$$\ell_o = C_s \Delta \tag{13.52}$$

in which the value of Smagorinski constant C_s is found to be somewhat dependent on mean flow shear, stability, proximity to the surface, and possibly other factors. For example, the optimum values of $C_s = 0.13$ and 0.21 were found by Deardorff (1972a) for the neutral and unstable PBLs, respectively, using his sharp cutoff filter.

A different analysis by Mason (1994) with a spherical filter of scale ℓ_f yields the following expression for ℓ_o:

$$\ell_o = C_f \ell_f \tag{13.53}$$

where $C_f \cong 0.17$. The precise value of C_f, in general, must depend on the filter shape. It is clear from Equations (13.52) and (13.53) that when the filter scale is equal to the average grid size, $C_s = C_f$. Mason has shown that $\ell_f = \Delta$ with $C_s = C_f \cong 0.2$ is indeed the optimum choice of filter scale.

In large-eddy simulations of the atmospheric boundary layer one should also be concerned about the possible effects of buoyancy or stability on the SGS model. If the filter scale or the mean grid size actually falls within the locally isotropic range of eddy sizes, then one would expect a negligible influence of buoyancy on the SGS model. The SGS model should also include a parametric relation for the SGS heat flux similar to that of the Reynolds stress. The corresponding SGS diffusivity of heat K_h may differ from that of momentum. The commonly used assumption is $K_h = \alpha K_m$, in which $\alpha = 1$ to 3, depending on the PBL stability and the type of filter used in the LES (Deardorff, 1972a; Mason, 1994). This procedure appears to be well justified for simulating unstable and convective boundary layers (for these $\alpha > 1$), but may not be adequate in the stably stratified transition layer lying above the convective mixed layer.

In the LES of stably stratified flows including the SBL, the theoretical requirement of a small grid size, so that all SGS motions are locally isotropic, is unlikely to be achieved, and some direct effect of stability must be incorporated in the SGS model. For example, Deardorff (1973) used an equation of the subgrid turbulence kinetic energy and derived an expression for the SGS eddy viscosity that is dependent on the flux Richardson number. However, this amounts to a higher-order TKE closure in which the subgrid-scale TKE equation is included in the model, with the usual parameterizations of $\varepsilon \sim E^{3/2}/\ell$, $K_m \sim E^{1/2}\ell$, etc. The simpler approach of using a stability-dependent SGS eddy viscosity and mixing length seems equally successful and computationally less expensive. The most widely used relation is

$$K = \ell^2 S(1 - \mathrm{Rf})^{1/2} \tag{13.54}$$

which may be considered a generalization of Equation (13.49). The additional influence of stability on the mixing length ℓ should also be considered so that the ratio ℓ/ℓ_o decreases with increasing stability. The ratio of SGS diffusivities of heat and momentum may also depend on Rf or Ri. Such a stability-dependent SGS model is considered to be crucial to the large-eddy simulation of the SBL.

Further refinements in SGS modeling have been proposed, especially for the satisfactory LES of the stratified surface layer. As the surface is approached, the relative contribution of subgrid motions to the total fluxes and TKE increases and must be accounted for by the SGS model. The simple Smagorinski model is found to be unsatisfactory and considerably overestimates the mean velocity gradients. The split-eddy viscosity model of Sullivan *et al.* (1994) appears to be quite promising (Andren, 1995), as more complex dynamic and stochastic subfilter models (Mason, 1994) require much larger computational resources. One may also consider using a TKE closure model in which the SGS eddy viscosity and the rate of energy dissipation are parameterized as

$$K = C\ell E^{1/2} \tag{13.55}$$

$$\varepsilon = C_\varepsilon E^{3/2}/\ell \tag{13.56}$$

Note that the estimated values of $C \cong 0.1$ and $C_\varepsilon \cong 0.93$ by Moeng and Wyngaard (1988) are quite different from the similar constants in the expressions (13.39) and (13.40) involving the total (overall) eddy viscosity and TKE. In the surface layer, ℓ is specified as an increasing function of height, such that $\ell = kz$ close to the surface and it approaches its constant value ℓ_o in the interior flow. The following interpolation formula is frequently used for this purpose:

$$\ell^{-n} = (kz)^{-n} + \ell_o^{-n} \tag{13.57}$$

in which an optimum value of $n \cong 2$ has been determined (Mason, 1994). A refinement of the vertical mesh near the surface, so that Δz is sufficiently small close to the surface and gradually increases to its constant interior value has also been recommended.

LES studies of the PBL The numerical solution of the grid-volume-averaged equations of motion, potential temperature, and scalar concentrations with an appropriate SGS model also depends on the initial and boundary conditions. In LES studies of the PBL, the initial conditions are usually specified assuming a simple reference state of the atmosphere in geostrophic balance, using actual sounding data, or results of an analytical model. The initial velocity profile may also contain small random fluctuations at all grid points. It is hoped that after some spinup time, model results essentially become independent of the somewhat arbitrary initial conditions and depend only on the relevant external forcings, such as large-scale pressure gradients, the surface roughness, and the surface heat flux or temperature, that are also specified directly as surface and upper boundary conditions.

The LES approach has most successfully been utilized for studying the mean structure as well as turbulent transport and diffusion processes in unstable and convective boundary layers. Deardorff (1970a, 1972a) was the first to use this approach (he called it three-dimensional numerical modeling) for studying neutral and unstable atmospheric boundary layers with specified surface roughness, surface heat flux, and horizontal pressure gradients (representing a constant geostrophic wind). A rigid lid was imposed at the top of the model domain for a crude simulation of an inversion at the top of the mixed layer. Thus, only simple cases of barotropic, stationary, unstable and neutral PBLs with no entrainment from the top were simulated in this early study. Still, the results of numerical simulations led to an important discovery of mixed-layer similarity scaling by Deardorff (1970b). They also confirmed that the gradient-transport hypothesis is not generally valid in unstable and convective mixed layers in which large scale convective motions (updrafts and downdrafts) dominate turbulent transport and diffusion processes. The most remarkable of his LES results were the displays of the instaneous fields showing large-eddy structures and their evolution in time and spacc. Ensemble or horizontal spatial averages of simulated fields including turbulence statistics provided further justifications for utilizing the proposed similarity theories and scaling for neutral and convective boundary layers.

More realistic simulations of the evolving unstable and convective boundary layers were obtained by Deardorff (1974) by replacing the rigid lid at the top by an inversion layer, allowing for penetrative convection and entrainment of the free atmospheric air, and using more sophisticated higher-order SGS closure models. Increased computer power also permitted increased domain size and grid resolution in the model. Deardorff's (1974) LES results compared well with observations during convective conditions. They were also used in formulating better models of the evolution of the mixed-layer height and entrainment velocity. Later, Deardorff (1980) extended the LES approach to study strato-cumulus-capped mixed layers. His pioneering LES studies at the National Center for Atmospheric Research were followed by other LES studies of turbulence and diffusion in the CBL (Moeng, 1984; Wyngaard and Brost, 1984; Moeng and Wyngaard, 1988; Nieuwstadt *et al.*, 1992).

An intercomparison of four different SGS parameterizations, by Nieuwstadt *et al.* (1992) shows that the simulated mean flow and turbulent structure of the convective mixed layer are not very sensitive to the details of SGS parameterization. However, the mean gradients and turbulence in the shear-dominated surface layer are found to be strongly dependent on the SGS parameterization or model. In particular, the simple Smagorinski model has been shown to be inadequate and more sophisticated SGS models with better treatment of near-surface turbulent transfer processes are recommended for more realistic simulations of the surface layer of the CBL as well as other shear flows, such as neutral and stably stratified PBLs (Mason, 1994; Sullivan *et al.*, 1994).

Although the neutral PBL was included in Deardorff's (1970a, 1972a) early LES studies, more recent simulations of the same have revealed a much stronger sensitivity of results to the details of SGS parameterization and much slower approach to steady state or equilibrium condition than previously realized (Andren *et al.*, 1994). In particular, the mean fields (e.g., the velocity components U and V) show large inertial oscillations whose amplitudes decrease with increasing dimensionless time tf. But, it would take more than ten inertial periods ($tf > 10$) to reach a near steady state. The mean velocity gradients vary only slowly in time and higher moments (e.g., turbulent variances and fluxes) are less sensitive to the inertial oscillation. For example, the layer-averaged TKE reaches its equilibrium value only after one or two inertial periods.

In the intercomparison study by Andren *et al.* (1994), a shallow neutral PBL was simulated using four different LES models and a rather coarse grid so that results might be sensitive to the details of SGS parameterization or model. It is found that mean gradients and turbulence structure are indeed sensitive to the magnitude of the SGS eddy diffusivities. Resolved-scale motions are generally more intense in the lower part of the PBL in the models with smaller SGS eddy diffusivities. For the mean fields, the noted failure of the Smagorinski SGS model in producing the logarithmic mean wind and scalar profiles cannot be corrected by changing the value of the Smagorinski coefficient C_S. Differences between models with and without stochastic backscatter are, in general, largest close to the surface, but also noticeable in the lower one-third of the PBL. Sullivan *et al.* (1994) have proposed an alternative two-part eddy viscosity model for the shear-dominated lower (near-surface) layer. Their model utilized the TKE closure, but can be adopted in a first-order SGS closure model as well (Ding *et al.*, 2001a, b).

Although the LES approach has successfully been used for modeling neutral, unstable and convective boundary layers for more than thirty years, its application to the SBL started only recently when increased computer speed and memory permitted the use of a small enough grid size for most of the TKE to be resolved. Mason and Derbyshire (1990) used a modified version of the Smagorinski SGS model in which SGS eddy viscosity was related to flux Richardson number. They conducted a series of simulations of slightly stable boundary layers and showed their results to be broadly consistent with observations and the local similarity theory of Nieuwstadt (1984). Later Brown *et al.* (1994) used a revised version of Mason and Thomson's (1992) stochastic backscatter model and extended the stability range to include moderately stable conditions ($L = 31$ m to 517 m and $h/L = 1.4$ to 5.4). The use of stochastic backscatter in their SGS model resulted in better resolution of turbulence and more realistic velocity and temperature profiles in the surface layer, which are consistent with the empirical Monin–Obukhov similarity functions. Their LES results are found to be quite sensitive to the SGS model with or without the stochastic backscatter.

The importance of using a good SGS model in the LES of the SBL has also been demonstrated by Andren (1995) who compared the results of simulations with two different SGS models, viz., Moeng's (1984) TKE closure model with a standard SGS mixing length formulation and that with a more refined two-part eddy viscosity formulation of Sullivan *et al.* (1994). Mean velocity and scalar gradients in the surface layer are much better simulated by the two-part eddy viscosity model which is found to be comparable to the computationally more expensive stochastic backscatter model of Brown *et al.* (1994). It yields more vigorous resolved-scale fields. The variance and covariance profiles in the simulated slightly stable boundary layer are well represented by their similarity forms (13.12)–(13.14). Eddy diffusivities of momentum, heat, and any passive scalar show similar distributions with normalized height in the SBL with their maximum values occurring around $z/h \cong 0.25$. Scalar diffusivities are 20–30% larger than eddy viscosity in the bulk of the SBL, but their ratios (K_h/K_m, etc.) decrease with sharply increasing Richardson number (Ri) near the top of the SBL.

Kosovic and Curry (2000) used a more complicated and computer intensive, nonlinear SGS model which is capable of reproducing the effects of backscatter of energy and of the SGS anisotropies that are characteristic of shear-driven flows. They explored the effects of the changes in surface cooling rate, geostrophic wind, inversion strength, surface roughness, and latitude on the mean flow and turbulence structure of the simulated SBL and carried out 21 LES cases with three different grid resolutions. The simulation times varied from 12 h to 24 h, depending on the latitude and corresponded to approximately six times the inertial period (f^{-1}). In order to test the SGS model effects, Kosovic and Curry (2000) also ran two simulations with the linear SGS models used by Andren (1995). A comparison of the normalized total (domain-averaged) TKE for different LES cases showed that the Smagorinski-type model is too dissipative, resulting in unrealistically low TKE levels. But, the linear two-part eddy viscosity model of Sullivan *et al.* (1994) produced much higher TKE levels which are 2–3 times larger than those produced by Kosovic's nonlinear model for the same grid resolution. The simulated TKE levels with the nonlinear SGS model also showed strong sensitivity to grid resolution. Longer simulation time resulted in the development of stronger inversion at the top of the SBL, which presented an obstacle to further SBL growth. But this may be due to imposing a constant surface cooling rate, rather than the constant surface heat flux.

More recently, Ding *et al.* (2001a, b) have studied weakly unstable to moderately stable atmospheric boundary layers using the LES version of the Terminal Area Simulation System (TASS-LES) with a new SGS model. Their proposed SGS model incorporates some aspects of the two-part eddy viscosity model of Sullivan *et al.* (1994) and further refinements including the stability dependence of SGS mixing length and two-part separation of eddy diffusivity of heat similar to that of eddy viscosity. The strong sensitivity of the simulated mean velocity and potential temperature gradients in the lower parts of neutral and weakly unstable boundary layers to different SGS parameterizations or models is evident from Figure 13.9. The proposed two-part eddy viscosity and diffusivity model is found to be quite effective in simulating the empirical Monin–Obukhov similarity functions ϕ_m and ϕ_h in all the weakly unstable to moderately stable cases (see also Figure 13.10).

For their weakly unstable and neutral cases, Ding *et al.* (2001a) used a 2000 m × 2000 m × 750 m domain with 50 × 50 × 75 grid points in x, y, and z

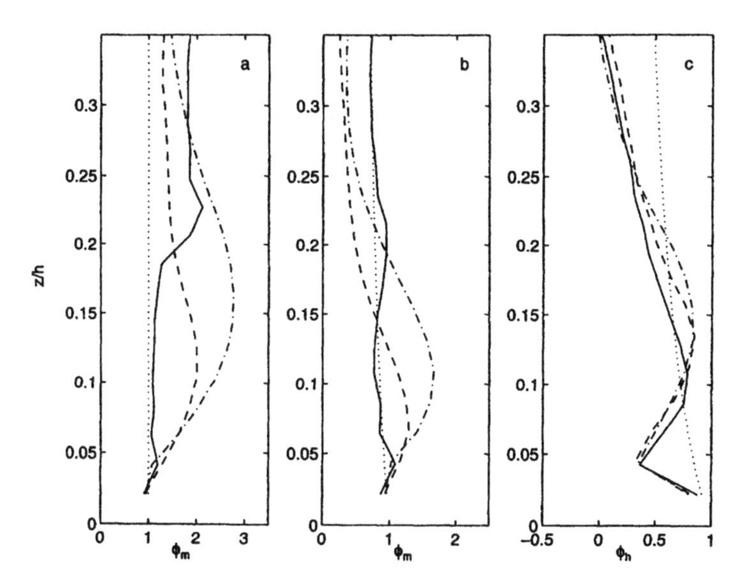

Figure 13.9 Comparison of simulated and empirical (measured) dimensionless gradients of mean velocity (a, b) and potential temperature (c) or Monin–Obukhov similarity functions in the lower part of the PBL for (a) the neutral case and (b) weakly unstable case. Dotted lines represent the empirical similarity functions, dashed-dotted lines the Smagorinski model, dashed lines the simplified two-part eddy viscosity model, and solid lines the proposed SGS model. [From Ding *et al.* (2001a).]

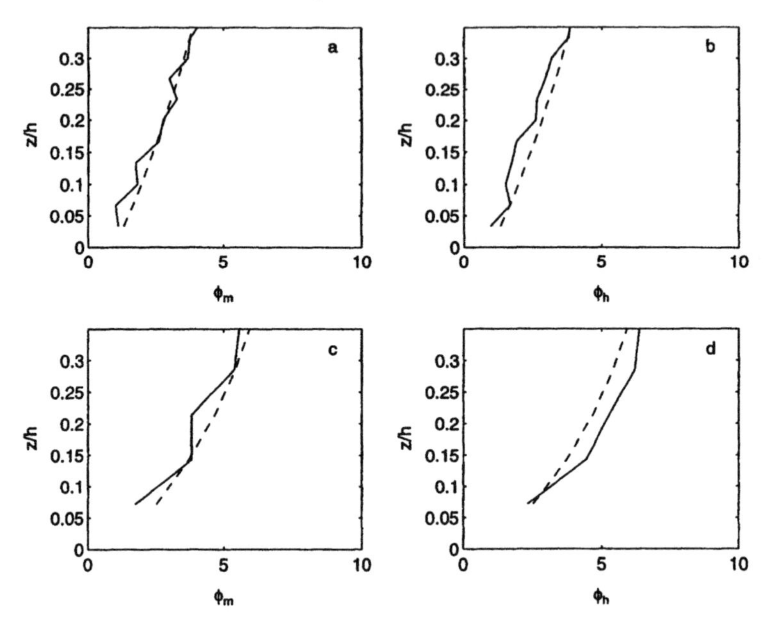

Figure 13.10 Comparison of simulated and empirical (measured) Monin–Obukhov similarity functions ϕ_m (a, c) and ϕ_h (b, d) in the lower part of the SBL. (a, c) The weakly stable WS1 case and (b, d) the moderately stable MS1 case. [From Ding *et al.* (2001b).]

directions, respectively. The initial potential temperature is 300 K below 450 m; it increases by 8 K across the 60 m thick inversion layer from 450 m to 510 m, and further increases with a small gradient of 0.003 K m^{-1} above 510 m. Initially, velocity field is in geostrophic balance with a constant geostrophic wind of 15 m s^{-1} in the x direction. The surface roughness parameter $z_0 = 0.16$ m and the surface heat flux is specified as 0.02 K m^{-1}. The simulation time for the weakly unstable case is 3.5 h of which the last 1 h is used for obtaining the average statistics. The simulation of the neutral case is built on the weakly unstable flow simulation by setting the surface heat flux to zero and running an additional 2 h simulation of which the last 1 h is used for obtaining averages. For these short simulation times, mean flow is not expected to attain a steady state, but turbulence structure reaches a quasi-steady state. Figure 13.11 shows the simulated momentum flux and TKE profiles for the neutral case. Note that the total momentum flux profile is close to linear with significant contributions from the proposed SGS model in the lower one-fourth of the PBL. The normalized TKE profile is characterized by large gradients in the surface layer with a maximum value of TKE$/u_*^2 \cong 5.8$ near the surface. Although this value is consistent with observations of turbulence (velocity

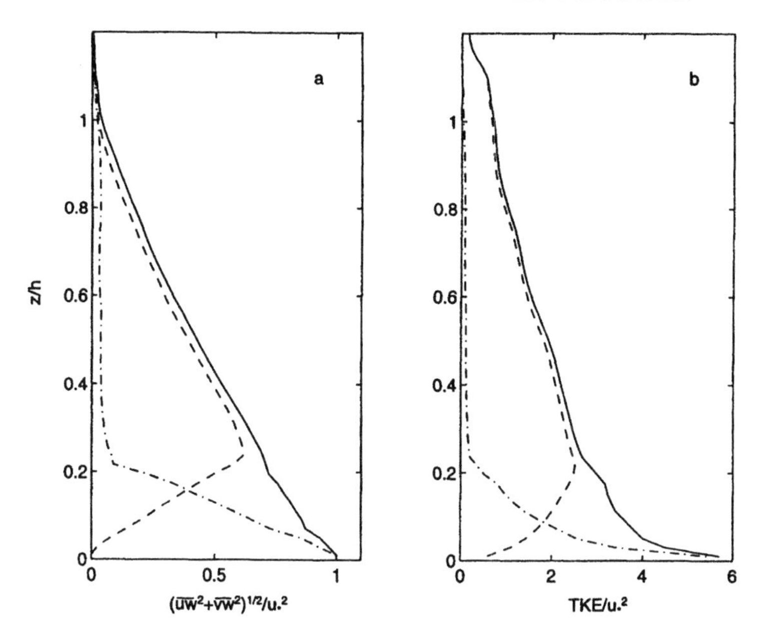

Figure 13.11 Ensemble-averaged profiles of the normalized (a) horizontal momentum flux and (b) TKE from the large-eddy simulation of the neutral PBL. Dashed lines represent resolved parts, dashed-dotted lines subgrid parts, and solid lines total flux and TKE. [From Ding *et al.* (2001a).]

variances and TKE) in the near-neutral surface layer, the rapid decrease of TKE with height is not consistent with the neutral surface-layer similarity theory.

The large-eddy simulations of weakly and moderately stable boundary layers by Ding *et al.* (2001b) are successively built from slightly unstable and neutral simulations by gradually reducing the surface heat flux in several steps. Most of the simulations had the same initial conditions but different geostrophic winds assuming a barotropic environment. Periodic boundary conditions are applied in horizontal (x and y) directions, while in the vertical direction, a sponge layer comprising of three grid intervals has been added on top of the model domain. Both the heat and momentum fluxes are assumed to vanish at the top boundary. The lower boundary employs a no-slip condition and the surface heat flux is specified as a function of time. Most simulations of the SBL are performed using small domains (e.g. 500 m × 500 m × 500 m) and finer grid resolutions (e.g. $\Delta x = \Delta y - \Delta z = 10$ m), although domain and grid sizes are also varied to examine their possible influences on LES results. Of the total simulation time of 6.5 h, a small positive (upward) surface heat flux is maintained during the first 1 h, followed by zero heat flux during the next 2 h, a decreasing negative heat flux during the next 2 h, and a constant negative (downward) heat flux in the final 1.5 h of simulation. The final values of the specified surface heat flux ranged between -0.02 and $-0.05\,\mathrm{K\,m\,s^{-1}}$, while the specified geostrophic wind speed ranged between 7.5 and 15 $\mathrm{m\,s^{-1}}$. With different combinations of these external forcing parameters and an initial potential temperature profile with a 30 m thick inversion layer between 320 m and 350 m, several weakly and moderately stable cases could be simulated. Some of the LES results of a weakly stable (WS1) case and a moderately stable (MS1) case are presented here as illustrative examples.

Figure 13.12 compares the ensemble-averaged velocity and potential temperature profiles for the above two cases. These are similar to those typically observed in the SBL. In both cases, Θ profiles show a three-layer structure: a stable surface layer of strong gradients, a middle layer of small or vanishing gradients, and an elevated inversion layer on the top. However, in the moderately stable case, the lower layer has much stronger gradients, while the middle layer has almost zero gradient and is totally detached from the underlying shallow SBL. Wind speed profiles show the formation and evolution of the low-level jet whose height decreases and intensity increases with increasing time of simulation, especially in the weakly stable (WS1) case. A rather broad

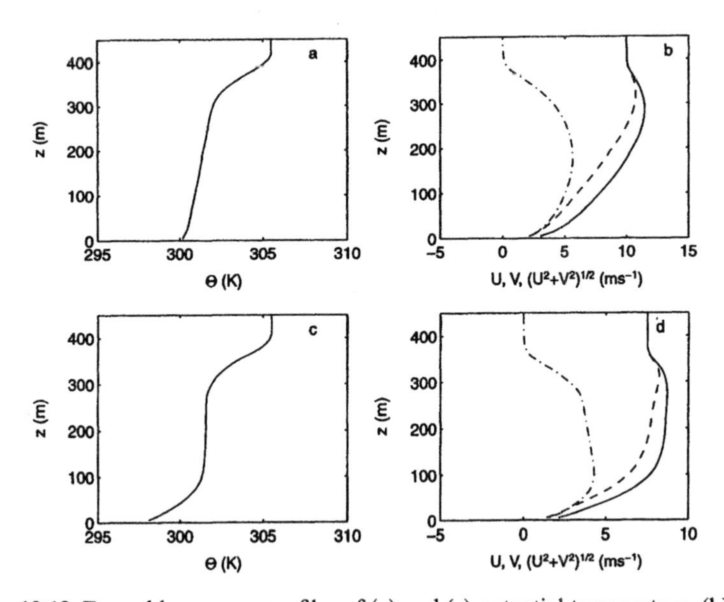

Figure 13.12 Ensemble-average profiles of (a) and (c) potential temperature, (b) and (d) wind speed, and horizontal velocity components in the simulated SBL. In (b) and (d) solid line represents wind speed, dashed line U-component, and dash-dotted line V-component. (a) and (b) the weakly stable (WS1) case; and (c) and (d) the moderately stable (MS1) case. [From Ding *et al.* (2001b).]

maximum in wind speed spans the entire depth of the residual layer in the moderately stable (MS1) case.

The simulated vertical distributions (profiles) of ensemble-averaged momentum flux, TKE, and heat flux are shown in Figures 13.13 and 13.14 for the two (WS1 and MS1) cases. They appear to be similar, except for the large difference in the SBL heights for the two cases. Both the momentum and heat fluxes decrease almost linearly with height in the lower part of SBL. The near-surface values of TKE/u_*^2 differ only slightly between weakly and moderately stable cases.

Much larger differences are found in the time histories (evolution) of TKE and Ri in weakly and moderately stable cases, as revealed by Figures 13.15 and 13.16. These figures show spatial distributions of instantaneous Ri and TKE in a horizontal plane in the middle of the SBL at four different times during the simulated evolution of the SBL: 200, 250, 300, and 350 minutes. Note that in the weakly stable case, TKE contours show large areas of continuous turbulence in which Ri < 0.25. In the moderately stable case at $t = 350$ min, however, Ri values are greater than 0.25 in almost the entire domain and TKE contours reveal very weak and patchy turbulence. Figure 13.17 shows the time evolution of the domain-averaged TKE for the two cases. Note that, following the evening

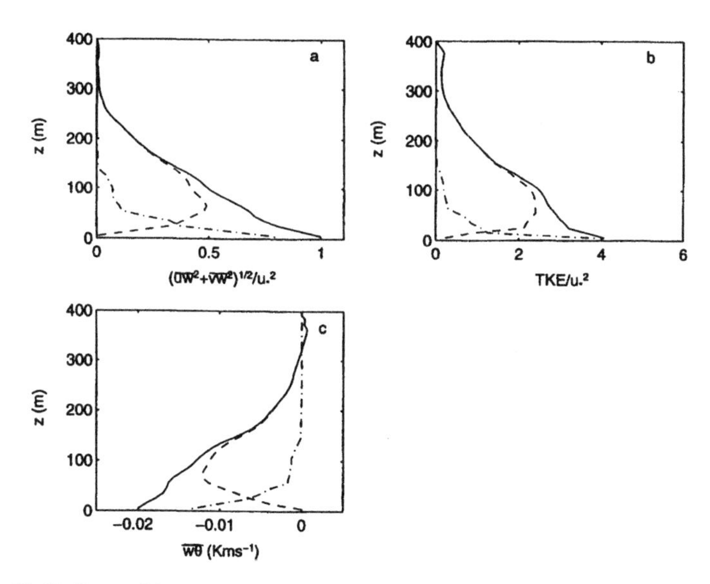

Figure 13.13 Ensemble-averaged profiles of (a) normalized momentum flux, (b) normalized TKE, and (c) heat flux from the simulation of the weakly stable case WS1. Dashed lines represent resolved parts, dash-dotted lines subgrid parts, and solid lines total fluxes and TKE. [From Ding *et al.* (2001b).]

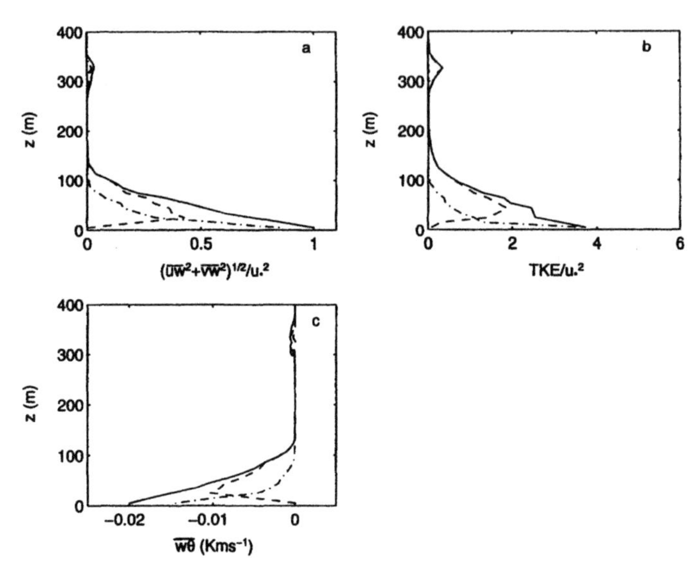

Figure 13.14 Same as Figure 13.13. but for the moderately stable case MS1. [From Ding *et al.* (2001b).]

transition period when the surface heat flux is zero, turbulence in the SBL decays in response to decreasing surface heat flux with time. A quasistationary state with nearly constant value of the domain-averaged TKE is reached only after 300 min when the surface heat flux is held constant.

13.4 Parameterization of the PBL

The governing equations of motion, thermodynamic energy, specific humidity, and other scalars used in large-scale models of the atmosphere contain subgrid scale variances and covariances (fluxes) of subgrid scale perturbations of velocity, temperature, specific humidity, etc. (Pielke, 1984). A perturbation is defined as the deviation of a variable from its grid-averaged value where averaging is implied over the grid volume ($\Delta x \Delta y \Delta z$) as well as over the finite time increment (Δt). The subgrid scale perturbation fluxes must be parameterized in terms of grid-averaged variables in order to form a closed set of equations for the latter. An accurate or satisfactory parameterization of subgrid scale fluxes is crucial to the satisfactory performance of any mesoscale or large-scale model, because the terms involving subgrid scale perturbations are usually of the same order or even larger than the corresponding terms involving grid volume-averaged resolved variables (Pielke, 1984).

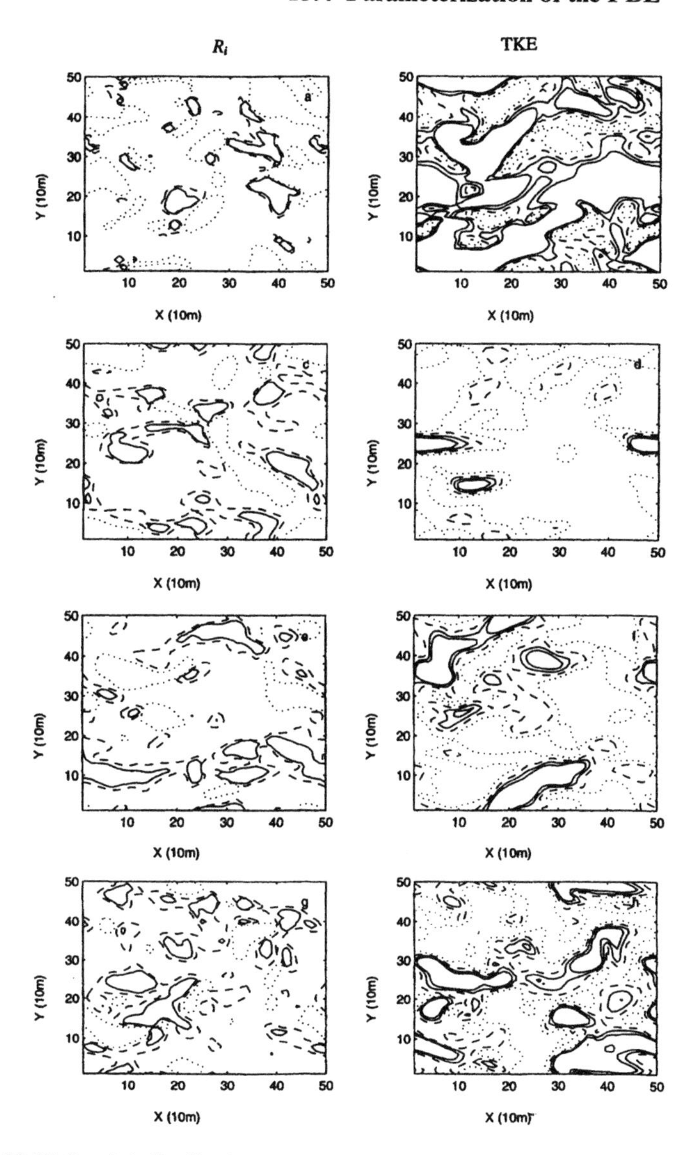

Figure 13.15 Spatial distribution of instantaneous R_i and TKE in a horizontal (x–y) plane near the middle of the SBL at four different times during the time evolution of the SBL in response to decreasing surface heat flux for the weakly stable case WS1: (a) and (b) $t = 200$ min.; (c) and (d) $t = 250$ min.; (e) and (f) $t = 300$ min.; (g) and (h) $t = 350$ min. (a), (c), (e) and (g) R_i contours: dotted lines represent 0, dashed lines 0.25, and solid lines 0.50; (b), (d), (f) and (h) TKE contours: dotted lines represent 0.05 $m^2\,s^{-2}$, dashed lines 0.10 $m^2\,s^{-2}$, and solid lines 0.15 $m^2\,s^{-2}$ or 0.20 $m^2\,s^{-2}$. [From Ding *et al.* (2001b).]

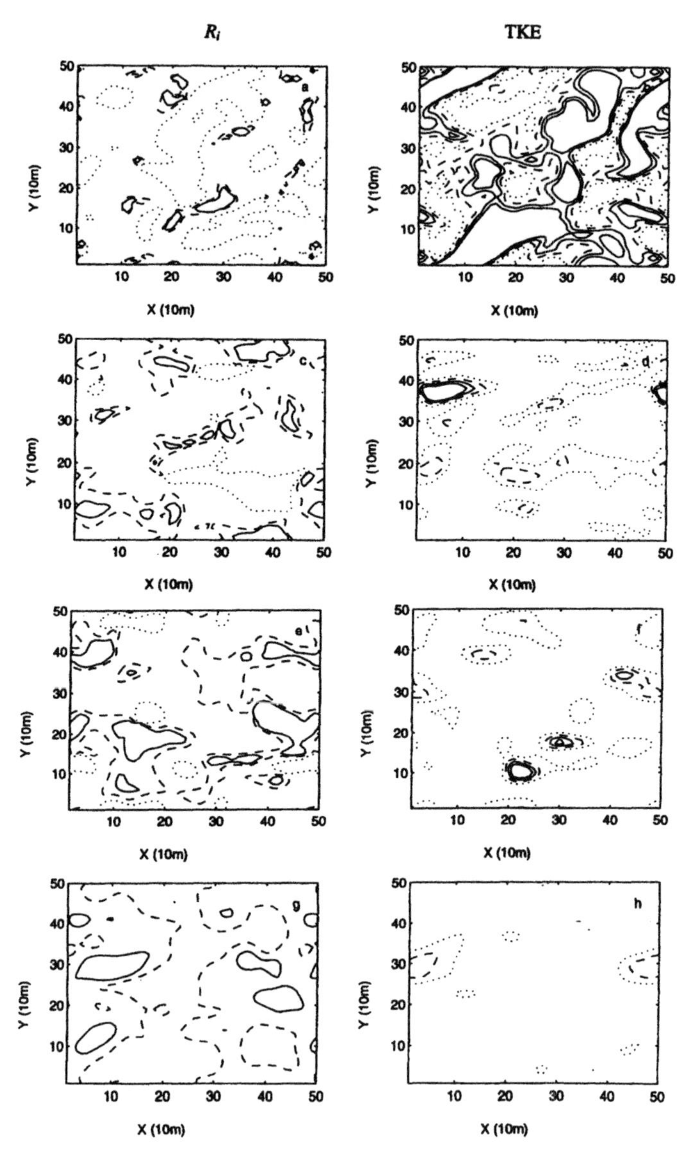

Figure 13.16 Same as Figure 13.15, but for the moderately stable case MS1. [From Ding *et al.* (2001b).]

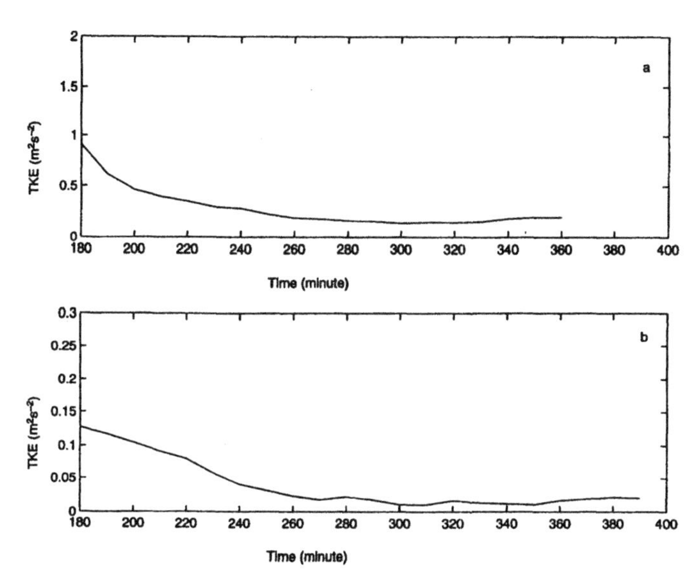

Figure 13.17 Time evolution of the domain-averaged TKE for (a) the weakly stable case WS1 and (b) the moderately stable case MS1 during the last 3.5 h simulation. [From Ding *et al.* (2001b).]

In developing parameterizations for subgrid-scale averaged quantities, it needs to be recognized that the preferred representation is an ensemble average over the grid volume rather than simply the grid volume average. The ensemble average represents the most likely value of the subgrid-scale quantity (variance or covariance) whereas the grid volume average represents just one realization (Pielke, 1984). Thus, most parameterizations discussed here may represent only the most likely (ensemble-averaged) estimates of subgrid-scale quantities. Another distinction that needs to be made, but is rarely recognized or even mentioned, is that between the subgrid-scale perturbation fluxes and the corresponding turbulent fluxes in the PBL. The former obviously depend on the grid spacings and may not be equal to the corresponding turbulent fluxes, except for the rare case when grid spacings are equal to the dimensions of the largest turbulent eddies. In large-domain mesoscale and large-scale models with horizontal grid sizes of tens or hundreds of kilometers, perturbation fluxes are likely to be larger than turbulent fluxes, because the former also contain the contributions of the subgrid mesoscale perturbations in addition to the contribution of turbulence. In mesoscale models of small domains with horizontal grid sizes of less than 2 km (i.e., the size of most energy-containing turbulent eddies), on the other hand, subgrid-scale perturbation fluxes might be smaller than the total turbulent fluxes. Since PBL parameterizations in mesoscale models attempt to represent total ensemble-averaged fluxes due to turbulence only, these are most appropriate for mesoscale models with horizontal grid sizes of the order of 1 km.

13.4.1 Parameterization of turbulent fluxes

Most mesoscale atmospheric circulation and dispersion models attempt to resolve the PBL by placing several discrete grid levels within the lowest 1 km of the model domain. Ideally, the lowest grid level should lie within the surface layer. Anthes *et al.* (1980) have shown that a detailed multi-level resolution of the PBL is essential in a mesoscale model of flow over complex terrain with differential heating (e.g., across land–water boundaries), because significant vertical gradients of meteorological variables occur within the PBL. Simpler, mixed-layer representations of the PBL might be adequate in large-scale or general circulation models. Different types of parameterizations are used to represent the surface fluxes and turbulence variances, kinetic energy, fluxes, and scales at other discrete levels in the PBL, as these are usually not resolved by most widely used first-order closure models.

Surface fluxes The surface fluxes of momentum, heat, and moisture are usually parameterized by using the following bulk transfer relations:

$$\begin{aligned} \tau_0 &= \rho C_D V_1^2 \\ H_0 &= \rho c_p C_H V_1 (\Theta_0 - \Theta_1) \\ E_0 &= \rho C_E V_1 (Q_0 - Q_1) \end{aligned} \tag{13.58}$$

where V_1, Θ_1, and Q_1 are the mean wind speed, potential temperature, and specific humidity, respectively, at the first grid level (z_1), Θ_0 and Q_0 are the values at $z = z_0$, and C_D, C_H and C_E are the bulk transfer coefficients of momentum (C_D is also called the drag coefficient), heat and water vapor, respectively. Using a viscous or molecular sublayer parameterization, Θ_0 and Q_0 can be related to their values at the ground surface (Pielke, 1984).

In some large-scale models, using only crude parameterization of surface fluxes, constant values of bulk-transfer coefficients are specified with some distinction made between land and water surfaces and also between stable and unstable conditions. In most cases with large variations in surface roughness over the model domain and strong diurnal variations of stability in the surface layer, however, it is inappropriate to treat C_D and C_H (usually $C_E = C_H$) as constants. These could easily be specified as continuous functions of surface roughness and stability, using the Monin–Obukhov similarity theory (Arya, 1984; Pielke, 1984). When z_1 lies in the surface layer, C_D and C_H can be represented as functions of z_1/z_0 and the bulk Richardson number

$$\mathrm{Ri_b} = \frac{g}{T_{v0}} \frac{(\Theta_{v1} - \Theta_{v0})z_1}{V_1^2} \tag{13.59}$$

Computations show that C_D and C_H can vary over very wide ranges which depend on ranges of z_0 and $\mathrm{Ri_b}$ values encountered (Arya, 1977).

Fluxes and variances at interior levels Observations of heat and water vapor fluxes in unstable and convective boundary layers suggest that the fluxes vary more or less linearly from their maximum values at the surface to zero at the top of the PBL. For the total momentum flux, a linear profile is also expected under barotropic conditions. However, the presence of baroclinicity can cause large deviations from the linearity (Kaimal *et al.*, 1976). Observations in the SBL also suggest simple profiles of fluxes decreasing with height, as represented by Equations (13.12)–(13.14). The use of these similarity-scaled profiles of fluxes, variances and TKE should probably constitute the simplest parameterizations of these at interior levels in the PBL (Arya, 1984). The scaling parameters essentially depend on the surface fluxes, whose parameterization is discussed in the previous section, and the PBL height. These similarity-scaling-based parameterizations may not be applicable, however, in the vicinity of steep topographical features or sharp changes in surface characteristics.

A more widely used scheme for parameterizing the interior turbulent fluxes is based on the gradient-transport relations in which eddy diffusivities or mixing lengths are specified as functions of height and stability (Businger and Arya, 1974; Brost and Wyngaard, 1978; Pielke, 1984). An implicit specification of eddy diffusivities in terms of mixing lengths might be preferable to an explicit eddy diffusivity profile, especially in mesoscale models with fine vertical grid resolution in which vertical gradients are well resolved. The dependence of subgrid-scale mixing length on the vertical grid spacing can also be considered in such parameterizations.

In the above-mentioned gradient-transport scheme, fluxes are always down-gradient. Simple modifications to gradient-transport relations can also allow for some countergradient fluxes as often observed in the convective mixed layer (Deardorff, 1972a). More satisfactory parameterizations of subgrid-scale fluxes in the CBL would be those based on nonlocal mixing schemes, such as the asymmetric convective mixing (ACM) model (Pleim and Chang, 1992). These have already been discussed in Section 13.3.1.

13.4.2 Height of the PBL or inversion

In most of the parametric relations discussed here, the height of the PBL (h) or the capping inversion (z_i) is an important length scale which must be calculated or parameterized. A variety of theoretical models have been proposed in the literature for predicting h or z_i during the daytime unstable and convective conditions (Deardorff, 1974; Driedonks, 1982; Arya and Byun, 1987; Stull, 1988). Differences in the predictions of different types of models generally fall within the uncertainties of estimating the mean large-scale vertical velocity (W_h) at the top of the PBL.

The commonly used prognostic equation for the PBL height is Equation (13.29) in which both the entrainment velocity (w_e) and the mean vertical velocity (W_h) must be specified or parameterized. For example, the simple thermodynamic model of the mixed-layer growth described earlier in Equation (13.31) implies that

$$w_e = \frac{(1+C)}{\gamma h}\frac{H_0}{\rho c_p} \cong \frac{1.2}{\gamma h}\frac{H_0}{\rho c_p} \tag{13.60}$$

where $\gamma = \partial\Theta/\partial z$ just above the top of the mixed layer.

Deardorff (1974) has pointed out the various limitations of the above model and proposed the following interpolation expression based on his large-eddy simulations of the PBL:

$$w_e = \frac{1.8(W_*^3 + 1.1u_*^3 - 3.3u_*^2 fh)}{\gamma h^2(g/\Theta_0) + 9W_*^2 + 7.2u_*^2} \tag{13.61}$$

Note that in both Equations (13.60) and (13.61), the rate of growth of h is directly proportional to the surface heat flux and inversely proportional to the stability of the overlying inversion layer. Unlike Equation (13.60), Equation (13.61) remains well behaved and approximately valid during each of the limiting cases of vanishing γ, u_*, or W_*.

The evolution of the nocturnal stable boundary layer (SBL) is rather slow, except immediately following the evening transition period, and it is not clear whether a simple diagnostic relation of h, such as Equation (13.7), might be adequate or one should carry an appropriate rate equation for the same. Since the SBL constitutes only a shallow surface inversion layer of weak turbulent transports and small TKE, its detailed parameterization in a mesoscale or larger-scale model may not be that important.

Other aspects and details of the PBL parameterizations used in air pollution and dispersion applications have been described elsewhere (Arya, 1999, Chapter 4).

13.5 Applications

The material presented in this chapter describes the theories, models, and measurements of mean flow and turbulence in the homogeneous atmospheric boundary layer. It has practical applications in the following contexts or areas:

- boundary-layer meteorology;
- air pollution meteorology;
- similarity scales used for presenting the atmospheric boundary layer data in the dimensionless form;
- turbulence closure models of the PBL;
- large-eddy simulations of the PBL;
- mean wind and thermal structure of the PBL;
- parameterization of the PBL in large scale atmospheric models;
- parameterization of mean and turbulent transports and mixing height in atmospheric dispersion and air quality models.

Problems and Exercises

1. Compare and contrast between the daytime convective boundary layer and the nighttime stable boundary layer over a homogeneous land surface, considering the following PBL characteristics:
(a) mean potential temperature profile or thermal structure;
(b) mean wind speed and direction profiles;
(c) the PBL height and its evolution with time.

2.
(a) Distinguish between the mixed-layer similarity theory and scaling and the local free convection similarity theory and scaling. What are their limitations?
(b) Show that in the free convective surface layer, $\sigma_w/W_* = C_w(z/h)^{1/3}$, where C_w is an empirical constant.

3. The following observations were made at a suburban site during a summer afternoon:

Mean wind speed at 10 m height = 1.5 m s^{-1}.
Mean temperature at 10 m height = 27.0°C.
Sensible heat flux near the surface = 500 W m^{-2}.
Height of the inversion base = 2000 m.

Using the mixed-layer similarity relations, estimate the following:
(a) Deardorff's convective velocity;
(b) standard deviation of horizontal velocity fluctuations in the mixed layer;
(c) turbulence kinetic energy at 10, 100, 1000 m;
(d) the rate of energy dissipation at 100, 1000, and 2000 m.

4.
(a) Discuss the implied assumptions in the derivation of the SBL height relation (13.7), as well as its implications and possible limitations when $L/f \to \infty$.
(b) What are the other height scales that might limit the SBL height under slightly stable conditions and even moderately stable conditions in low latitudes?
(c) Compare the SBL height from Equation (13.7) with the neutral PBL height $h = cu_*/|f|$ and estimate the range of stability parameter μ_* in which the former is larger than the latter. Which of the two height expressions should be used under such slightly stable conditions and why?

5. Use the mean wind and temperature data from the Cabauw mast, given in Problem 3 of Chapter 6, and estimate the following scaling parameters using the appropriate similarity relations:
(a) u_* and L from the surface layer (5 m and 10 m) data;
(b) the SBL height from Equation (13.7);
(c) the heights where wind speed and Ri are maximum.
(d) Compare the different estimates of the SBL height with the measured value of 100 m by a colocated sodar, and comment on the accuracy or uncertainty of such estimates.

6. Discuss the relative merits and limitations of the Rossby number and generalized similarity theories of the PBL. Which one would you prefer for representing the turbulence structure (e.g. the vertical profiles of fluxes, variances, and TKE) of (a) the SBL and (b) the CBL and why?

7.
(a) Simplify the Equations (13.16) of mean motion for the stationary and horizontally homogeneous PBL and integrate them with respect to height z from zero to the top of the PBL to obtain the expressions for the momentum fluxes $(\overline{uw})_0$ and $(\overline{vw})_0$ at the surface.

(b) Show that, if the surface-layer coordinate system is used with the x axis oriented along the mean wind direction near the surface

$$\int_0^h (U - U_g)dz = 0$$

Discuss the implication of this integral constraint on the U profile under barotropic and baroclinic conditions with $\partial U_g/\partial z > 0$ and $\partial U_g/\partial z < 0$. Under what conditions would you expect a stronger low-level jet?

(c) Similarly, obtain the expressions for $\overline{uw}$ and $\overline{vw}$ at any height z and describe an indirect method of determining the stress profiles from the vertical sounding of mean winds and geostrophic winds.

8.

(a) What are the merits and limitations of the first-order local closure models of the PBL?

(b) Describe a simple nonlocal closure model that might be more appropriate for the convective boundary layer.

9.

(a) Using the simple thermodynamic integral model Equation (13.33), obtain an expression for the time evolution of the mixed-layer height for a sinusoidal variation of the surface heat flux with time during the daytime period P_d between sunrise and sunset.

(b) Using the above expression of the mixed-layer height with $P_d = 14h$ and $h_0 = 200$ m examine the sensitivity of the evolution of the mixed-layer height to $(\overline{w\theta})_{max}/\gamma$ within the expected range of this parameter (say, 10^4–10^6).

10.

(a) In the TKE closure relations (13.39)–(13.40) with parameterized length scales ℓ and ℓ_ε, show how you would estimate the two empirical constants from the well-known empirical similarity relations in the neutral surface layer.

(b) Are the same values of the constants also applicable to stratified surface layers and the PBL as a whole? Give reasons for your answer.

11.

(a) Discuss the uncertainties in the estimates of the empirical constants C_k, C_1 and C_2 that appear in the E–ε model equations. Are the so-called 'standard' values of these constants, based on laboratory experiments, also applicable to the atmospheric boundary layer and why not?

(b) Discuss the limitations of the TKE closure models for their application to the ABL, especially under unstable and convective conditions.

12.

(a) Discuss the merits and limitations of second- and higher-order closure models.

(b) Compare and contrast the large-eddy simulation approach with higher-order closure modeling.

(c) Distinguish between the first-order subgrid-scale closure model and a similar ensemble-averaged closure model, and show that the former may not have the same limitations as the latter.

13. Of all the PBL and turbulence modeling approaches discussed in Chapters 9 and 13, which one would be the most appropriate to use and why, for the following situations:

(a) Evolution of the atmospheric boundary layers over a period of several days.

(b) Modeling the morning and evenings transitions between stable and unstable boundary layers and vice versa.

(c) Simulating the turbulence generation and decay processes is a very stable, shallow PBL.

(d) Modeling and parameterizing the turbulent exchange processes in the PBL in a mesoscale atmospheric model.

Chapter 14 | Nonhomogeneous Boundary Layers

14.1 Types of Surface Inhomogeneities

Micrometeorological theories, observations, and methods discussed in the preceding chapters are strictly applicable to the atmospheric boundary layer flow over a flat, uniform, and homogeneous terrain. Over such an ideal surface, the PBL is also horizontally homogeneous and in equilibrium with the local surface characteristics, especially when synoptic conditions do not change rapidly. Such a simple state of affairs may exist frequently over open oceans, seas, and large lakes, as well as over extensive desert, ice, snow, prairie, and forest areas of the world. More often, though, land surfaces are characterized by surface inhomogeneities, which make the atmospheric boundary layers over them also nonhomogeneous.

Surface inhomogeneities, which have important effects on atmospheric flows, include boundaries between land and water surfaces (coastlines); the transitions between urban and rural areas or between different types of vegetation; mesoscale oceanic eddies and other regions of varying sea-surface temperature; and hills and valleys. In going over these terrain inhomogeneities, the flow encounters sudden or gradual changes in surface roughness, temperature, wetness, or elevation. Quite often, changes in several surface characteristics occur together, and a complete understanding of modifications in PBL properties (velocity, temperature, and humidity) requires that changes in the surface roughness, temperature, wetness, and elevation be treated simultaneously. For the sake of simplicity and convenience, however, micrometeorologists have studied the effects of changes in the surface roughness, temperature, etc., separately. Due to nonlinearity of the system, however, the individual effects may not simply be additive; the overall effect of a combination of changes in surface characteristics may differ substantially from the sum of individual effects. There are only a few studies of the combined effects of two or more types of surface inhomogeneities occurring simultaneously.

14.2 Step Changes in Surface Roughness

Modification of the boundary layer following a step change in surface roughness normal to the direction of flow has been studied extensively experimentally as well as theoretically. A schematic of the approach flow and the modified flow due to change in the surface roughness is shown in Figure 14.1 for the case of neutral stability. Note that the approach wind profile $U_1(z)$ is a function of the upwind friction velocity u_{*1} and the surface roughness z_{01}. Following the roughness changes from z_{01} to z_{02}, the friction velocity is modified and so is the wind profile near the new surface. Modifications to mean wind profile and turbulence as indicated by computed variances and fluxes are found to be confined to a layer whose thickness increases with distance from the line of discontinuity in the surface. The modified layer is commonly referred to as an internal boundary layer (IBL), because it grows within another boundary layer associated with the approach flow.

Above the IBL, flow characteristics are the same as in the approach flow at the same height above the surface. These are essentially determined by upwind surface characteristics. The influence of the upstream surface may be expected to disappear at a sufficiently large distance from the roughness discontinuity, where the IBL has grown to the equilibrium PBL depth for the new surface, or it is otherwise limited by a strong low-level inversion.

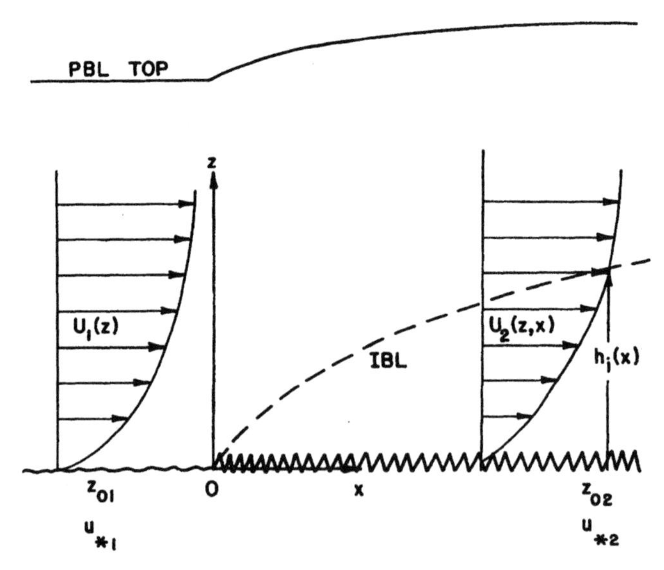

Figure 14.1 Schematic of the internal boundary layer development and wind-profile modification following a step change in the surface roughness.

In some of the earliest experimental studies of flow over a step change in surface roughness, hundreds of bushel baskets or Christmas trees were placed on a frozen lake (Lake Mendota, Wisconsin) in winter and modifications to the near-surface wind and temperature profiles due to the change in roughness were studied (Kutzback, 1961; Stearns and Lettau, 1963). The observed wind profiles did not extend beyond a height of 3.2 m and no independent shear stress or drag measurements were made in these experiments.

A more comprehensive field study was conducted by Bradley (1968), who measured changes in the surface shear stress and wind profiles with distance from the roughness discontinuity in going from a comparatively 'smooth' (but aerodynamically rough) tarmac surface ($z_0 = 0.02$ mm) to a moderately rough surface of wire mesh with spikes ($z_0 = 2.5$ mm), and vice versa. Figure 14.2

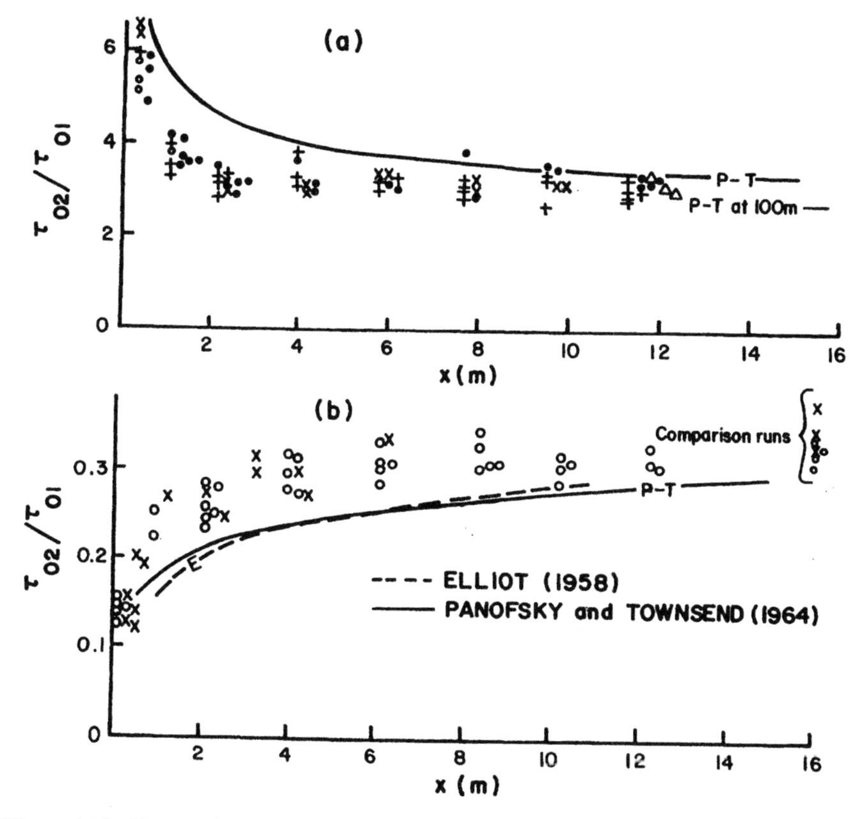

Figure 14.2 Observed variations of the normalized surface shear stress downwind of (a) smooth to rough (from $z_0 = 0.02$ mm to $z_0 = 2.5$ mm) and (b) rough to smooth (from $z_0 = 2.5$ mm to $z_0 = 0.02$ mm) transitions compared with certain theories. [After Bradley (1968).]

shows the variations in surface stress with distance for the two cases (smooth to rough and rough to smooth) of the internal boundary layer flow under near-neutral stability conditions. Here the local stress is normalized by the magnitude of the equilibrium stress for the upwind surface.

Note that the observations of Figure 14.2 indicate that there is a rather sharp change in the surface stress near the roughness discontinuity, with a considerable amount of overshoot above or undershoot below its expected equilibrium value for the new (rougher or smoother) surface. Relaxation to the equilibrium value occurs very rapidly in the first few meters and gradually thereafter. No detectable changes in the surface stress are apparent from observations beyond a distance of 10–15 m from the roughness discontinuity. But a number of theoretical and numerical model studies predict slow adjustment occurring over much longer distances (Elliot, 1958; Panofsky and Townsend, 1964; Rao *et al.*, 1974); some of these theoretical results are shown in Figure 14.2.

Observed changes in the wind profile following step changes in the surface roughness from smooth to rough and vice versa, are shown in Figure 14.3. At any downwind distance from the surface discontinuity, the lower part of the wind profile is representative of the local surface, while the upper part represents the upwind surface.

Within the lowest 10–15% of the PBL the velocity profile in neutral stability may be approximated as (Elliot, 1958)

$$\begin{aligned} U &= (u_{*2}/k)\ln(z/z_{02}), \quad \text{for } z \leq h_i \\ U &= (u_{*1}/k)\ln(z/z_{01}), \quad \text{for } z > h_i \end{aligned} \tag{14.1}$$

In reality, the IBL is not as sharply defined as is implied by the discontinuity (kink) in the velocity profile [Equation (14.1)], and there is a finite blending region around $z = h_i$ in which the velocity profile gradually changes from one form to another. There is also a question about the validity of the log law in the lower layer, especially in the region where the friction velocity might be changing rapidly and has not reached its equilibrium value u_{*2}. The more appropriate equations of mean motion in the lower part of the IBL, where the Coriolis effects can be neglected, are

$$\begin{aligned} U(\partial U/\partial x) + W(\partial U/\partial z) &= \partial \tau_{zx}/\partial z \\ \partial U/\partial x + \partial W/\partial z &= 0 \end{aligned} \tag{14.2}$$

which form the basis for simpler theoretical models. Note that eliminating the pressure gradient and Coriolis acceleration terms from Equation (14.2) would limit their validity to rather short distances from the surface discontinuity.

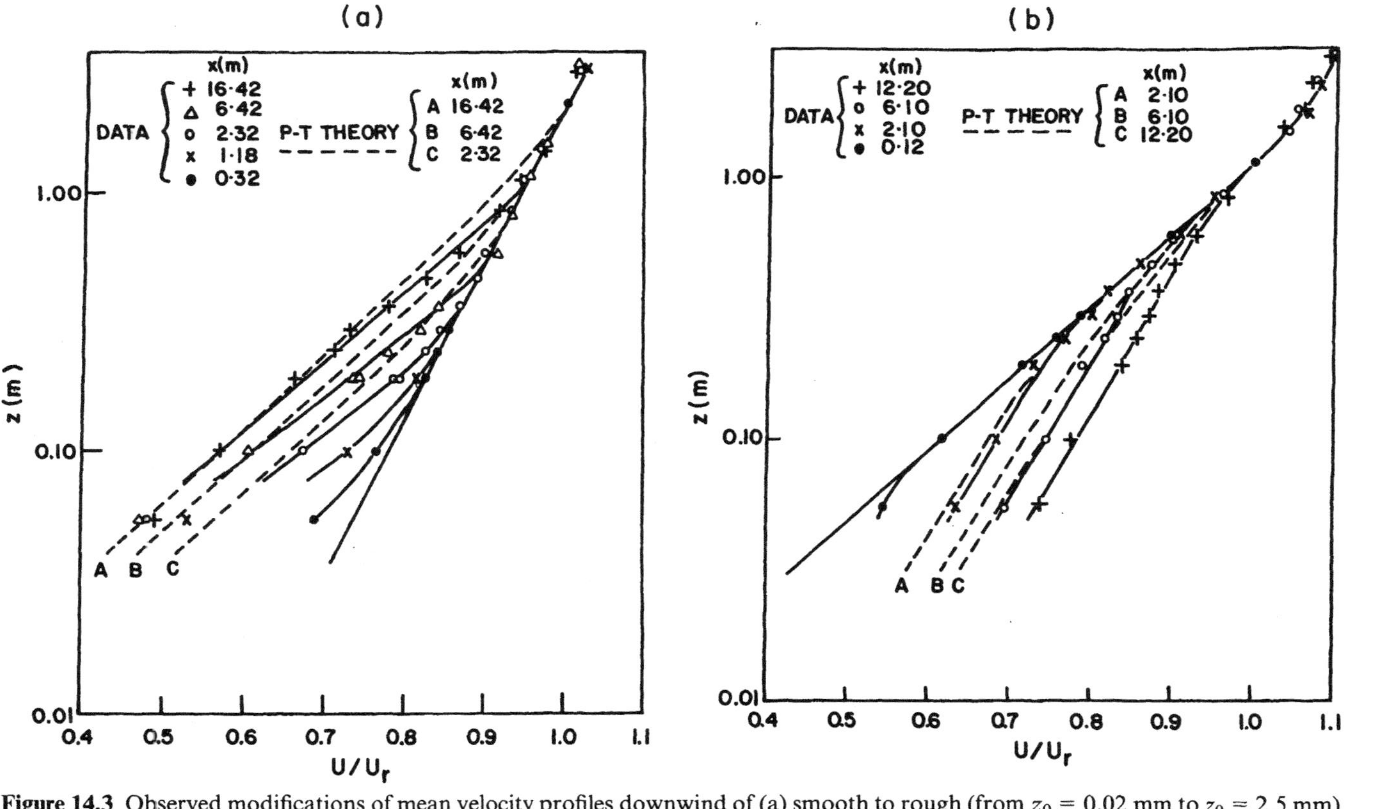

Figure 14.3 Observed modifications of mean velocity profiles downwind of (a) smooth to rough (from $z_0 = 0.02$ mm to $z_0 = 2.5$ mm) and (b) rough to smooth (from $z_0 = 2.5$ mm to $z_0 = 0.02$ mm) transitions compared with Panofsky and Townsend's (1964) theory. [After Bradley (1968).]

According to a number of field and laboratory observations taken under neutral stability conditions, the growth of the internal boundary layer thickness follows an approximate power law

$$h_i/z_{02} = a_i(x/z_{02})^{0.8} \tag{14.3}$$

with estimated values of the empirical constant a_i between 0.35 and 0.75. Actually, a_i depends also on the definition of the IBL thickness. The top of the IBL may be defined as the level where mean wind speed, turbulent momentum flux, or one of the velocity variances reaches a specified fraction (0.90–0.99) of its upstream equilibrium value. An experimental verification of Equation (14.3) is shown in Figure 14.4, using Bradley's (1968) field data and those from a wind tunnel simulation. Interestingly, the value of the exponent in Equation (14.3) is the same as that in the approximate classical relationship for the growth of a turbulent boundary layer over a flat plate parallel to the flow (Schlichting, 1960).

More recent modeling studies of the mean flow and turbulence structure in a neutral boundary layer over a change in surface roughness have also indicated small modifications in the flow upstream of the roughness change or discontinuity (Claussen, 1987). The models based on the TKE closure are found to give much better agreement with experimental data than those based on the first-order (eddy viscosity or mixing length) closure (Claussen, 1988).

Effects of stability on the development of internal boundary layers over warmer and colder than air surfaces have not been investigated systematically. A few experimental and theoretical studies of the IBL under unstable conditions suggest that Equation (14.3) may also be applicable to these conditions, perhaps with a larger value of a_i than for the neutral case. The IBL growth in a stably stratified atmosphere, on the other hand, may be expected to be slower, suggesting smaller values of the coefficient a_i and possibly also of the exponent in Equation (14.3). There is a definite need for further studies of the growth and structure of the IBL under stably stratified conditions.

There have been a few experimental and theoretical studies of the wind-profile modification and the IBL development to large distances of the order of tens of kilometers from the roughness discontinuity, where the Coriolis effects cannot be ignored. It is found that there is a gradual shifting of surface wind direction in a cyclonic or anticyclonic sense in going from smooth to rough or rough to smooth surface, respectively. For example, wind direction shifts of 10–20° have been observed in air flowing from rural to urban areas and vice versa (Oke, 1974). Such changes in the surface wind direction can be explained in terms of the observed dependence of the surface cross-isobar angle (α_0) on surface roughness (α_0 increases with surface roughness, as discussed in Chapter 6). These wind direction changes have important implications for plume trajectories crossing urban and rural boundaries, as well as shorelines. They

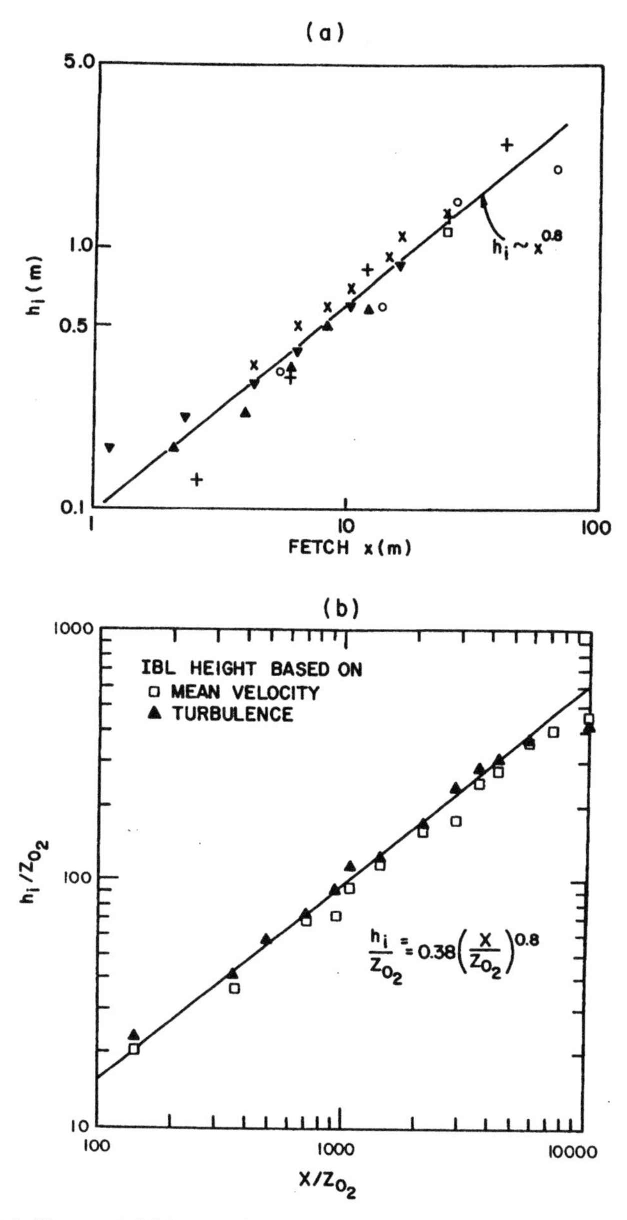

Figure 14.4 Observed thickness of the internal boundary layer as a function of distance downwind of a step change in the surface roughness based on (a) Bradley's (1968) field data and (b) Pendergrass and Arya's (1984) wind tunnel data. [(b) Reprinted with permission from *Atmospheric Environment*, Copyright © (1984), Pergamon Journals Ltd.]

may also result in distribution of horizontal divergence and convergence and associated distributions of clouds and precipitation.

Many natural surfaces actually present a series of step changes in surface roughness corresponding to different land uses. In principle, one can think of an IBL associated with each step change in surface roughness, so that after a few steps the PBL would be composed of several IBLs. The wind profile is not simply related to the local roughness, except in a shallow surface layer, but represents the integrated effects of several upwind surfaces, depending on the layer (height range) of interest. Observed wind profiles under such conditions often show kinks (discontinuities in slope) when U is plotted against log z; each kink may be associated with an interface between the adjacent IBLs. Such a fine tuning of the wind profile and identification of IBL interfaces may not be possible, however, unless the wind profile is adequately averaged, well resolved, and taken under strictly neutral conditions (most routine wind soundings do not meet these requirements).

14.3 Step Changes in Surface Temperature

Differences in albedos, emissivities, and other thermal properties of natural surfaces lead to their different surface temperatures, even under a fixed synoptic weather setting. Consequently, inhomogeneities in surface temperature may occur independently of changes in surface roughness. Most dramatic changes in surface temperature occur across lake shorelines and sea coastlines. In response to these abrupt changes in surface characteristics, thermal internal boundary layers develop over land during onshore flows and over water during offshore flows. Thermal boundary layers also develop in air flow from rural to urban areas and vice versa. On still smaller scales, thermal internal boundary layers develop near the boundaries of different crops, as well as across roads and rivers (Rider *et al.*, 1963).

When temperature differences between two or more adjacent surfaces are large enough and the ambient (geostrophic) flow is weak, thermally induced local or mesoscale circulations are likely to develop on both sides of surface temperature discontinuity. The best known examples of such circulations are the land and sea breezes. Internal boundary layers are often embedded in such thermally driven mesoscale circulations.

14.3.1 Thermal IBL growing over a warmer surface

Perhaps the most dramatic changes in the atmospheric boundary layer occur when stably stratified air flowing over a cold surface encounters a much warmer (relative to air) surface. This frequently occurs in coastal areas during midday and afternoon periods; the resulting flow is called a sea breeze or lake breeze. A similar situation on a large scale occurs during periods of cold-air outbreaks over relatively warm waters (e.g., Great Lakes, Gulf Stream, and Kuroshio) in late fall or winter.

If the temperature difference between downwind and upwind surfaces is fairly large (say, $\Theta_{02} - \Theta_{01} > 5$ K), the thermal internal boundary layer (TIBL) developing over the warmer surface would most likely be very unstable or convective (Figure 14.5). The growth of such a TIBL can be described by a simple mixed-layer model based on the mean thermodynamic energy equation, under stationary conditions

$$U(\partial\Theta/\partial x) = -(1/\rho c_p)(\partial H/\partial z) \tag{14.4}$$

Integrating Equation (14.4) with respect to z from 0 to h_i gives

$$h_i U_m(\partial\Theta_m/\partial x) = (1/\rho c_p)(H_0 - H_i) \tag{14.5}$$

in which U_m and Θ_m are the mixed-layer averaged wind speed and potential temperature and H_0 and H_i are the heat fluxes at the surface and at the top of the TIBL or mixed layer ($z = h_i$), respectively. An equation for the growth of the TIBL with distance x from the temperature discontinuity can be obtained after making reasonable assumptions about the potential temperature profile and the heat flux profile, as shown in Figure 14.5. From these it is clear that

$$\begin{aligned} \Theta_m &= \Theta_{01} + \gamma h_i - \Delta\Theta \\ \partial\Theta_m/\partial x &= \gamma(\partial h_i/\partial x) \end{aligned} \tag{14.6}$$

where $\gamma \equiv (\partial\Theta/\partial z)_0$ is the potential temperature gradient in the approach flow, which is also the gradient above the TIBL. Here, we have assumed that the jump in potential temperature ($\Delta\Theta$) at the top of the TIBL remains constant. If one also assumes that the downward heat flux at the top of the TIBL is a constant fraction of the surface heat flux, i.e., $H_i = -AH_0$, Equations (14.5) and (14.6) yield an expression for the height of the TIBL

$$h_i = a_1(H_0 x/\rho c_p U_m \gamma)^{1/2} \tag{14.7}$$

in which $a_1 = (2 + 2A)^{1/2} \cong 1.5$ is an empirical constant. The use of Equation (14.7) requires the knowledge of the surface heat flux and the mixed-layer wind speed over the downwind heated surface, in addition to that of γ.

An alternative, simpler formula is obtained by expressing or parameterizing the surface heat flux as $H_0 = \alpha\rho c_p U_m(\Theta_{02} - \Theta_{01})$, where $\alpha \cong 2.1 \times 10^{-3}$ is an

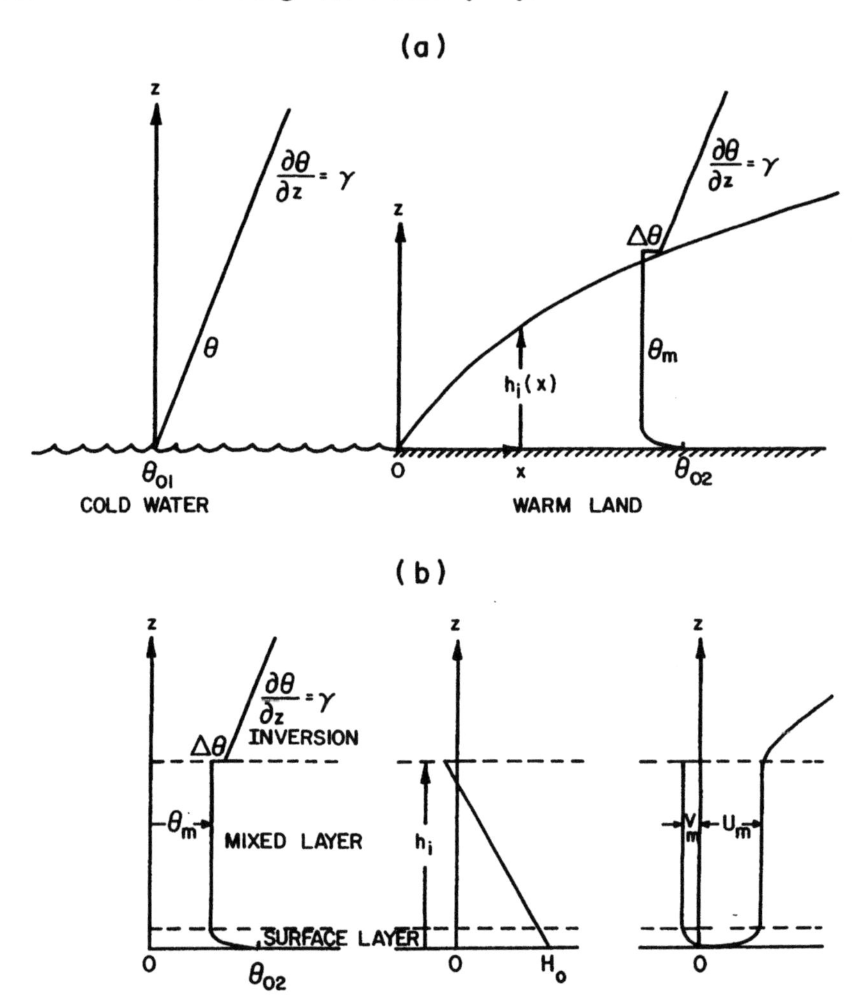

Figure 14.5 (a) Schematic of a thermal internal boundary layer developing when a stable air flow encounters a much warmer surface, and (b) idealized potential temperature, heat flux, and wind profiles in the same.

empirical coefficient (Vugts and Businger, 1977). Then, Equation (14.7) can be expressed as

$$h_i = a_2[(\Theta_{02} - \Theta_{01})x/\gamma]^{1/2} \tag{14.8}$$

where $a_2 = [2(1 + A)\alpha]^{1/2} \cong 0.1$. This does not require the knowledge of surface heat flux or mixed-layer wind speed. The latter is introduced in the TIBL height expression if one uses the more conventional bulk transfer formula for the surface heat flux

$$H_0 = C_H \rho c_p U_m (\Theta_{02} - \Theta_m)$$

with $C_H \approx C_D = (u_*/U_m)^2$. This type of parameterization leads to the TIBL height formula

$$h_i = a_3 (u_*/U_m)[(\Theta_{02} - \Theta_m)x/\gamma]^{1/2} \tag{14.9}$$

with $a_3 \cong [2(1 + A)C_H/C_D]^{1/2} \cong 2.0$. The above expression has the disadvantage of containing the x-dependent variables U_m and Θ_m of the modified mixed layer. Implicit or explicit relations for the variation of Θ_m with x have been given (Fleagle and Businger, 1980). Using some additional assumptions, the TIBL height can also be expressed as (Venkatram, 1977)

$$h_i = a_4 (u_*/U_m)[(\Theta_{02} - \Theta_{01})x/\gamma]^{1/2} \tag{14.10}$$

where $a_4 \cong 1.7$. Some of the above expressions for h_i have been compared and verified against observations (Raynor *et al.*, 1979; Stunder and SethuRaman, 1985). Equation (14.7) appears to be the best, because its derivation involves the least number of assumptions.

The above-mentioned formulations of the TIBL growth are based on mixed-layer assumptions for the TIBL and neglect of the near-surface layer, in which velocity and potential temperature generally have strong gradients. These are not expected to be valid for short (less than 1 km) distances from the temperature discontinuity, where modifications are likely to be confined to the shallow surface layer. Detailed experimental investigations of air modification in the surface layer in the immediate vicinity of a step change in surface temperature have been reported by Rider *et al.* (1963) and Vugts and Businger (1977). Here, the growth of the TIBL is more rapid and more closely follows the $x^{0.8}$ behavior, similar to that of an IBL following a step change in roughness (Elliot, 1958). Turbulent characteristics of a shallow convective internal boundary layer have been studied by Smedman and Hogstrom (1983).

A number of experimental and numerical modeling studies have been conducted on large lake/sea breezes and thermal internal boundary layers associated with such breezes (Pielke, 1984; Arritt, 1987). It is found that water surface temperature has little effect on lake breeze development, as long as the water is cold enough that the boundary layer over the lake is stably stratified. If the lake surface is warmer than air, the more vigorous fluxes due to convective turbulence produce major changes in the lake breeze (Arritt, 1987). Lake and sea breezes are more commonly observed during late spring and summer seasons when land surface gets rapidly heated up during mid-day and afternoon hours.

14.3.2 Thermal IBL growing over a colder surface

As compared to the rapidly growing and highly energetic unstable or convective TIBL developing in cold air advecting over a much warmer surface, the reverse situation of comparatively shallow and smooth stable TIBL that develops in warm air advection over a colder surface has received very little attention. An early observational study was made by Taylor (1915) on the ice-scout ship SS Scotia near the coast of Newfoundland. More recently, Raynor *et al.* (1975) have reported on some measurements of the depth of the stable TIBL at the southern shore of Long Island, New York, as a function of fetch. The TIBL resulted from modification of warm continental air following a southwesterly trajectory over colder oceanic waters; its depth could be represented by an empirical relationship

$$h_i = a_5(u_*/U)[(\Theta_{01} - \Theta_{02})x/|\partial T/\partial z|]^{1/2} \tag{14.11}$$

which is similar to Equation (14.10) for the unstable TIBL. Here, $|\partial T/\partial z|$ is the absolute value of the lapse rate or temperature gradient over the source region or above the inversion in the modified air, U is the mean wind speed at a reference level (say, 10 m) near the surface, and a_5 is an empirical constant of the order of unity which may depend on the choice of the reference level for measuring or specifying wind speed.

Note that despite the apparent similarities of the TIBL height equations [Equations (14.10) and (14.11)] in different situations of cold- and warm-air advection, respectively, for a given fetch the magnitudes of h_i and $\partial h_i/\partial x$ are found to be much smaller in the latter case. These and other IBL height equations showing unrestricted growth of internal boundary layers with fetch or distance from the discontinuity in the surface roughness, temperature, etc., must cease to be valid beyond a certain distance where the IBL has grown to the equilibrium depth of the PBL for the given surface and external conditions. Observations indicate that this might take tens to hundreds of kilometers, depending on the equilibrium PBL depth.

Example Problem 1

Calculate and compare the development of thermal internal boundary layers in the following two different situations:

(a) A stably stratified boundary-layer flow with $\partial\Theta/\partial z = 0.05$ K m^{-1} overland advecting over a 10°C warmer sea.

(b) An unstable continental boundary-layer flow with an average lapse rate of 0.015 K m^{-1} in the surface layer advecting over a 10°C colder sea surface. Assume a typical value of $C_D = 0.5 \times 10^{-3}$ for the stably stratified TIBL.

Solution

(a) The appropriate expression for the TIBL height in this case is Equation (14.8)

$$h_i = a_2[(\Theta_{02} - \Theta_{01})x/\gamma]^{1/2}$$

in which $\Theta_{02} - \Theta_{01} = 10$ K, $\gamma = 0.05$ K m^{-1}, and $a_2 \cong 0.1$. Substituting these values in the above expression, we get

$$h_i \cong 1.414x^{1/2}$$

in which x is in meters. The computed TIBL heights at various distances offshore are:

x (m):	100	200	500	1000	2000	5000	10 000	20 000
h_i (m):	14.1	20.0	31.6	44.7	63.2	100	141	200

(b) The appropriate expression for the TIBL height growing over the colder sea surface is Equation (14.11)

$$h_i = a_5 \frac{u_*}{U} \left[\frac{(\Theta_{01} - \Theta_{02})x}{|\partial T/\partial z|} \right]^{1/2}$$

where, $a_5 \cong 1.0$, $u_*/U = C_D^{1/2} = 0.02236$, $\Theta_{01} - \Theta_{02} = 10$ K, and $\partial T/\partial z = -0.015$ K m^{-1}.
Substituting these values in the above expression, we obtain

$$h_i \cong 0.577x^{1/2}$$

in which x is in meters. The computed heights of the stable TIBL at various distances offshore are:

x (m):	100	200	500	1000	2000	5000	10 000	20 000
h_i (m):	5.7	8.2	12.9	18.2	25.8	40.8	57.7	81.6

Note that the stable TIBL growing over the colder sea surface is much shallower than the unstable TIBL growing over the warmer sea.

14.4 Air Modifications over Water Surfaces

When continental air flows over large bodies of water, such as large lakes, bays, sounds, seas, and oceans, it usually encounters dramatic changes in surface roughness and temperature. Consequently, significant modifications in air temperature, specific humidity, cloudiness, winds, and turbulence occur in the developing IBLs with distance from the shoreline. Most dramatic changes in air properties and ensuing weather phenomena occur when cold air flows over a much warmer water surface. Intense heat and water vapor exchanges between the water surface and the atmosphere lead to vigorous convection, formation of clouds, and, sometimes, precipitation. Some examples of this phenomenon are frequent cold-air outbreaks over the warmer Great Lakes in late fall and early winter and the resulting precipitation (often, heavy snow) over immediate downwind locations. It has been the focus of several field experiments, including the 1972 International Field Year of the Great Lakes (IFYGL). Figure 14.6 presents the observed changes in potential temperature and cloudiness across Lake Michigan during a cold-air outbreak when the water–air temperature difference was about 12.5 K (Lenschow, 1973). Note the rapid warming and

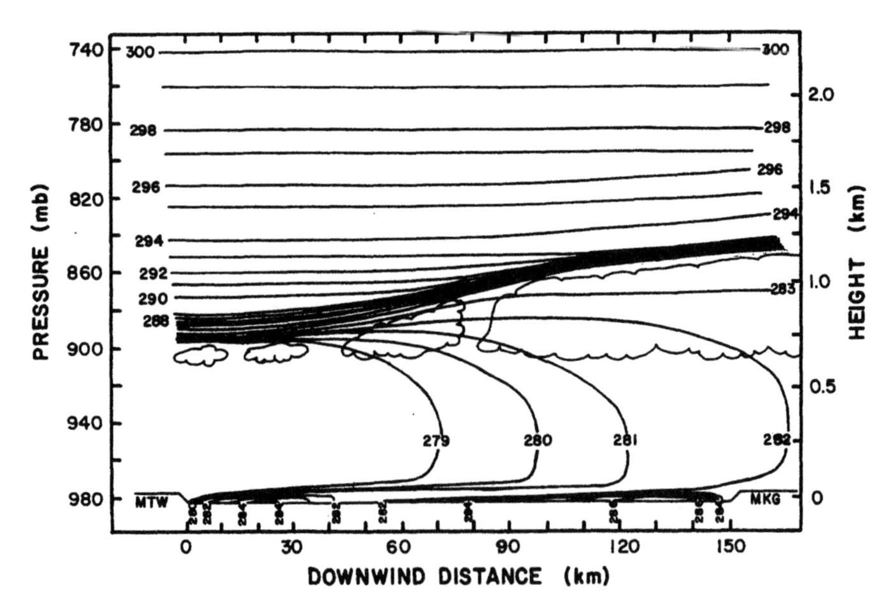

Figure 14.6 Observed modifications in potentials temperature and cloudiness across Lake Michigan during a cold-air outbreak on November 5, 1970. Mean wind is from left to right. [After Lenschow (1973).]

moistening of the approaching cold, dry air mass by the warmer lake surface water.

Other major investigations of air modification during cold-air outbreaks over warm waters have been conducted during the 1974 and 1975 Air Mass Transformation Experiments (AMTEX) over the Sea of Japan and a more recent (1986) Genesis of Atlantic Lows Experiment (GALE). In the latter experiment was observed one of the most intense cold-air outbreaks with air–sea temperature differences of 20–25 K over the Gulf Stream just off the North Carolina coast. The ocean-surface temperature here varies considerably with distance from the coast, with several step changes between the coast and the eastern edge of the Gulf Stream. Several research aircraft, ships, and buoys were used to study air-mass modifications in the boundary layer at different locations. The cold-air outbreak of January 28, 1986, was most spectacular in that the ocean surface was enshrouded in a steamlike fog with abundant number of steam devils, water spouts, and other vortexlike filaments visible from low-flying aircraft. The PBL depth increased from about 900 m at 35 km offshore to 2300 m at the eastern edge of the Gulf Stream (284 km offshore). At the same time, the cloudiness increased from zero to a completely overcast, thick deck of stratocumulus with snow and rainshowers developing over the eastern edge of the Gulf Stream. The subcloud mixed layer was characterized by intense convective turbulence with strong updrafts and downdrafts (Wayland and Raman, 1989).

A simple theory of predicting changes in air temperature and specific humidity as a function of fetch for the case of cold-air advection over a warm sea, but in the absence of condensation and precipitation processes in the modified layer, is given by Fleagle and Businger (1980). When the top of the TIBL reaches above the lifting condensation level, however, the structure of the TIBL changes dramatically. Complex interactions take place between cloud particles, radiation, and entrainment at the top of the TIBL. These and complex topographical effects can be considered in numerical mesoscale PBL models (Huang and Raman, 1990).

Modifications of warm continental air advecting over a cold lake or sea surface are equally dramatic, but the modified layer is much shallower. Turbulence is considerably reduced in this shallow, stably stratified TIBL, while visibility may be reduced due to fog formation under appropriate conditions. Cases of strong warm-air advection commonly occur in springtime over the Great Lakes and along the east coast of the United States and Canada. Figure 14.7 presents some observations of air modification under such conditions prevailing over Lake Michigan (Wylie and Young, 1979). These were made from a ship traveling in the downwind direction across the lake, with instruments attached to a tethered balloon. Note that the surface-based inversion that formed over water rose to about 140 m after a fetch of 50 km

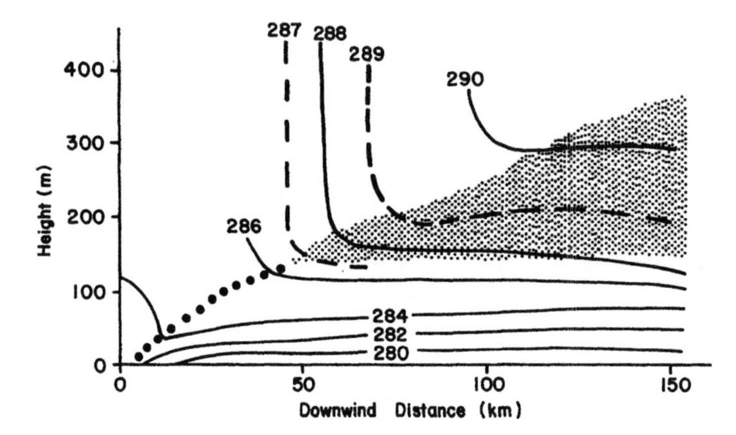

Figure 14.7 Observed modifications in potential temperature over Lake Michigan during warm-air advection. Surface-based inversion is unshaded area below the dotted line; shading denotes stable isothermal layer. [After Wylie and Young (1979). Copyright © (1979) by D. Reidel Publishing Company. Reprinted by permission.]

and remained at a constant level for the rest of the fetch along the lake. A growing isothermal layer separated the surface inversion layer from the adiabatic mixed layer advected from upwind land areas. Large variations in wind speed and direction were observed within the inversion layer. Note that the thermal internal boundary layer, including the isothermal layer, kept growing over the whole fetch of more than 150 km across the lake.

14.5 Air Modifications over Urban Areas

Urbanization, which includes residential, commercial, and industrial developments, produces radical changes in radiative, thermodynamic, and aerodynamic characteristics of the surface from those of the surrounding rural areas. Therefore, it is not surprising that as urbanization proceeds the associated weather and climate often are modified substantially. Such modifications are largely confined to the so-called urban boundary layer, although the urban 'plume' of pollutants originating from the city may extend hundreds of kilometers downwind. Many studies of urban influences have identified significant changes in surface and air temperatures, humidity, precipitation, fog, visibility, air quality, surface energy fluxes, mixed-layer height, boundary layer winds, and turbulence between the urban and rural areas. Here, we give a brief description of some of the better known urban-induced phenomena.

14.5.1 The urban heat island

The most frequently observed (or felt) and best documented climatic effect of urbanization is the increase in surface and air temperatures over the urban area, as compared to rural surroundings. Closed isotherms over the urban and suburban areas generally separate these areas from the rural environs. This condition or phenomenon has come to be known as the urban heat island, because of the apparent similarity of isotherms to contours of elevation for a small, isolated island in the ocean. This analogy is carried further in the schematic representation of near-surface temperature in Figure 14.8 for a large city on a clear and calm evening, as one travels from the countryside to the city center. This shows a typical 'cliff' of steep rise in temperature near the rural/suburban boundary, following by a 'plateau' over much of the suburban area, and then a 'peak' over the city center (Oke, 1987, 1995). The maximum difference in the urban peak temperature and the background rural temperature defines the urban heat-island intensity (ΔT_{u-r}). Over large metropolitan areas, there may be several plateaus and peaks in the surface temperature trace. Also, there are likely to be many small-scale variations in response to distinct intraurban land uses, such as parks, recreation areas, and commercial and industrial developments, as well as topographical features, such as lakes, rivers, and hills. Strictly urban influences on the local weather and climate can be studied in an isolated manner only in certain interior (as opposed to coastal) cities in relatively flat terrain. For this reason, St Louis, Missouri, was chosen to be the arena for several large field studies, viz., Metropolitan Meteorological Experiment (METROMEX, 1971–1976) and Regional Air Pollution Study (RAPS, 1973–1977), of urban influences on local weather and climate.

Figure 14.9 shows a map of the St Louis metropolitan area representing major land-use types. The principal urban area is located about 16 km south of the confluence of the Mississippi and Missouri rivers. The primary topographic

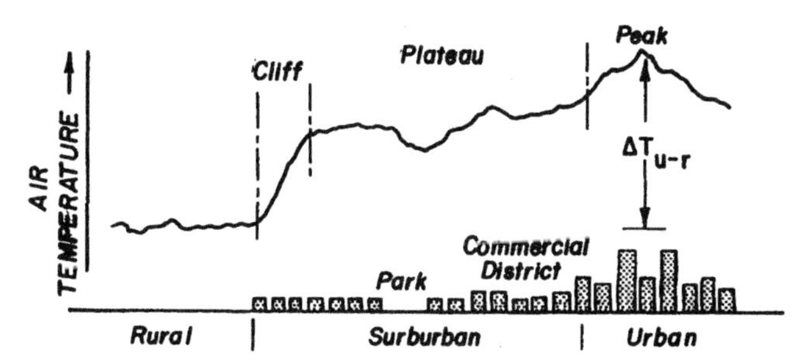

Figure 14.8 Schematic representation of variation in air temperature in going from a rural to an urban area. [After Oke (1987).]

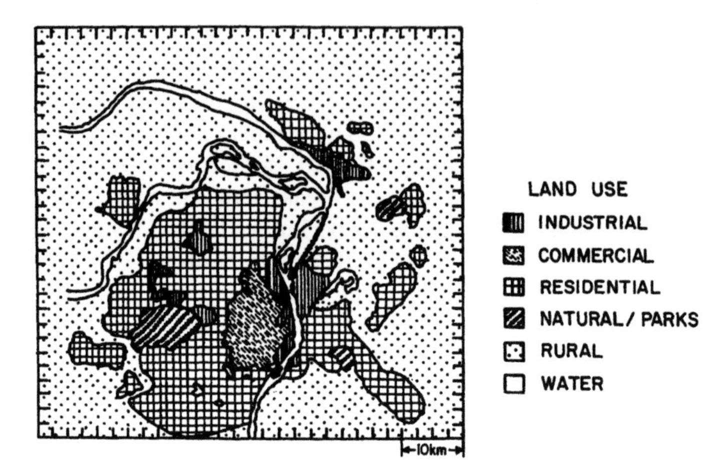

Figure 14.9 Land use map of the Saint Louis Metropolitan area during RAPS. [After Byun (1987).]

features in the area are the river valley with a prominent bluff rising 20–60 m above the flood plain east of the merged river and heavily wooded (70–90 m in height) hills to the southwest. A slightly smoothed pattern of aircraft-observed (with an infrared radiometer) ground-surface temperatures on a clear, summer afternoon is shown in Figure 14.10. Note that it clearly shows the urban heat-island phenomenon with an intensity of $\Delta T_{u-r} \cong 8.4$ K at that particular time. The difference in air temperatures at a 10 m height is likely to be much smaller.

Other observations of the urban heat island over small and large cities have been reviewed by Landsberg (1981) and Oke (1974, 1995). The intensity of an urban heat island depends on many factors, such as the size of city and its energy consumption, geographical location, month or season, time of day, and synoptic weather conditions. The maximum intensity for a given city occurs in clear and calm conditions (e.g., under a stationary high), a few hours after sunset. The intensity becomes minimal or zero under highly disturbed (stormy), windy weather conditions.

Under reasonably stationary synoptic weather conditions, the heat-island intensity shows a pronounced diurnal variation with a minimum value around midday and maximum value around or before midnight (Oke, 1987, Chapter 8). There are likely to be marked differences in the diurnal variations of ΔT_{u-r} between winter and summer related to the anthropogenic heat release.

Some attempts have been made to correlate the maximum nocturnal heat-island intensity (based on air temperatures at 10 m above ground level) with population (considered as a surrogate for the city size and energy consumption) and near-surface wind speed. Figure 14.11 shows different relationships of

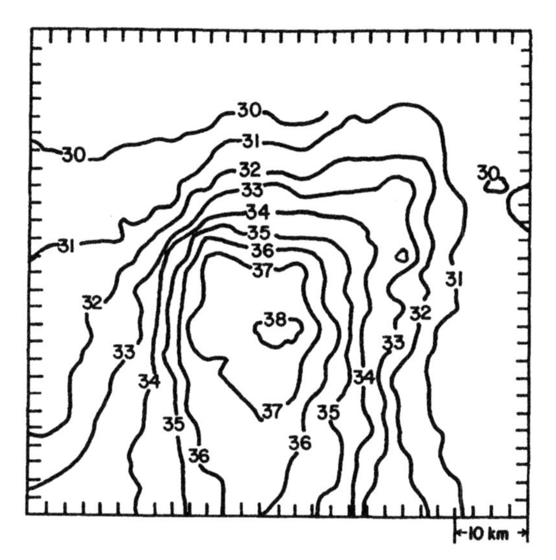

Figure 14.10 Observed patterns of ground-surface temperatures showing the heat-island phenomenon over the Saint Louis Metropolitan area during RAPS. [After Byun (1987).]

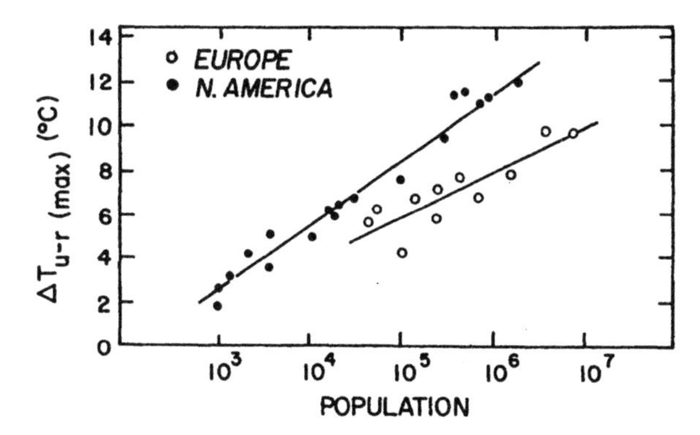

Figure 14.11 Empirical relation between the observed heat-island intensity and city population for North American and European cities. [After Oke (1987).]

$\Delta T_{(u-r)\max}$ with population (P) for some of the cities of Europe and North America with minimal topographical influences. Different curves probably reflect differences in the per capita energy consumption and in population density between European and North American cities. The data and regression lines in Figure 14.11 are for calm and cloudless conditions. For moderate wind speeds, ΔT_{u-r} is found to decrease in inverse proportion to the square root of regional (nonurban) wind speed, and vanishes at a 'critical' wind speed of about

$9\ \text{m s}^{-1}$ (measured at a height of 10 m at a rural site) for large cities (Oke, 1987, Chapter 8).

The use of city population alone is not very satisfactory in explaining the maximum intensity of urban heat island. A much stronger correlation and apparently unique relationship is obtained between the maximum heat-island intensity and the geometry of the street canyons in the city center, as characterized by the average height to width (H/W) ratio. The following regression relationship has been obtained from the data from 31 cities in North America, Europe, and Australia (Oke, 1987, Chapter 8):

$$\Delta T_{\text{u-r(max)}} = 7.54 + 3.97 \ln(H/W) \tag{14.12}$$

This relationship indicates that urban canyon geometry exerts a fundamental control on the urban heat island. Urban geometry and the density of development influence physical processes such as the trapping of both incoming solar and outgoing longwave radiations, the amount of anthropogenic heat released in the urban area, and turbulent transports within urban canyons.

The extent and intensity of the urban heat island is also greatly influenced by topography and the presence of large bodies of water within or around the urban area. In coastal cities, land and sea breezes have a profound influence, while in hilly areas nocturnal drainage and valley flows may dominate over the strictly urban influences.

What are the causes for the urban heat-island phenomenon? A number of causes have been hypothesized and most of them are verified by observations. The leading candidates are (Oke, 1987, Chapter 8) as follows:

- Increased incoming longwave radiation ($R_{L\downarrow}$) due to absorption of outgoing longwave radiation and re-emission by polluted urban atmosphere.
- Decreased outgoing longwave radiation loss ($R_{L\uparrow}$) from street canyons due to a reduction in their sky view factor by buildings.
- Increased shortwave radiation (R_S) absorption by the urban canopy due to the effect of street canyons on albedo.
- Greater daytime heat storage (ΔH_s) due to the thermal properties of urban materials and heat release at nighttime.
- Addition of anthropogenic heat (H_a) in the urban area in process emission (heating and cooling), transportation, and industrial operations.
- Decreased evaporation and, hence, the latent heat flux (H_L) due to the removal of vegetation and surface waterproofing of the city.

In short, all the components of the energy balance in the urban canopy are modified in such a way that they add to the heat-island effect in the positive sense. Some of these are effective only during daytime when they contribute to the storage of heat in the urban canopy; heat release at nighttime keeps the urban air warmer. The relative role of each component is likely to differ from one season to another and also from one city to another.

14.5.2 The urban boundary layer

The urban boundary layer (UBL) is modified by both the urban heat island and the increased roughness (one can also think of an urban area as a 'roughness island'). In the absence of any topography, the roughness elements of a city are its buildings, whose heights (h_0) generally increase toward the city center. Practical considerations make it difficult to determine the value of the roughness length or parameter (z_0) directly from wind-profile measurements over a large city center or commercial area; such observations should be made above the prevailing height of the buildings. But rough estimates can be made by extrapolating the z_0/h_0 and d_0/h_0 relationships discussed in Chapter 10 (as in the case of tall vegetation, the roughness parameterization of an urban canopy must include a zero-plane displacement). In an urban area, the roughness parameter may vary from fractions of a meter in suburbs to several meters over the city center (Grimmond and Oke, 1999).

In near-calm or weak-wind and clear-sky conditions, thermal modification of the boundary layer in response to the urban heat island is likely to dominate over increased roughness effects. The thermally induced circulation that is superimposed on any weak background flow is radially inward toward the city center at lower levels and outward from the city center at upper levels, with a rising motion over the center and subsidence over the surrounding environs. During the day this circulation may extend up to the base of the lowest inversion, which may then acquire a dome shape in response to the induced circulation (rising and subsiding motions). Since mixing is likely to be restricted to the mixed layer below the inversion, it is also referred to as the 'mixed-layer dome,' or 'dust dome,' in which dust, smoke, and haze from urban emissions accumulate during stagnant conditions. Figure 14.12 depicts a schematic of thermally induced circulation and dust dome over an urban area. Figure 14.13 shows the same (mixed-layer dome) with aircraft-observed profiles of potential temperature and specific humidity in the UBL of St Louis, Missouri (Spangler and Dirks, 1974). At that time, winds were nearly uniform in the lowest 500 m, variable above this height up to the inversion base, and again uniform at 5 m s^{-1} above the inversion. Note that a rural-to-urban variation in mixed-layer height of at least 400 m was observed even when the urban heat-island intensity was not particularly strong ($\Delta T_{u-r} < 1$ K).

Aircraft measurements of turbulent fluxes and variances in the daytime UBL also indicate considerable urban-scale variations of these parameters (Ching,

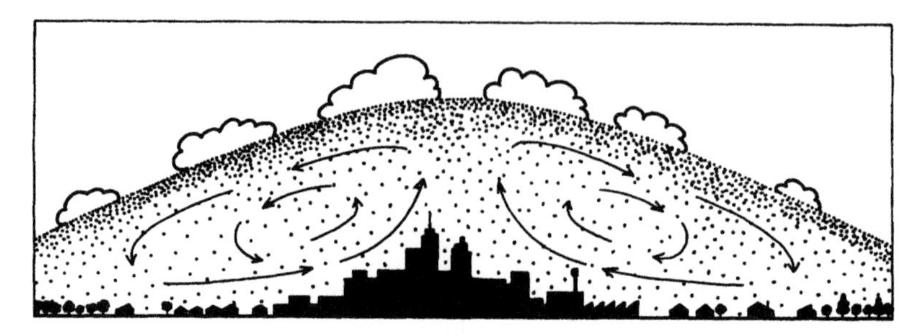

Figure 14.12 Schematic of dust dome and thermally induced circulations over a large city under calm or light winds. [After Lowry (1967). Copyright © (1967) by Scientific American, Inc. All rights reserved.]

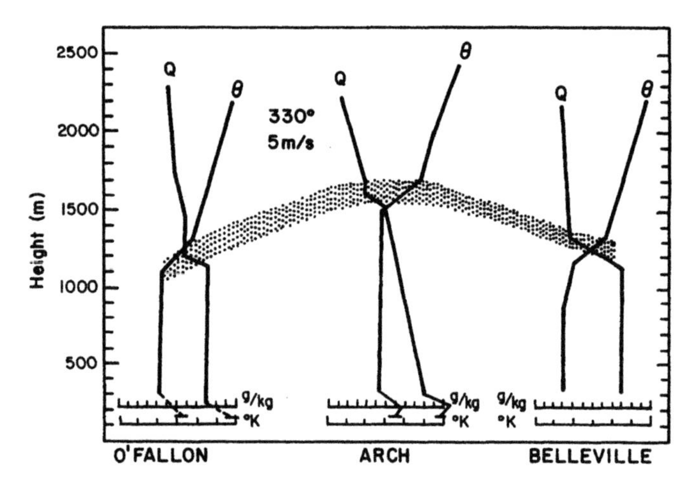

Figure 14.13 Observed mixed-layer dome and potential temperature and specific humidity profiles over Saint Louis on 12 August 1971. [After Spangler and Dirks (1974). Copyright © (1974) by D. Reidel Publishing Company. Reprinted by permission.]

1985). For example, observations over St Louis show that the sensible heat flux varied by a factor of 2 to 4, with the largest values over the city center. The latent heat flux also varied by a factor of 4, but with smallest values over the city center. Consequently, the Bowen ratio exhibited even larger spatial variability, with a maximum value of about 1.8 over the city center and a minimum value of less than 0.2 over some nonurban areas. Spatial patterns of turbulent velocity variances are similar to those of the sensible heat flux. There are also large differences observed in average summertime heat fluxes in suburbs of North American cities located in different geographic regions (Grimmond and Oke, 1995).

The structure and dynamics of the daytime urban boundary layer are similar to the convective boundary layer over the surrounding rural area, except for the UBL being more turbulent, warm, dry, and polluted. The potential temperature profiles above the roof level are characterized by negative gradients in the shallow surface layer, very small (nearly zero) gradients in the much deeper urban mixed layer, and large positive gradients in the capping inversion layer. The urban heat-island intensity is relatively small in magnitude but extends through a significant portion of the mixed layer and up to tens of kilometers downwind of the city (Oke, 1995). Mean wind profiles also have large gradients in the surface layer including the urban canopy layer, relatively small gradients in the mixed layer, and much larger gradients in the capping inversion.

At night the urban boundary layer shrinks to a depth of a few hundred meters because strong stability of the approach flow suppresses turbulent mixing in the vertical direction. Still, the UBL over the city center is much thicker than the nocturnal stable boundary layer upwind of the city. The combination of increased temperatures (heat island) and increased roughness over a moderate-size and a large-size city can easily destroy the nocturnal surface inversion and dramatically modify the nocturnal boundary layer as it advects over the city. This process or phenomenon is illustrated in Figure 14.14, which shows a helicopter-measured along-wind vertical temperature cross-section, as well as the vertical potential temperature profiles at different locations in Montreal, Canada, on a winter morning. Other measurements of air temperature from automobile traverses in the city indicated an urban heat-island intensity of about 4.5 K at that time (Oke and East, 1971).

Figure 14.14 also shows that air is progressively warmed as it traverses the urban area (the weak flow from the N/NE was along the main helicopter traverse route), and the depth of the modified urban boundary layer increases downwind. At the rural site (sounding location 7) the atmosphere is very stable from the surface up to at least 600 m. In moving over the urban area, the bottom (urban canopy) layer becomes unstable and the layers above become neutral or weakly stable. Above the top of the UBL, the capping inversion layer retains the characteristics of the 'rural' approach flow. The UBL attains its maximum depth (300 m) over the city center and appears to become shallower farther downwind of the city center. A surface inversion may also reform over the downwind rural area. The newly developing rural boundary layer downwind of the city is, thus, a stably stratified internal boundary layer which grows slowly with distance. The modified urban air advecting above this shallow surface inversion layer is called the urban plume.

In comparison to the strong diurnal changes in stability of the lower atmosphere in surrounding rural areas, the urban atmosphere experiences only small diurnal variations in stability. The UBL remains well mixed both by day and by night, although the mixing depth usually undergoes a large diurnal oscillation.

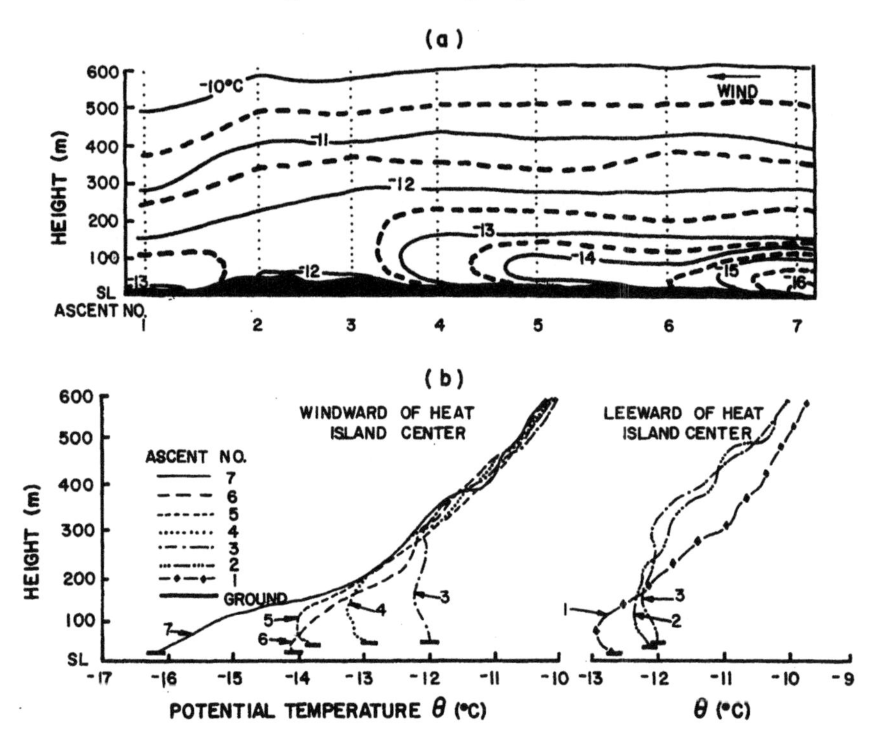

Figure 14.14 Modifications of potential temperature in the UBL due to the urban heat island over Montreal on a winter morning. (a) x-z cross-sectional distribution of Θ, and (b) vertical profiles of Θ at the various locations along wind. [After Oke and East (1971). Copyright © (1971) by D. Reidel Publishing Company. Reprinted by permission.]

The difference in the stabilities of the rural and urban boundary layers at nighttime explains why surface winds are often greater in the latter. Strong inversion reduces the near-surface winds in the rural area and essentially decouples them from stronger winds aloft. More efficient vertical mixing in the UBL, on the other hand, results in increased momentum (winds) in the surface layer.

Observational studies of the UBL have shown that the urban heat-island phenomenon, quantified by the urban–rural temperature difference, extends through the whole depth of the UBL. However, the heat-island intensity is maximum at the surface, decreases with height, and vanishes at the top of the UBL. There is also some evidence of the so-called cross-over effect, according to which the heat-island intensity becomes slightly negative above the top of the UBL (Oke and East, 1971; Oke, 1995). This effect is observed only at low wind speeds ($U < 3$ m s^{-1}).

A simple but appropriate relationship between the intensity of a nocturnal urban heat island and the maximum mixing depth over a city is

$$h_u = (\Theta_u - \Theta_r)/(\partial\Theta_r/\partial z) \cong (T_u - T_r)/[(\partial T_r/\partial z) + \Gamma] \qquad (14.13)$$

in which T_u and T_r are urban and rural air temperatures near the surface. It is based on the assumptions that the rural temperature sounding can be characterized by a constant gradient, at least above the shallow surface inversion layer, and there is no large-scale cold- or warm-air advection. Good agreement has been found between the predicted and observed values of h over New York, Montreal, and other cities. More sophisticated models will be needed to predict the growth of the UBL in approaching the city center from the edge of the city.

The urban heat island exerts strong influences on both the mean flow and turbulence structure of the UBL. The thermal anomaly represented by the city affects the approaching airflow by altering the local pressure field, by modifying the stability, and by increasing turbulence. The magnitudes of such influences depend upon the strengths of the gradient flow and the urban heat island (Oke, 1995).

With calm or very weak winds and a strong urban heat island, a thermally induced circulation system forms over the urban area and its rural environs. This circulation is essentially caused by the rising of warm air over the city and sinking of cold air over the surrounding countryside. The near-surface air from the rural surroundings converges toward the thermal low that forms over the city, while at the upper levels the airflow diverges away from the city. During the daytime, this circulation may extend up to the base of the lowest inversion, which may then acquire a dome shape with the maximum height near the city center. A radially symmetrical circulation would be expected to occur only over a circular city in the absence of gradient wind. More frequently, even light gradient winds over irregularly shaped urban heat islands cause more complex circulation patterns whose centers may not coincide with city centers. The low-level flow tends to accelerate toward the center of circulation and takes an anticyclonic curvature (Oke, 1995). Presence of topography and large green spaces (parks) in the urban area may also greatly influence these thermal circulations. Many observational, experimental, and numerical modeling studies of urban heat-island-induced circulations and urban boundary layers have been reported in the literature (Oke, 1987, 1995).

With stronger winds and therefore weaker urban heat island, the mechanical effects of increased surface roughness usually dominate over the thermal effects. Consequently, the mean wind speed decreases in going toward the city center. Even at a height of 100 m, the reduction in wind speed may be more than 20%. On the other hand, the surface drag or shear stress increases by more than 40%. Turbulence intensities are expected to increase even more, because σ_u, σ_v and σ_w

usually increase in proportion to u_* while mean wind speed decreases in going toward the city center. Wind direction shifts have also been observed in the near-surface air passing over the city. Wind first turns to the left or cyclonically in passing over the city and, then, to the right anticyclonically as the flow recovers its original direction downwind of the city. Wind direction shifts of 10–20° may occur over large urban areas. Such changes in the direction and magnitude of transport winds, as well as in turbulent intensities over and around urban areas have important consequences for the dispersion of pollutants from urban sources.

It has been commonly observed that the UBL is characterized by stronger turbulence as compared to the rural PBL upwind of the urban area. The increased turbulence is a result of the increased production of turbulence kinetic energy over an urban area due to increases in both the shear and buoyancy production terms in the TKE equation. Turbulent variances, fluxes and other statistics in the above canopy surface layer and the UBL, however, follow the same similarity scaling and display similar structure as turbulence in the homogeneous PBL for the same values of stability and other similarity parameters (Oke, 1995).

14.6 Building Wakes and Street Canyon Effects

In the preceding section we discussed the urban effects on the atmospheric boundary layer well above the urban canopy, and details of three-dimensional, complex flows around buildings were glossed over. For urban inhabitants, the local environment on streets around buildings, i.e., within the urban canopy, is perhaps more important than that in the above-canopy boundary layer. The canopy flow is much more complex, however, and is not amenable to any simple, generalized, quantitative treatment. Therefore, we give here only a brief qualitative description of some of the observed features of flow around buildings and other urban structures. These observations have largely come from physical simulations of flow around isolated and somewhat idealized buildings, as well as from clusters of buildings in environmental wind tunnels. Most of our conceptual understanding of such flows is derived from fundamental studies in bluff-body aerodynamics and environmental fluid mechanics.

14.6.1 Characteristic flow zones around an isolated building

There have been many experimental studies of flow distortion (modification) around isolated bluff bodies (e.g., circular and rectangular cylinders, flat plates, etc.) placed in a uniform, streamlined (laminar) approach flow. Comparatively fewer studies have been made of the modified flow fields around surface-mounted structures immersed in thick boundary layers, as in the atmosphere. Although details of the modified flow vary with the characteristics of the approach flow, as well as with the size, shape, and orientation of the bluff body, certain flow phenomena and characteristic zones are commonly observed.

Flow separation and recirculating cavity Flow separation is said to occur when fluid initially moving parallel to a solid surface suddenly leaves the surface, because it can no longer follow the surface curvature or because of a break (discontinuity) in the surface slope. Flow separation is essentially a viscous flow phenomenon, as an inviscid or ideal fluid can, in principle, go around any bluff body without separation. The surface boundary layer developing in a real fluid flow, on the other hand, usually separates and moves out into the flow field as a free shear layer. The flow near the surface immediately downstream of the boundary layer separation is in the opposite direction. This is illustrated by the schematics of flow separation from a curved surface (Figure 14.15a) and a

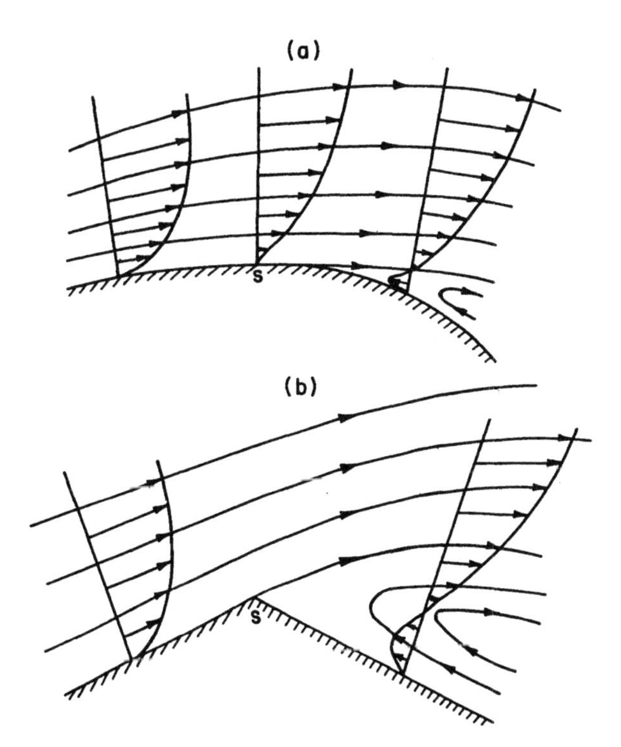

Figure 14.15 Schematic of flow separation from a (a) curved surface and (b) sharp-edged surface, and typical velocity profiles and streamline patterns.

sharp-edged surface (Figure 14.15b). Note that the surface streamline leaves off-tangentially or parallel to the initial slope from the separation point S.

The primary cause of flow separation is the loss of mean kinetic energy (e.g., in a diverging flow) in the boundary layer and consequent gain in the potential energy and, hence, in the surface pressure along the direction of flow, until the flow is no longer possible against the increasing adverse pressure gradient. All fluids have a natural tendency to flow in the direction of decreasing pressure, i.e., along a favorable pressure gradient. A significant adverse (positive) pressure gradient in the direction of flow can lead to abrupt and dramatic changes in the flow following its separation from the boundary.

The flow around a curved surface, as shown in Figure 14.15a, often leads to an unsteady separation in which the separation point S may not remain fixed but moves up and down the surface in response to perturbations to the flow, including the shedding of vortices by the separated shear layer. Even the average location of S is found to be quite sensitive to slight surface irregularities, as well as to the Reynolds number (Re), depending on the radius of curvature of the surface. On the other hand, the flow separation at a salient edge, as shown in Figure 14.15b, is steady, and its location is fixed, irrespective of the Reynolds number. This property is found to be quite useful in reduced-scale model simulations of atmospheric flows around buildings in fluid-modeling facilities, such as wind tunnels and water channels, in which Reynolds numbers are necessarily much smaller than those in the atmosphere. Gross features of flow around sharp-edged bluff bodies are found to be essentially Reynolds-number independent for all large enough Reynolds numbers (say, $Re > 10^4$).

An example of flow separation from the upwind edge of the roof of a rectangular building is shown in Figure 14.16, in which other flow zones are also depicted. Note that the separated streamline eventually reattaches to the ground (or the roof in the case of a long building) at the downwind stagnation point. Below this a relatively stagnant zone of recirculating flow, or 'cavity,' is formed. Contaminants released into this region or those entrained from above often lead to very high concentrations in the cavity region, inspite of some cross-streamline diffusion and mixing with the outer fluid. The cavity region is characterized by a recirculating mean flow with low speeds but large wind shears and high-turbulence intensities. Flow separation also occurs from the sides of the building (not shown in Figure 14.16), so that the three-dimensional cavity is a complex structure.

Dimensions of the cavity envelope depend on the building aspect ratios W/H and L/H, where L, W, and H are the building length, width, and height, respectively, as well as on the characteristics of the approach flow, such as the boundary layer depth relative to the building height (h/H) and stability. In a neutral boundary layer, the maximum cavity length may vary from H (for $W/H < 1$) to $15H$ (for $W/H > 100$, and $L/H < 1$), while the maximum depth of

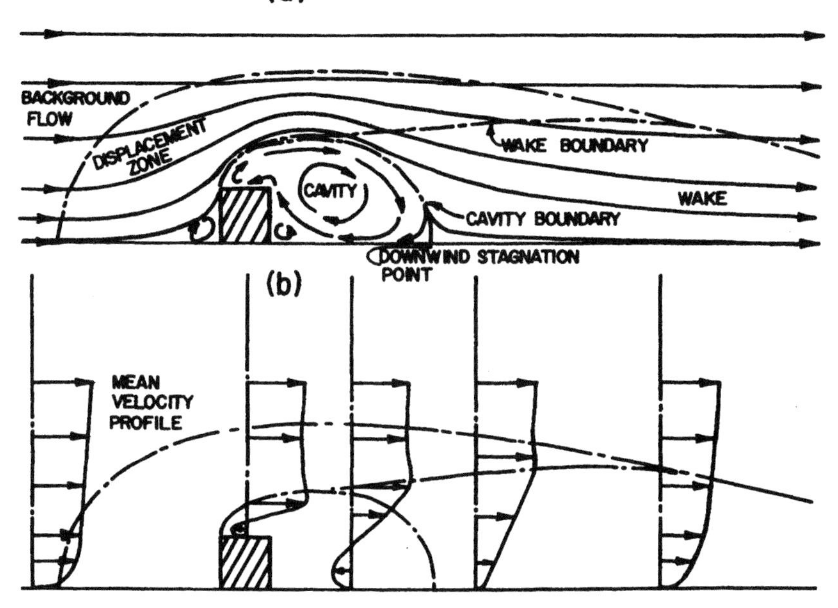

Figure 14.16 Schematic of displacement, cavity and wake flow zones around a two-dimensional sharp-edged building: (a) mean streamline pattern and (b) mean velocity profiles at the various locations along the flow. [After Halitski (1968).]

the cavity can vary between H and $2.5H$. For a more comprehensive review of wind tunnel and field data on cavity dimensions and empirical relations based on the same, the reader should refer to Hosker (1984).

Figure 14.16 also shows a small frontal cavity or vortex region confined to the upwind ground-level corner of the building. This is caused by the separation of the flow from the ground surface, resulting from an adverse pressure gradient immediately upwind of the building (note that the pressure is maximum at the stagnation point on the upwind face of the building).

Wake formation and relaxation The region of flow immediately surrounding and following the main recirculating cavity, but still affected by the building, is called the building 'wake.' According to some broader definitions, the wake also includes the lee-side cavity region, sometimes called the 'wake cavity,' and comprises the entire downstream region of the flow affected by the obstructing body (all bluff and streamline bodies have wakes, even though some may not have any recirculating cavity in their lee). Here we use the more restrictive definition excluding the cavity from the wake, so that there is no flow reversal in the latter.

Sometimes the wake is subdivided into the 'near wake' and the 'far wake.' The mean flow and turbulence structure in the former are strongly affected by separating shear layers, as well as by vortices shed from the building edges, as shown in Figure 14.17. The separated shear layers have large amounts of vorticity, generated when these layers were still attached to the building as a boundary layer. They also have strong wind shears across them. After separation the shear layers become unstable and generate a lot of turbulence and, frequently, also periodic vortices called Karman vortices, which grow as they travel downwind. Another prominent vortex system in flows around three-dimensional surface obstacles is the so-called horseshoe vortex. It is a standing horizontally oriented vortex generated near the ground upwind of the obstacle; it wraps around the obstacle and then trails off downwind as a counter-rotating vortex pair (Figure 14.17). This vortex resembles a horseshoe when viewed from above, hence its name. These large vortices produce oscillations in the wake boundary, while smaller eddies transport momentum across mean streamlines. As a consequence of these processes, mean velocity in the direction of flow decreases rapidly, while turbulence increases with downwind distance in the near wake, which may extend up to a few building heights from the upwind face of the building. Building-generated turbulence is found to be predominant in the near wake and cavity regions.

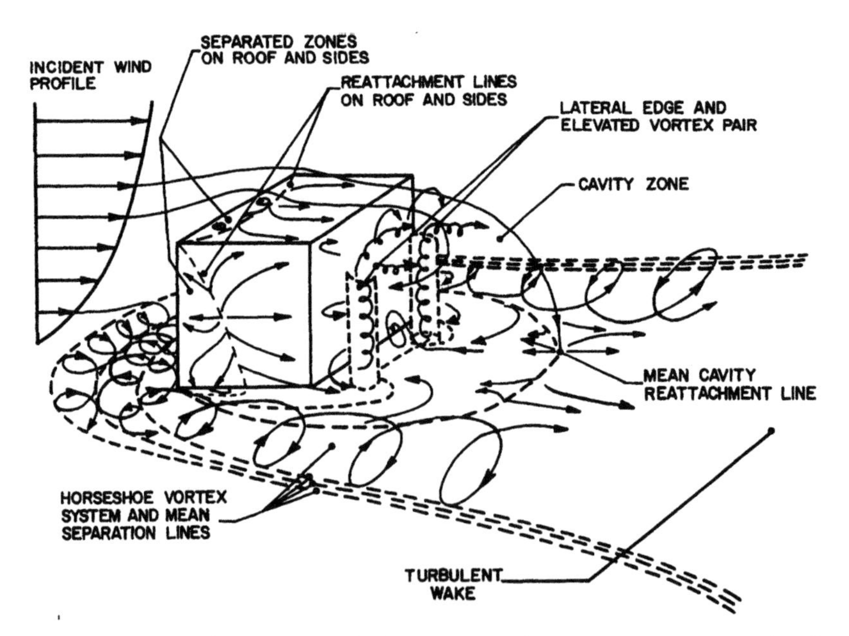

Figure 14.17 Schematic of separated flow zones and vortex systems around a three-dimensional sharp-edged building in a deep boundary layer. [After Woo *et al.* (1977).]

The near wake is followed by a more extensive far-wake region in which perturbations to mean flow and turbulence decay with distance, following an exponential or power law, and become insignificant only at large downwind distances. The longitudinal, lateral, and vertical extents of the wake depend on building dimensions and their aspect ratios. Both the width and the thickness (height) of the wake grow with distance behind the building, if the wake is defined as the region encompassing all building-induced perturbations to the background or approach flow, even if they become practically undetectable. This is the classical theoretical definition of wake, which extends to infinity along the flow direction as it grows in cross-wind directions. A more practical definition of the wake in which perturbations in the mean and turbulent flow are significant and measurable, i.e., they are larger than a certain specified percentage (say, 10%) of the background flow, would make the wake boundary a finite envelope with a maximum length, width, and height, which depend on building aspect ratios. Wind tunnel observations in simulated building wakes indicate that significant building-induced perturbations to flow may not extend beyond 10 to 20 building heights in the downwind direction and 2 to 3 building heights or widths in other directions. Sometimes, the far wake is further subdivided into an inner layer or a surface layer, and an outer mixing layer. The former is assumed developing downwind of the cavity reattachment point and has the character of an internal boundary layer in which flow adjusts to the local surface (see Plate, 1971, Chapter 4). The outer mixing layer has the character of a classical wake far from any bounding surface.

External region of streamline displacement A third region of relatively minor modifications to the flow also envelops the building and its wake and cavity regions. Streamlines in this so-called external or displacement region are at first deflected upward and laterally outside in response to the growing wake, and then downward and inside to their undisturbed pattern, as the building influence decays and disappears. These displacements and associated perturbations to the flow decrease with increasing distance outward from the wake boundary. Such perturbations to the flow in the external region are considered to be essentially inviscid and small, allowing for simplified theoretical treatments. Note that external zones of streamline displacement must also exist, to some extent, in all the nonhomogeneous boundary layers (Townsend, 1965). Outside the displacement zone is the region of undisturbed flow.

14.6.2 Clustering and street canyon effects

In urban settings, buildings are placed in clusters of individual houses, high-rise apartments, or commercial buildings. When the spacing between adjacent buildings is less than 10 to 20 building heights, which is generally the case, the wakes and cavities associated with individual buildings interact, producing a variety of complicated and often discomforting flow patterns. There have been only a few simulation studies of clustering effects of buildings in urban settings, and one or two field studies (Hosker, 1984). Theory and numerical simulations cannot cope with such complex and highly turbulent flows. Experimental efforts are also beset by serious instrumental and logistical difficulties. Our qualitative understanding of flow phenomena associated with building clusters is largely due to wind tunnel simulation studies (Jeram *et al.*, 1995).

When buildings are nearly of the same size and height, and are arranged in regular rows and columns parallel to straight streets in an otherwise flat area, air flow is accelerated just above the roof level outside of any roof cavities, as well as in any side streets parallel to the wind direction. Winds can become strong and steady, particularly in narrow and deep street canyons, due to direct blocking of flow by windward faces of the buildings and channeling in side-street canyons. Streets normal to the ambient wind direction are relatively sheltered when the ambient atmospheric boundary layer flow is weak, but strong standing vortices can develop in cross-streets at moderate ambient wind speeds. These are augmented forms of lee-side cavities where the recirculating flow is enhanced by deflection down the windward face of the adjacent downstream building. Winds are more gusty (turbulent) in the presence of these building-induced vortices. If the ambient wind flow is at an angle to the streets or buildings, lee-side vortices take on a 'corkscrew' motion with some along-street movement. The winds in the side streets are also reduced and become more gusty, reflecting intensification of corner vortices. All of the above-mentioned flow patterns strongly depend on building aspect ratios and the relative spacing between them. Irregular arrangements of building blocks along curved streets can result in an almost infinite variety of flow patterns. Some of the basic fluid flow phenomena underlying these complex flow patterns are schematically illustrated and discussed by Hosker (1984).

When adjacent buildings in a cluster differ considerably in their heights or aspect ratios, flow patterns around them become even more complex and highly asymmetrical. Although a wide variety of situations can be visualized and actually occur in urban settings, systematic experimental studies (primarily wind tunnel simulations) are so far confined to the so-called two-body problem in which a much taller building is placed downwind of a shorter building, or vice versa (Hosker, 1984). Here, too, the flow patterns depend strongly on the relative building heights, spacing between the two buildings, and the building aspect ratios, as well as on the characteristics of the approach flow. The presence of a much smaller building upwind of a tall building appears to affect only the upwind portion of the horseshoe vortex, which wraps around any isolated building, and the downwind cavity and wake of the taller building are less affected. These can be considerably modified, however, if the shorter building is placed downstream of the taller buildings. The taller building has a more dominating influence on the circulation around a smaller building in its wake than the other way around.

14.7 Other Topographical Effects

In this section we consider the effects of small hills, ridges, and escarpments on atmospheric boundary layer flows. Effects of large mountains extend well above the boundary layer and involve mesoscale and, sometimes, synoptic-scale motions, which are outside the scope of this book. Unlike the flows around sharp-edged buildings or groups of buildings in which mechanically generated turbulence dominates over the ambient turbulence in the approach flow, flows around gentle, low hills and ridges are quite sensitive to mean shear, stratification, and turbulence in the approach flow. Therefore, their description is presented here according to different approach flow conditions. This is largely based on laboratory fluid-modeling experiments, some theoretical and numerical model studies, and a few field experiments. For more comprehensive reviews of the literature, the reader should refer to Hunt and Simpson (1982), Taylor *et al.* (1987), and Bains (1995).

14.7.1 Neutral boundary layer approach flow

The characteristic flow zones shown in Figure 14.16 for the boundary layer modified by an isolated building can also be applied to neutral boundary layer flow around an isolated hill, ridge, or escarpment. However, flow separation and associated cavity regions are likely to be unsteady and much smaller, if they exist at all, around gentle topographical features. The flow may not separate even on the lee side, unless the maximum downwind slope is large enough (say, ≥ 0.2). Behind long, steep ridges, however, the cavity region may extend up to 10 hill heights in the downstream direction. The cavity is at most a few hill heights long and frequently smaller for thee-dimensional hills.

An important universal feature of the flow over a hill is its speeding up in going over the hill's top. For quantifying this, a speed-up factor $S_F = U(z)/U_0(z)$ is defined as the ratio of wind speed at some height above the hill to that at the same height above the flat surface in the approach flow. While S_F decreases with increasing height, its maximum values range from just over 1 to about 3, depending on the hill slope, and aspect ratio. The largest speed-up factors are observed over three-dimensional hills of moderate slope. In approaching an isolated hill or a two-dimensional ridge, the flow near the surface first decelerates slightly before reaching the upwind base of the hill and then accelerates in going over the hill. Flow also accelerates in going around the hillsides. These features, as well as deceleration in the near wake, are shown in Figure 14.18, based on measurements over an approximately circular hill (Brent Knoll).

Another feature of the air flow over a hilltop is that the wind speed increases with height much more rapidly than it does over a level ground. Above this shallow surface shear layer, the wind profile becomes nearly uniform; sometimes it is characterized by a maximum wind speed (low-level jet) near the surface. Wind profiles at the tops of escarpments are also found to have similar characteristics.

The lee-side hill wakes are similar to building wakes; these are characterized by reduced mean flow and enhanced turbulence. The maximum topographically induced perturbations to the flow in the near wake depend on the aspect ratio, slope, and shape of the hill. However, their decay with distance in the far wake appears to follow similar relations. In particular, the maximum velocity deficit $(\Delta U)_{max}$ is given by

$$(\Delta U)_{max}/U_0 = m(x/H)^{-1} \qquad (14.14)$$

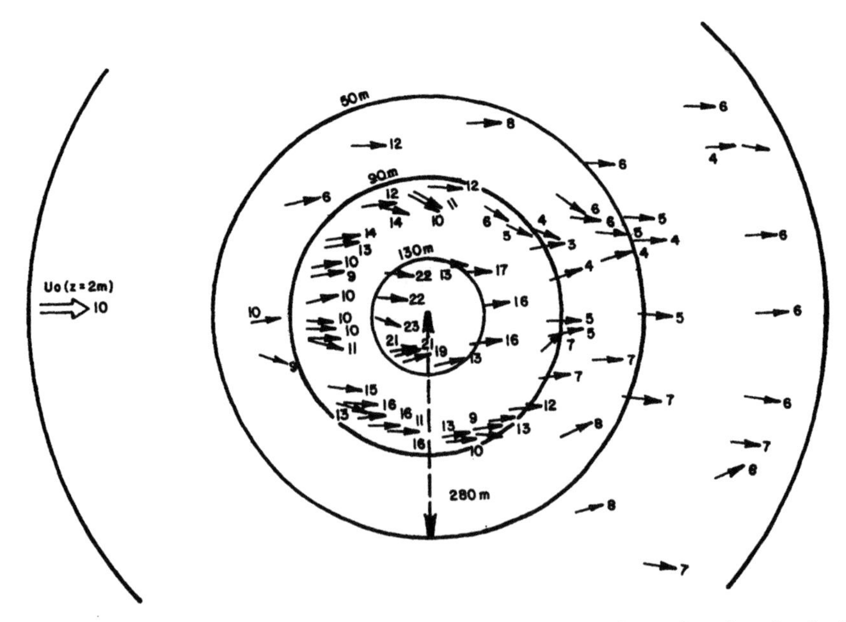

Figure 14.18 Observed mean wind speeds and directions near the surface ($z = 2$ m) of an approximately circular low hill (Brent Knoll). [After Mason and Sykes (1979).]

where U_0 is the mean velocity at hill height in the approach flow, and m is a coefficient depending on the hill and boundary layer parameters. The rate of growth of the wake also depends on the hill slope and aspect ratio (Arya *et al.*, 1987).

Flow at an oblique angle to a long ridge is often characterized by the generation of trailing vortices and a persistent swirling turbulent flow in the wake. Thus, the ridge acts like a vortex generator. These and other three-dimensional effects of hills are poorly understood, and so are the Coriolis effects of the earth's rotation on flow around hills.

Very few systematic studies have been made on groups of hills and valleys. The simplest arrangement is that of similar, nearly two-dimensional ridges and valleys located parallel to each other. Wind tunnel experiments indicate that the speed-up of flow over the ridge top and the flow distortion in the near wake are maximum for the first ridge and valley, as if downwind ridges did not exist. But these effects diminish in the downwind direction as subsequent ridges and valley interact with the flow. After the first few ridges and valleys, the speed-up factor over the tops of ridges approaches a constant value of only slightly above unity, and the summit velocity profile assumes a new logarithmic form with nearly the same roughness length as upwind. Turbulence intensities over the hill crests do not differ much from their undisturbed (flat terrain) values except for some enhancement of the lateral component σ_v/U for wind directions not normal to ridges and valleys. In the valleys between the ridges mean flow is reduced, with possible recirculation cavities forming over steeper lee-side slopes, and turbulence is considerably enhanced. These results have been confirmed in a recent field study of atmospheric flow over a succession of ridges and valleys under near-neutral conditions (Mason and King, 1984). When the flow is across the valley, wind speeds in the valley are about one- to two-tenths of that on the summit. The size of the separated region is very sensitive to flow direction and the presence of any sharp discontinuity in the terrain profile. When the flow is predominantly along the valley, wind speeds in the valley are higher, but still less than that on the summit, and no channeling effect is observed, at least in near-neutral conditions.

A grouping of three-dimensional hills can produce complex flow patterns, similar to those for groups of buildings. For example, channeling of flow between two round hills or in a gap between a long ridge often leads to strong, persistent winds. These effects are particularly pronounced in the presence of stable stratification, which forces the flow to go around rather than over the hills.

14.7.2 Stably stratified approach flow

Effects of topography on the flow are considerably modified in the presence of stable stratification. Some of the effects and the associated flow patterns are qualitatively similar to those for the neutral flow, although stratification does influence their intensity. Other effects are peculiar to stably stratified flows over and around hills, e.g., the generation of lee waves, formation of rotors and hydraulic jump, inability of the low-level fluid to go over the hills, and upstream blocking.

Stratification effects on flow around topography are generally described in terms of a Froude number (F or F_L) based on the characteristic height (H) or length scale (L_1) of the topographical feature in the direction of flow

$$F = U_0/NH \tag{14.15}$$

$$F_L = U_0/NL_1 \tag{14.16}$$

Where U_0 is the characteristic velocity of approach flow (say, at $z = H$), and N is the Brunt–Baisala frequency

$$N \equiv \left(-\frac{g}{\rho_0}\frac{\partial \rho}{\partial z}\right)^{1/2} = \left(\frac{g}{\Theta_0}\frac{\partial \Theta}{\partial z}\right)^{1/2} \tag{14.17}$$

which is the natural frequency of internal gravity waves or lee waves; the corresponding wavelength is $\lambda = U_0/2\pi N$. Physically, the Froude number is the ratio of inertia to gravity forces determining the flow over topography. It has an inverse-square relationship to the bulk Richardson number, i.e., $F \sim \mathrm{Ri}_B^{-2}$. Note that if $F \ll 1$ the stratification is considered to be strong, while for $F \gg 1$ it is near neutral ($F = \infty$ for a strictly neutral stability). The Froude number also provides the criteria for the possible generation of lee waves and separation of flow in the lee of the hill. In particular, simple Froude number criteria have been developed from theory and experiments in the simple case of a uniform, inviscid approach flow with constant density or potential temperature gradient (Hunt and Simpson, 1982; Bains, 1995). These are schematically shown in Figure 14.19 for two-dimensional hills. Here, the critical Froude number (F_c) for separation represents the highest value of F at which the boundary layer separation is suppressed by lee waves. It depends on the stratification of the approach flow, as well as on the various hill parameters (e.g., height, aspect ratio, and shape).

Figure 14.19 also indicates that different types of flow patterns may develop in the lee of a two-dimensional hill, depending on the Froude number and the maximum hill slope. Similar and other types of flow phenomena are found to occur in the lee of three-dimensional hills and under different approach flow conditions (e.g., stratified shear flow and mixed layer capped by an elevated inversion). Perhaps the most spectacular flow phenomena related to topography are the severe downslope winds, locally known as Chinook, Föhn, or Bora, followed by a 'hydraulic jump.' These are analogous to the passage of water

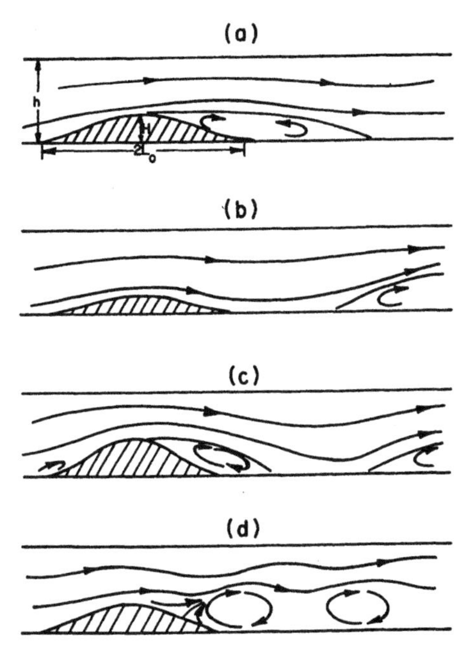

Figure 14.19 Stratified flows over two-dimensional hills in a channel or under a strong inversion, showing the effect of lee waves on flow separation. (a) Supercritical $F > F_c$, no waves possible; separation is boundary-layer controlled. (b) Hill with low slope; subcritical $F < F_c$, downstream separation caused by lee-wave rotor. (c) Hill with moderate slope; supercritical $F > F_c$, boundary-layer separation on lee slope. (d) Hill with moderate slope; subcritical $F < F_c$, lee-wave-induced separation on lee slope. [After Hunt and Simpson (1982).]

over a dam spillway as it is rushing down at very high speeds and creating a violent hydraulic jump at the base of the dam. The necessary approach flow conditions for the corresponding atmospheric flow are a strong, elevated inversion capping a high-speed mixed-layer flow of depth $h > H$, such that $U_0/N(h - H) \cong 1$, where N is the Brunt–Vaisala frequency for the inversion layer. Figure 14.20 shows the schematics of two widely different, but possible, flow patterns with an approach mixed-layer flow. Note that, here, the relevant Froude number determining the flow over the hill is $U_0/N(h - H)$.

An important aspect of stably stratified flows over and around three-dimensional topographical features is the increasing tendency of fluid parcels to go around rather than over the topography with increased stratification (decreasing Froude number). This is because such fluid parcels do not possess sufficient kinetic energy to overcome the potential energy required for lifting the parcel through a strong, stable density gradient. Whether a given parcel in the

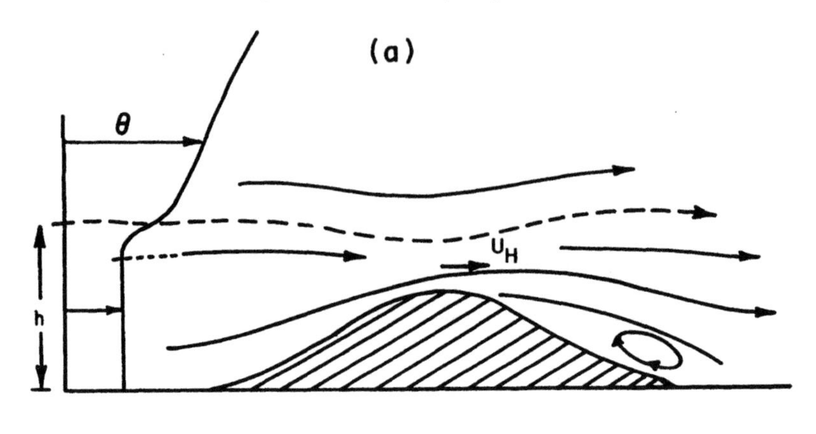

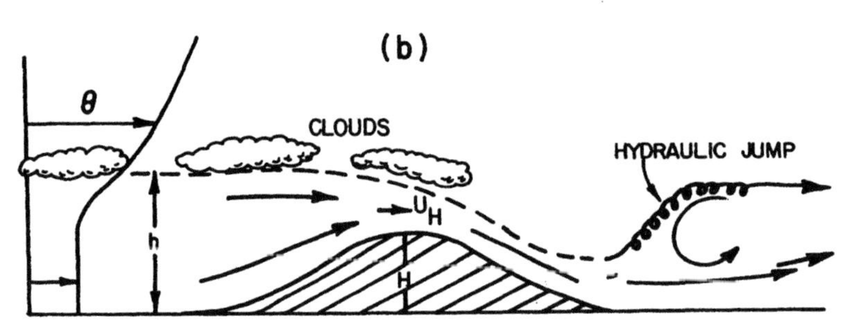

Figure 14.20 Schematics of air flow over a two-dimensional ridge with an elevated inversion upwind. (a) Low wind speed with $U_0/N(h - H) \ll 1$, flow separation on lee slope. (b) High wind speed with $U_0/N(h - H) \cong 1$, no flow separation on lee slope, hydraulic jump downwind. [After Hunt and Simpson (1982).]

approach flow would go over or around the hill depends on the height of the parcel relative to hill height, its initial lateral displacement from the central (stagnation) streamline, shear and stratification in the approach flow, and topographical parameters. Parcels approaching the hill at low heights are likely to go around, while those near the hilltop may go over the hill. One can define a dividing streamline that separates the flow passing around the sides of the hill from that passing over the hill. The height H_s of this dividing streamline can be estimated from the simple criterion that the kinetic energy of a fluid parcel following this streamline be equal to the potential energy associated with stratification that the parcel must overcome, i.e.,

$$\tfrac{1}{2}\rho U_0^2 = g\int_{H_s}^{H} (H - z)\left(-\frac{\partial \rho}{\partial z}\right) dz \tag{14.18}$$

where U_0 is the approach flow velocity at $z = H_s$. This integral equation, first suggested by Sheppard (1956), can be used for any approach flow and topographical arrangement. In practice, it must be solved for H_s iteratively, because the unknown appears as the lower limit of integration. For the particular case of a stratified approach flow with a constant density gradient, Equation (14.18) reduces to the simpler formula

$$H_s = H(1 - F) \tag{14.19}$$

which was suggested by Hunt and Snyder (1980) on other theoretical and experimental bases.

Laboratory experiments and a few observations of smoke plumes impinging on hillsides have confirmed the validity and usefulness of the dividing streamline concept (Snyder *et al.*, 1985). These also indicate that Equation (14.18) provides only a necessary (but not the sufficient) condition for the fluid above H_s to go over rather than around a hill. It can still pass around the sides and take a path requiring less potential energy to overcome than that given by Equation (14.18). Experiments also indicate that the lateral aspect ratio (W/H) of the hill is relatively unimportant, but the stratification and shear in the approach flow, upwind slope of the hill, and obliqueness of the flow to an elongated hill are all important in determining the flow over and around the hill (Bains, 1995).

In the case of strongly stratified ($F \ll 1$) flow approaching normal to a very long (nearly two-dimensional) ridge, there is a distinct possibility of upstream blocking of the fluid, because very little can go over the ridge. A region of nearly stagnant fluid occurs below the dividing streamline and far away from the edges where it can go around the ridge. Such a flow field may not acquire a steady state, because the upwind edge of the blocked flow propagates farther upstream as a density front and certain columnar disturbance modes and gravity waves generated near the ridge can also propagate upstream. Topographically blocked flows have important implications for local weather and for dispersion of pollutants from upwind sources.

In stably stratified flows over hills and mountain ridges, a variety of flow patterns may develop on the lee side, depending on the Froude number, hill slope and shape, presence of elevated inversions, and wind shear (Hunt and Simpson, 1982; Bains, 1995). Perhaps, the most spectacular of these are strong downslope winds as gravity flows, which are sometimes followed by a hydraulic jump. Similar phenomena have also been observed in laboratory experiments on stratified water flows over model hills (Bains, 1995). Cold downslope winds, called bora, have been observed in different parts of the world, but most notably along the coast of Yugoslavia (Atkinson, 1981, Chapter 3). Some of the boras start rather abruptly, while others develop gradually over several hours as the synoptic flow pattern changes to the appropriate direction (normal to the mountain ridge).

Strong downslope winds also occur when warm and moist winds are forced up the slope of a tall mountain ridge resulting in condensation and precipitation near the top of the mountain. With most of the moisture removed from the air and the latent heat of condensation added to the same, the drier and warmer air begins to descend on the lee side slope, where it becomes much warmer due to dry adiabatic compression. Such orographically forced warm and dry downslope winds have acquired local names such as chinook, foen, and Santa Anna, depending on their geographical location. Both cold and warm downslope winds are strong, gusty, dry and fairly constant in direction. The approach flows typically associated with these winds are normal to the mountain ridges and usually contain a stable layer at a critical height (Atkinson, 1981, Chapter 3).

When the approach flow is stably stratified, any lee-side separation of flow may be suppressed by stratification. But it can be induced by lee waves and their associated rotors under appropriate conditions (Hunt and Simpson, 1982). Since the flow is usually separated from the hillsides, the near-wake region is always turbulent even when the approach flow may not be turbulent, especially outside of the shallower stable boundary layer. This has important implications for the dispersion of pollutants at nighttime from sources near the downwind base of a hill.

Vortex streets are often shed by hill sides, especially under strongly stratified flows with a low-level inversion. Satellite photographs of vortex streets downwind of small, isolated mountain islands show their wakes extending up to several hundred kilometers (Atkinson, 1981, Chapter 4). Experimental and theoretical studies of obstacle wakes in stably stratified flows have been reviewed by Bains (1995). Most of these studies considered only uniform, laminar, linearly stratified flows over or around idealized two- or three-dimensional hills. For a given hill shape and size, wake phenomena are described primarily as a function of Froude number or Richardson number. The effects of wind shear, large surface roughness, and complex irregular topography, such as those commonly encountered in nature, are hardly understood. A few field studies of near-surface flow around small, isolated hills and ridges have been conducted, but measurements did not extend into the far-wake region.

Example Problem 2

A uniform stably stratified flow with a characteristic velocity of 10 m s^{-1} outside the shallow stable boundary layer of depth $h = 100$ m and temperature gradient of 0.02 K m^{-1} approaches an isolated hill of height $H = 400$ m. The characteristic surface air temperature is 12°C and the surface pressure is about 1000 mb at the upwind base of the hill.

(a) Calculate the Brunt–Vaisala frequency and the characteristic wavelength of the lee waves that might be produced in the lee of the hill.
(b) Estimate the height of the dividing streamline below which the air flow might be expected to go around rather than over the hill.

Solution

(a) The Brunt–Vaisala frequency is given by

$$N = \left(\frac{g}{\Theta_0} \frac{\partial \Theta}{\partial z} \right)^{1/2}$$

in which,

$$\Theta_0 = 273.2 + 12 = 285.2 \text{ K}$$

$$\frac{\partial \Theta}{\partial z} = \frac{\partial T}{\partial z} + \Gamma = 0.02 + 0.0098 \cong 0.03 \text{ K m}^{-1}$$

Thus,

$$N = \left(\frac{9.81}{285.2} \times 0.03 \right)^{1/2} \cong 0.0321 \text{ Hz}$$

The characteristic wavelength of hill-induced lee waves is given by

$$\lambda = \frac{U_0}{2\pi N} = \frac{10}{2\pi \times 0.0321} \cong 49.6 \text{ m}$$

(b) The height of dividing streamline is given by Equation (14.19)

$$H_s = H(1 - F)$$

where F is the Froude number based on the hill height, i.e.,

$$F = \frac{U_0}{NH} = \frac{10}{0.0321 \times 400} \cong 0.78$$

so that,

$$H_s = 400(1 - 0.78) \cong 88 \text{ m}$$

Thus, the shallow stable boundary layer flow below the height of 88 m is expected to go around the hill.

14.8 Applications

Spatial variations of surface roughness, temperature, wetness, and elevation frequently cause nonhomogeneous atmospheric boundary layers developing over these surface inhomogeneities. A basic understanding of horizontal and vertical variations of mean flow, thermodynamic variables, and turbulent exchanges in nonhomogeneous boundary layers is essential in micrometeorology and its applications in other disciplines and human activities. Specifically, the following applications may be listed for the material covered in this chapter.

- Determining the development of modified internal boundary layers following step changes in surface roughness and temperature.
- Determining the fetch required for air flow adjustment to the new surface following a discontinuity in surface elevation or other properties.
- Selecting an appropriate site and height for locating meteorological or micrometeorological instruments for representative observations for the local terrain.
- Determining wind and air mass modifications under situations of cold or warm air advection over a warmer or colder surface.
- Estimating urban influences on surface energy balance and surface and air temperatures.
- Determining possible modifications of mean flow, turbulence, and the PBL height over an urban area.
- Estimating transport and diffusion of pollutants in the urban boundary layer and the urban plume.
- Estimating pollutant dispersion around hills and buildings.
- Estimating wind loads on buildings and knowing the wind environment around buildings.
- Siting and designing wind-power generators in hilly terrain.
- Other wind-engineering applications in urban architecture and planning.

Problems and Exercises

1. A near-neutral atmospheric boundary layer 1500 m thick encounters a sudden change in the surface roughness in going from land ($z_0 = 0.1$ m) to a large lake ($z_0 = 0.0001$ m) of the same surface temperature. The observed wind speed at 10 m height in the approach flow is 10 m s^{-1}. Assuming that the overall PBL thickness ($h = 1500$ m) and the surface layer thickness ($h_s \cong 150$ m) remain unchanged over the lake, calculate and plot the following parameters, using Equations (14.1) and (14.3), with $a_i = 0.4$ for the internal boundary layer.

(a) The IBL height as a function of distance, and the distances where it reaches the top of the approach flow surface layer and the PBL.
(b) The friction velocity and wind speed at 10 m height over the lake as functions of distance from the shore.

2. Repeat the calculations in Problem 1 for the case of an onshore neutral boundary layer flow with the same wind speed (10 m s^{-1}) and depth (1500 m) over the lake, and compare the results for the two cases.

3. Calculate and graphically compare the development of thermal internal boundary layers in the following two situations, using appropriate expression for h_i:
(a) A stably stratified boundary layer flow with $\partial\Theta/\partial z = 0.02$ K m^{-1} over land advecting over a 15°C warmer sea.
(b) An unstable warm continental boundary layer with a lapse rate of 0.02 K m^{-1} advecting over a 15°C colder sea surface. Assume a typical drag coefficient of 0.9×10^{-3} for the stably stratified IBL.

4.
(a) What are the leading causes and mechanisms for the formation of an urban heat island?
(b) If the population of a midwestern United States city has increased from 100 000 to 1 000 000 in the past 20 years, estimate the expected change in its maximum heat-island intensity and the annual rate of urban warming over that period.

5.
(a) Give physical reasons for the urban mixed-layer dome phenomenon and list conditions under which it might be enhanced and those in which it would be suppressed.
(b) Calculate and compare the nocturnal urban mixed-layer height at the city center, where the near-surface temperature is 15°C, with that at a suburban location of near-surface temperature 12°C, when the upwind rural sounding indicates a temperature gradient of 0.02°C m^{-1} over a deep-surface inversion layer and a near-surface temperature of 9°C.

6.
(a) What are the causes and consequences of boundary layer separation from hills and buildings?
(b) In what respects do the cavity and near-wake flow regions behind an isolated square-shaped building differ from those behind a cylindrical building?

7.

(a) Why is the length of a recirculating cavity region behind a long ridge much larger than that behind an axisymmetrical hill of the same height and slope?

(b) In what respects does a three-dimensional hill wake differ from a building wake?

8.

(a) Show that, for constant density-gradient uniform or shear flow, Equation (14.18) reduces to Equation (14.19).

(b) Calculate the dividing streamline height for a constant gradient shear flow with $\partial U_0/\partial z = 0.1\ \text{s}^{-1}$, $\partial \Theta/\partial z = 0.05\ \text{K m}^{-1}$, and $\Theta_0 = 280$ K, approaching an axisymmetric hill 200 m high. What are the implications of this to flow over and around the hill?

Chapter 15 | Agricultural and Forest Micrometeorology

15.1 Flux-profile Relations above Plant Canopies

In this final chapter we discuss the particular applications of micrometeorological principles and methods to vegetative surfaces, such as agricultural crops, grasslands, and forests. Here, the micrometeorologist is interested in the atmospheric environment, both above and within the plant canopies, as well as in the complex interactions between vegetation and its environment. First, we discuss the vertical profiles of mean wind, temperature, specific humidity, etc., and their relationships to the vertical fluxes of momentum, heat, water vapor, etc., above plant canopies.

The semiempirical flux-profile relations developed in Chapters 10–12 should, in principle, apply to any homogeneous surface layer, well above the tops of roughness elements ($z > h_0 \gg z_0$). It should be kept in mind, however that for any tall vegetation the apparent reference level for measuring heights must be shifted above the true ground level by an amount equal to the zero-plane displacement (d_0), which is related to the mean height (h_0) of roughness elements and roughness density. Also, for flexible plant canopies, z_0 and d_0 may not have fixed ratios to h_0 but may vary with wind speed. Physically, one can argue that for a vegetative canopy acting as a sink of momentum and a source or sink of heat, water vapor, CO_2, etc., the effective source or sink height must lie between $0.5h_0$ and h_0. For exchange processes in the atmospheric boundary layer well above the canopy, it is found convenient to consider the canopy as an area source or sink of infinitesimal thickness located at an effective height d_0 above the ground level, and to consider it an effective reference plane. Then, the zero-plane displacement does not appear explicitly in theoretical or semiempirical flux-profile relations, such as those given in Chapters 10–12. In most experimental situations and practical applications, however, the ground level is the most convenient reference plane for measuring heights. It is also the appropriate reference plane for representing mean and turbulent quantities within the canopy layer. For these reasons, in this chapter, we will use the height z above the local ground level and account for the zero-plane displacement, wherever appropriate (e.g., in the above-canopy flux-profile relations).

Applying the Monin–Obukhov similarity theory to the fully-developed part of the constant-flux surface layer, also called the 'inertial sublayer', in which the flux-profile relations are expected to have the same forms as in the idealized surface layer over low roughness, we can rewrite Equation (11.2) as

$$
\begin{aligned}
\frac{k(z-d_0)}{u_*}\frac{\partial U}{\partial z} &= \phi_{\mathrm{m}}(\zeta) \\
\frac{k(z-d_0)}{\theta_*}\frac{\partial \Theta}{\partial z} &= \phi_{\mathrm{h}}(\zeta) \\
\frac{k(z-d_0)}{q_*}\frac{\partial Q}{\partial z} &= \phi_{\mathrm{w}}(\zeta)
\end{aligned}
\tag{15.1}
$$

where $\zeta = (z-d_0)/L$ is the M–O stability parameter. As shown earlier in Chapters 11 and 12 the integral forms of Equation (15.1) represent the mean wind speed, potential temperature, and specific humidity profiles in the similarity forms as

$$
\begin{aligned}
\frac{U}{u_*} &= \frac{1}{k}\left[\ln\left(\frac{z-d_0}{z_0}\right) - \psi_{\mathrm{m}}\left(\frac{z-d_0}{L}\right)\right] \\
\frac{\Theta-\Theta_0}{\theta_*} &= \frac{1}{k}\left[\ln\left(\frac{z-d_0}{z_0}\right) - \psi_{\mathrm{h}}\left(\frac{z-d_0}{L}\right)\right] \\
\frac{Q-Q_0}{q_*} &= \frac{1}{k}\left[\ln\left(\frac{z-d_0}{z_0}\right) - \psi_{\mathrm{w}}\left(\frac{z-d_0}{L}\right)\right]
\end{aligned}
\tag{15.2}
$$

where Θ_0 and Q_0 are the values at $z = d_0 + z_0$, and θ_* and q_* are the scaling parameters, each representing the ratio of the appropriate vertical flux (in kinematic units) to the friction velocity u_* [Equation (12.12)]. The M–O similarity functions ψ_{h} and ψ_{w} are presumably equal, but may differ from the momentum function ψ_{m}, especially in unstable and free convective conditions. The empirical forms of these functions over homogeneous land surfaces with short roughness including vegetation, as well as water surfaces, have already been discussed in Chapters 11 and 12.

The gradient and bulk parameterizations of the vertical fluxes of momentum, heat, etc., in an idealized (constant-flux, horizontally homogeneous, stationary) surface layer have also been discussed in Chapters 11 and 12. These can be applied to the above-canopy inertial sublayer, especially in the height range $0.1h > z > 2h_0$. The range of validity of the surface layer similarity relations becomes narrower for taller canopies and for shallow boundary layers. It may disappear completely when the PBL thickness is less than 10–15 times the vegetation height, a distinct possibility over mature forest canopies.

15.1.1 Profile relations in the roughness sublayer

Even when allowance is made for the zero-plane displacement, significant deviations from the log law under neutral conditions and the more general M–O similarity profiles given in Equation (15.2) under stratified conditions have been observed to occur in the so-called roughness sublayer which is a transition layer between the canopy layer below and the constant-flux inertial sublayer above (see also Garratt, 1992, Chapter 3). This roughness transition layer is characterized by relatively small values of $(z - d_0)/z_0$, ranging between 10 and 150. It has been found from micrometeorological measurements over tall crops and forests that the dimensionless similarity functions ϕ_m, ϕ_h, etc., attain their expected surface-layer similarity forms only above a certain minimum height z_* which is estimated to be 1.5 to 2.5 times the average height of roughness elements (i.e., $z_*/h_0 = 1.5$–2.5). In terms of the roughness length/ parameter, $(z_* - d)/z_0$ has been estimated to vary between 10 and 150. Thus, the roughness-sublayer thickness may vary from less than a meter over short vegetation to several tens of meters over forest canopies.

Due to enhanced turbulent mixing in the roughness sublayer, dimensionless wind shear and vertical gradients of temperature and specific humidity are found to be significantly reduced from their equilibrium values in the inertial sublayer or the idealized surface layer over low roughness for the same value of $(z - d_0)/L$ (Raupach, 1979; Garratt, 1980; Hogstrom *et al.*, 1989). In particular, the large (up to 75%) reduction in the dimensionless gradient of potential temperature seems to be a universal finding over forest canopies. The reduction in the dimensionless wind shear appears to be related in some way to the density of canopy (Hogstrom *et al.*, 1989). Significant reductions (up to 65%) of ϕ_m have been observed over sparse Savannah and other forests, but not over dense forests. The dimensionless specific humidity gradient (ϕ_w) also shows strong reduction in the roughness sublayer, particularly just above the treetops. The ratio ϕ_w/ϕ_h is found to be about one under unstable conditions, but somewhat larger (about 1.4) under near-neutral conditions. It should be recognized, however, that it is not easy to measure ϕ_h accurately enough under near-neutral conditions when both $\partial\Theta/\partial z$ and θ_* tend to zero.

A somewhat controversial finding by some investigators is that the ratios between the measured dimensionless gradients in the roughness sublayer and their expected values in the idealized surface layer over low roughness are more or less independent of stability. The suggested modified similarity relations that account for the observed reductions of dimensionless gradients in the roughness sublayer are

$$\frac{k(z-d_0)}{u_*}\frac{\partial U}{\partial z} = \phi_m(\zeta)\phi^*_m(\eta)$$
$$\frac{k(z-d_0)}{\theta_*}\frac{\partial \Theta}{\partial z} = \phi_h(\zeta)\phi^*_h(\eta) \quad (15.3)$$
$$\frac{k(z-d_0)}{q_*}\frac{\partial Q}{\partial z} = \phi_w(\zeta)\phi^*_w(\eta)$$

in which the reduction factors $\phi^*_m(\eta)$, etc., are assumed to be functions of the normalized height parameter $\eta = (z-d)/(z_* - d)$ only, independent of stability. For sparse forest canopies, Garratt (1992) has suggested a simple empirical form for all the reduction factors $\phi^*_m \cong \phi^*_h \cong \phi^*_w = \phi^*$:

$$\phi^*(\eta) = \exp[-0.7(1-\eta)], \quad \text{for } \eta \leq 1 \quad (15.4)$$

It should be recognized, however, that over dense forest canopies, the empirical constants may be different for different reduction functions when they are represented by similar exponential expressions as Equation (15.4). The top of the roughness sublayer (height z_*) can be estimated in terms of the average height of the dominating trees (e.g., $z_* = 1.5h_0$ to $2.5h_0$, depending on the roughness density).

The similarity functions ψ_m, ψ_h, etc., in the profile equations (15.2) should also be modified in the roughness sublayer in accordance with the flux-gradient relations (15.3). Other parameters most directly affected by the reductions in ϕ_m, ϕ_h, etc., are the eddy diffusivities K_m, K_h, etc., which have been observed to increase in the roughness sublayer. The enhancement of eddy diffusivities is expected to be inversely proportional to $\phi^*(\eta)$. The suggested mechanisms for the enhancement of turbulent exchanges in the roughness sublayer are the wakes generated by roughness elements and thermals rising between individual elements (Monteith and Unsworth, 1990, Chapter 14).

15.1.2 Transfer coefficients and resistances

The flux parameterizations that employ the drag coefficient and other transfer coefficients are commonly used in micrometeorology and its applications to air–sea interaction, air pollution, hydrology, large-scale atmospheric modeling, etc. In these, the various fluxes are linearly related to the products of wind speed and the difference between the values of variables at some reference height in the surface layer and at the surface itself. For example

$$\begin{aligned} \tau_0 &= \rho C_D U^2 \\ H_0 &= \rho c_p C_H U(\Theta_0 - \Theta) \\ E_0 &= \rho C_W U(Q_0 - Q) \end{aligned} \tag{15.5}$$

in which the drag coefficient (C_D) and other transfer coefficients can be prescribed as functions of the surface roughness (more appropriately, $(z - d)/z_0$) and stability (e.g., Ri_B) parameters, as discussed in Chapter 11. On the basis of Reynolds analogy (similarity of transfer mechanisms of heat, water vapor, etc.), one expects that $C_H = C_W$.

Another commonly used approach or method of parameterizing the vertical fluxes in terms of the difference in mean property values at two heights is the 'aerodynamic resistance' approach, which is based on the analogy to electrical resistance and Ohm's law:

$$\text{current} = \frac{\text{potential difference}}{\text{electrical resistance}}$$

The analogous relations for turbulent fluxes are

$$\begin{aligned} \tau_0 &= \rho r_M^{-1} U \\ H_0 &= \rho c_p r_H^{-1}(\Theta_0 - \Theta) \\ E_0 &= \rho r_w^{-1}(Q_0 - Q) \end{aligned} \tag{15.6}$$

in which r_M, r_H, etc., are the aerodynamic or aerial resistances to the transfer of momentum, heat, etc. Note that a flux will increase as the appropriate aerodynamic resistance decreases. In engineering literature, transfer coefficients having the dimensions of velocity are used, instead of inverse resistances, in empirical parameterizations of fluxes.

The resistance approach is more frequently used in micrometeorological applications in agriculture and forestry. This is largely because of the simple additive property of resistances in series, which can be used to express the total aerodynamic resistance in terms of resistances of its components (bare soil surface, leaves, branches, and stems of plants). A disadvantage is that, unlike the dimensionless transfer coefficients, aerodynamic resistances are dimensional properties, having dimensions of the inverse of velocity, which depend not only on the canopy structure, but also on the wind speed at the appropriate height and atmospheric stability. Also, in practice, it is not easy to determine and specify the component resistances for estimating the overall canopy resistance. A comparison of Equations (15.5) and (15.6) clearly shows that the bulk transfer method is superior to the aerodynamic resistance method, but the two sets of coefficients are interrelated as

$$r_M^{-1} = UC_D; \quad r_H^{-1} = UC_H; \quad r_W^{-1} = UC_W \tag{15.7}$$

Thus, for a given canopy, aerial resistances are inversely related to wind speed and are not unique properties of canopy and the surface. The reciprocals of resistances may also be called transfer velocities or conductances. The deposition velocity of a gaseous pollutant is often parameterized in terms of the aerodynamic and other resistances (Arya, 1999, Chapter 10).

15.2 Radiation Balance within Plant Canopies

Solar or shortwave radiation is the main source of energy for the vegetation. Longwave radiation is also a major component of energy balance in daytime and the primary component at nighttime. In plant canopies radiation has important thermal and photosynthetic effects; it also plays a major role in plant growth and development processes. The visible part of the shortwave radiation (0.38–0.71 μm) is called the photosynthetically active radiation (PAR).

Radiative transfer in a plant canopy is a very complicated problem for which no satisfactory general solution has been found. There are significant amounts of internal radiative absorption, reflection, transmission, and emission within the canopy elements. Complications also arise from the large variability and inhomogeneity of the canopy architecture, as well as of individual plants. The canopy architecture also determines, to a large extent, the turbulent exchanges of momentum, heat, water vapor, CO_2, etc., within the canopy and is discussed briefly in the following section.

15.2.1 Canopy architecture

The size and shape of canopy elements, as well as their distributions in space and time, determine the physical characteristics of the canopy structure. A detailed description of a nonhomogeneous or nonuniform plant canopy in a complex terrain would necessarily involve many parameters. A commonly accepted simplification is made by assuming that the plant stand is horizontally uniform and its average characteristics may vary only with height above the ground.

The simplest characteristic of the canopy structure is its average height or thickness h_0. Along with the area density of the plants, h_0 determines the roughness parameter and the effective heights of the sources or sinks of momentum, heat, moisture, etc. Another widely used characteristic of the canopy architecture is the cumulative leaf area index (L_{AI}), defined as the area of the upper sides of leaves within a vertical cylinder of unit cross-section and height h_0. The leaf area index is a dimensionless parameter which is related to the foliage area density function $A(z)$ as (Ross, 1975)

$$L_{AI} = \int_0^{h_0} A(z)dz \tag{15.8}$$

Here, $A(z)$ is a local characteristic of the canopy architecture which represents the one-sided leaf area per unit volume of the canopy at the height z. A more convenient dimensionless local characteristic is the local cumulative leaf area index

$$L_{AI}(z) = \int_z^{h_0} A(z)dz \tag{15.9}$$

which represents the leaf area per unit horizontal area of the canopy above the height z. Note that $L_{AI}(0) = L_{AI}$.

In most crop and grass canopies leaves are the most active and dominant elements of interaction with the atmosphere, and the fractional area represented by plant stems and branches can be neglected. Exceptions are certain forest canopies in which woody elements may not be ignored. Particularly for a deciduous forest in late fall and winter, the woody-element silhouette area index (W_{AI}), which can be defined in a manner similar to L_{AI}, is an important architectural characteristic. The sum of the two indices is called the plant area index

$$P_{AI} = L_{AI} + W_{AI} \tag{15.10}$$

Both the leaf area density and the leaf area index are found to vary over a wide range for the various crops and forests. For a particular crop, they also depend on the density of plants and their stage of growth. For illustration purposes, Figures 15.1 and 15.2 show their observed distributions in a deciduous forest canopy in summer months. Note that for this particular canopy, $L_{AI} \cong 4.9$, $W_{AI} \cong 0.6$, and $P_{AI} \cong 5.5$. Observations of the same parameters for other forest and crop canopies are reviewed by Monteith (1975, 1976).

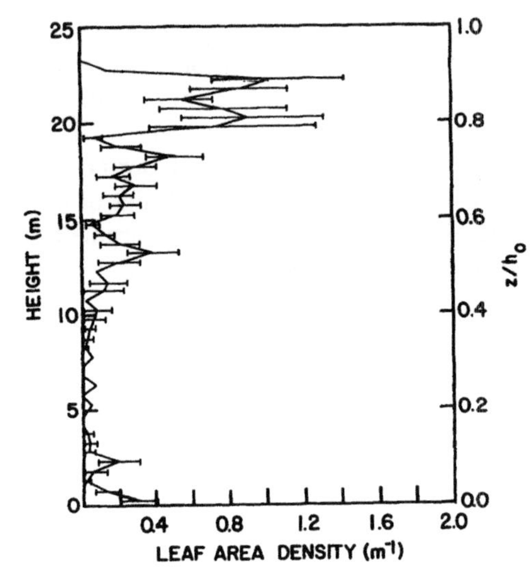

Figure 15.1 Vertical distribution of mean leaf area density in a deciduous forest in eastern Tennessee. [After Hutchinson *et al.* (1986).]

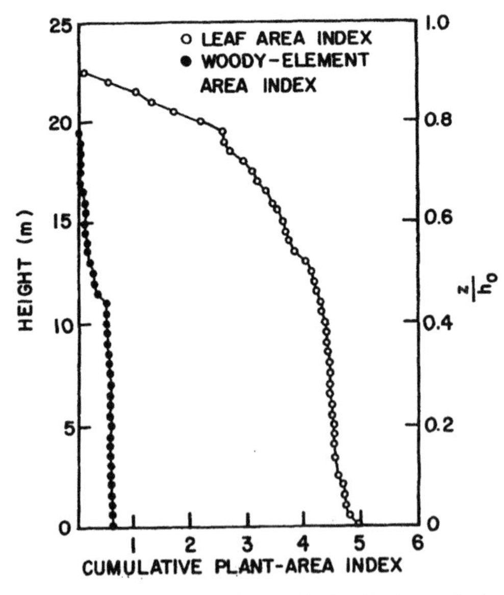

Figure 15.2 Vertical distribution of local cumulative leaf area index and woody element area index in a deciduous forest in eastern Tennessee. [After Hutchinson *et al.* (1986).]

The orientation of leaves is characterized by the mean leaf inclination angle from the horizontal. One can also measure the cumulative frequency distribution of the inclination angle. The foliage of many plant stands are more or less uniformly oriented, but some may display a narrow range of inclination angles around the horizontal, vertical, or somewhere in between (these are called panophile, erectophile and plagiophile canopies). A useful inclination index of the foliage is defined by Ross (1975). This index also depends on plant species and has a height and seasonal dependency.

15.2.2 Penetration of shortwave radiation

For a given incident solar radiation at the top of the canopy, the radiation regime in a plant canopy is determined by the canopy architecture, as well as by radiative properties of the various canopy elements and the ground surface. At any particular level in the canopy, the incoming solar radiation (both direct and diffuse) may vary considerably in the horizontal due to the presence of sunflecks and shadow areas with transitional areas, called penumbra, in between. Therefore, appropriate spatial averaging is necessary when we represent the radiative flux density as a function of height alone.

The transmission of incoming shortwave radiation into a plant canopy shows an approximately exponential decay with depth of penetration (more appropriately, the depth-dependent leaf area index), following the Beer–Bouguer law

$$R_{S\downarrow}(z) = R_{S\downarrow}(h_0)\exp[-\alpha L_{AI}(z)] \qquad (15.11)$$

in which α is an extinction coefficient. The same law, perhaps with a slightly different value of extinction coefficient, applies to the net shortwave radiation. A comparison of observed profiles of incoming and net shortwave radiations in a grass canopy with Equation (15.11) is shown in Figure 15.3. The diurnal variation of the extinction coefficient for the same canopy is shown in Figure 15.4. Radiation measurements in forest canopies also show the approximate validity of Equation (15.11), with P_{AI} replacing L_{AI}.

Foliage in a canopy not only attenuates the radiative flux density but it also changes its spectral composition. Leaves absorb visible (PAR) radiation more strongly than they absorb the near infrared radiation (NIR). This selective absorption reduces the photosynthetic value of the solar radiation as it penetrates through greater depths in the canopy. The spectral composition of total radiation also differs in sunflecks and shaded areas; in sunflecks the spectrum is similar to that of the incident total radiation, while in shaded areas the NIR dominates.

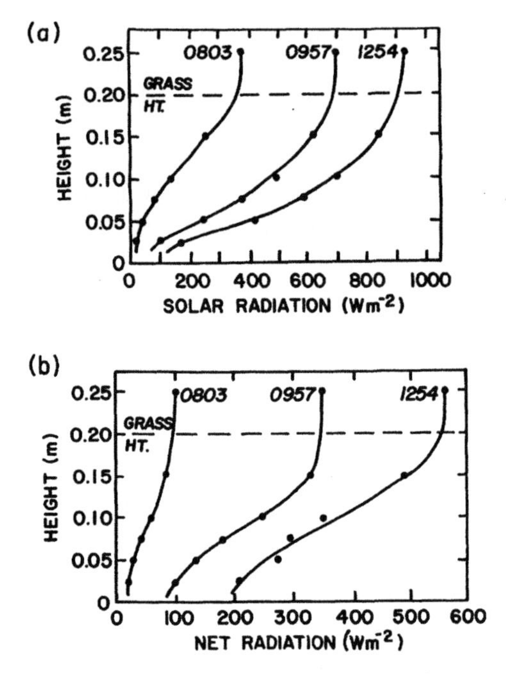

Figure 15.3 Observed profiles of (a) incoming solar radiation (R_S) and (b) net all-wave radiation (R_N) in a 0.2 m stand of native grass at Matador, Saskatchewan, on a clear summer day. [From Oke (1987); after Ripley and Redmann (1976).]

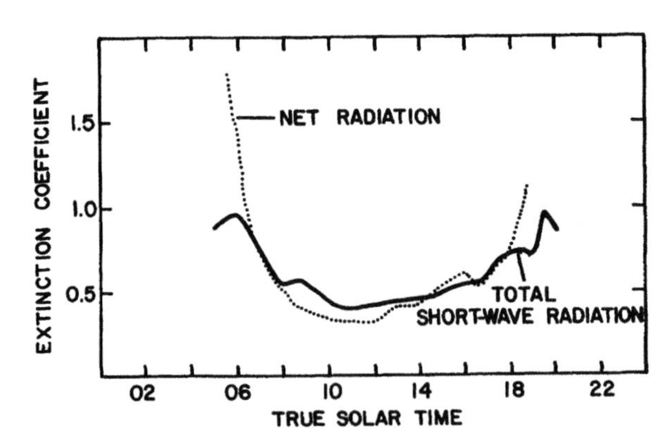

Figure 15.4 Diurnal variations of incoming solar and net radiation extinction coefficients for the native grass at Matador, Saskatchewan. [After Ripley and Redmann (1976).]

The shortwave reflectivity or albedo of a plant canopy depends on the average albedo of individual leaves, the canopy architecture (in particular, the canopy height, the leaf area index, and the leaf inclination index), and the solar altitude angle. The latter two factors determine the amount of penetration, radiation trapping, and mutual shading within the canopy. The longer the path of attenuation, the greater the amount of radiation to be trapped and the less to be reflected. Consequently, the albedo of most plant canopies is a decreasing function of both the canopy height and the solar altitude (Oke, 1987, Chapter 4).

15.2.3 Longwave radiation

The total or net longwave radiation (R_L) at any height within a plant canopy has four components: (1) the longwave radiation from the atmosphere which has penetrated through the upper layer of the canopy without interception; (2) the longwave radiation from the upper leaves and other canopy elements; (3) the radiation from lower canopy elements; and (4) the part of outgoing longwave radiation from the ground surface which is not intercepted by lower canopy elements. The net longwave radiation at the top of a plant canopy is almost always a loss, as with most other surfaces. However, the net loss usually diminishes toward the ground due to the reduction in the sky view factor (SVF), as the cold sky radiative sink is increasingly being replaced by relatively warmer vegetation surfaces. Consequently, in the lower part of a tall and dense canopy the role of sky becomes insignificant and R_L depends largely on the ground-surface temperature, the vertical distribution of mean foliage temperature, and ground and leaf surface emissivities. If the foliage temperature distribution is nearly uniform, which may be the usual case, R_L is also approximately constant, independent of height. The variation of R_L in the upper parts of the canopy is much more complicated (Ross, 1975).

15.3 Wind Distribution in Plant Canopies

Wind profiles above plant canopies ($z \geq 1.5h_0$) have been observed to follow the log law or the modified log law, depending on the atmospheric stability. Just above and within the canopy, however, the observed profiles deviate systematically (toward higher wind speeds) from the theoretical (modified log law) profile. The actual flow around roughness elements being three-dimensional, we can think of some generalized vertical profiles of wind and momentum flux only in a spatially averaged sense. With mechanical mixing processes dominating within the canopy, the stability effects can safely be ignored. If the large-scale pressure gradients can also be ignored, as is commonly done in the above-canopy surface layer, the equations of motion degenerate to the usual condition that the divergence of the total vertical momentum flux must be zero. The main difference between the flow inside the canopy and that above it is that considerable momentum is absorbed over the depth of the canopy in the form of drag due to the various canopy elements.

An appropriate equation for horizontally averaged wind flow within a canopy is (Plate, 1971; Businger, 1975)

$$\partial \overline{uw}/\partial z = -(1/\rho)(\partial P/\partial x) - \tfrac{1}{2}C_d A U^2 \tag{15.12}$$

where C_d is the average drag coefficient of the plant elements and A is the effective aerodynamic surface area of the vegetation per unit volume. Both C_d and A are likely to be complicated functions of height and the canopy structure. When the pressure term is negligible, e.g., in the upper part of the canopy, Equation (15.12) reduces to

$$\partial \overline{uw}/\partial z = -\tfrac{1}{2}C_d A U^2 \tag{15.13}$$

Practical models of wind flow in canopy are based on the solution to Equation (15.13) with appropriate assumptions for C_d, A, and the relation between shear stress and mean velocity gradient (e.g., an eddy viscosity or a mixing-length hypothesis). Using the simplest assumption that C_d is independent of height and the mixing-length relation [Equation (9.21)] is valid with a constant l_m, the solution to Equation (15.13) is an exponential wind profile (Inoue, 1963)

$$U(z)/U(h_0) = \exp[-n(1 - z/h_0)] \tag{15.14}$$

where $U(h_0)$ is the wind speed at the top of the canopy and $n \equiv (C_d A/4l_m^2)^{1/3} h_0$.

An exponential profile similar to Equation (15.14) can also be derived from an eddy viscosity relation for momentum flux with an exponential variation of K_m which is consistent with a constant l_m (Thom, 1975). If, on the contrary, K_m is assumed constant, the resulting canopy wind profile is of the form

$$U(z)/U(h_0) = [1 + m(1 - z/h)]^{-2} \tag{15.15}$$

where m is a numerical coefficient characterizing the plant canopy. The actual difference in the profile shapes given by Equations (15.14) and (15.15) is small for the appropriate choice of profile parameters n and m (compare the two, for example, for $n = m = 3$), despite the strong contrast in the associated K_m profiles. This seems to indicate that the predicted canopy wind profile is not very sensitive to the assumed K_m or l_m distribution within reasonable limits.

The simplifications and assumptions used in the derivation of the exponential profile [Equation (15.14)] are such that one may wonder if it could represent the observed wind profiles in plant canopies. Surprisingly, however, a number of observed profiles do agree with Equation (15.14), especially in the upper half of canopies. This is partially because the parameter n is only weakly dependent on C_d and A. Figure 15.5 shows a comparison of some observed profiles in wheat and corn canopies with the exponential profile Equation (15.14). The data suggest that a unique (similarity) profile shape may exist for each canopy type. Cionco (1972) has summarized the best estimated empirical values of n for various canopies; they usually lie between 2 and 4. This simple theoretical or semiempirical model may represent only a part of the canopy wind profile. The model can be refined further by considering variations with height of C_d, A, and l_m (Cionco, 1965).

In a tall and well-developed canopy such as a mature forest, most of the momentum is absorbed in the upper part of the canopy where leaves are concentrated. Farther down, as the leaf area density decreases to zero, the pressure gradient term in Equation (15.12) is likely to become more significant and perhaps dominate over the flux-divergence term. Consequently, wind speed may increase with decreasing height above the ground, until surface friction reverses this trend close to the ground level. This suggests that a low-level wind maximum (jet) may appear in the lower part of certain forest canopies. This is indeed confirmed by many observations, a sample of which is shown in Figure 15.6. Note that a variety of wind profile shapes are found in forest canopies.

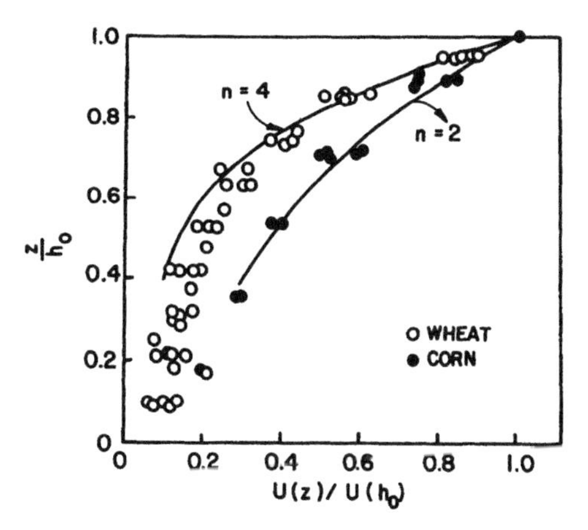

Figure 15.5 Comparison of observed mean velocity profiles in wheat and corn canopies with the exponential profile Equation (15.14) with $n = 4$ and 2, respectively. [After Plate (1971) and Businger (1975).]

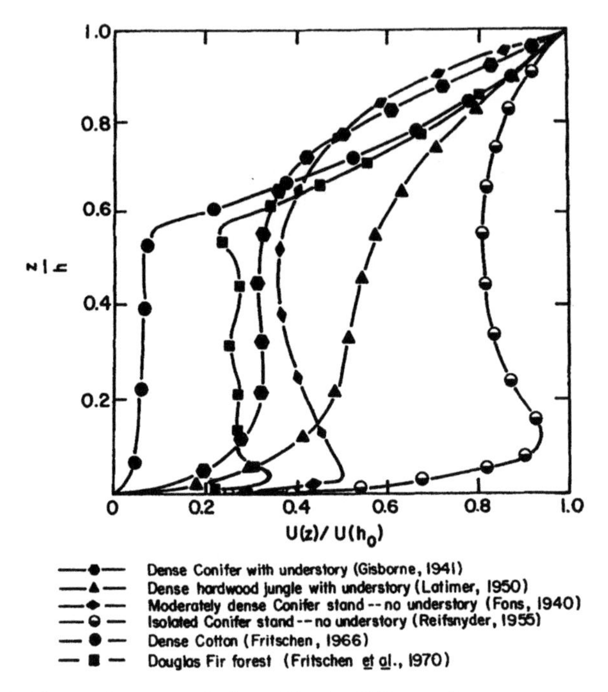

Figure 15.6 Observed mean wind profiles in different forest canopies. [From Businger (1975); after Fritschen *et al.* (1970).]

15.4 Temperature and Moisture Fields

Temperature profiles have been measured in various crop and forest canopies; these are quite different from those above the canopies. There are differences in canopy temperature profiles due to different canopy architectures. There are also significant diurnal variations in these due to variations of dominant energy fluxes. Here, we shall discuss only some typical observations and point out some of the common features of canopy temperature and humidity profiles. For a more comprehensive review of the literature the reader should refer to Monteith (1975, 1976).

Figure 15.7 shows the observed hourly averaged air temperature profiles within and above an irrigated soybean crop over the course of a day. Often, during daytime, there is a temperature maximum near middle to upper levels of the canopy. Since the sensible heat exchange in the canopy is expected to be down the temperature gradient, the heat flux is downward (negative) below the level of temperature maximum and upward (positive) above it. Thus, the sensible heat diverges away from the source region which coincides with the

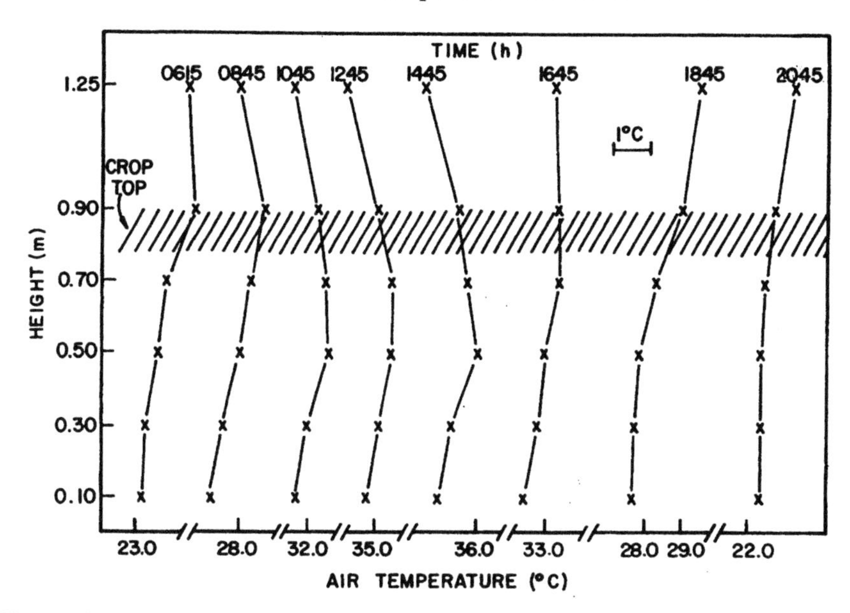

Figure 15.7 Mean air temperature profiles within and above an irrigated soybean crop on a summer day at Mead, Nebraska. [After Rosenberg *et al.* (1983).]

level of maximum leaf area density where most of the solar radiation is absorbed. The temperature inversion in the lower part of the canopy is a typical feature of daytime temperature profiles in tall crop and forest canopies. At night, temperature profiles in lower parts of canopies are close to isothermal, because canopies trap most of the outgoing longwave radiation. Often, the minimum temperature occurs not at the ground level but just below the crown. The nocturnal temperature inversion extends from above this level. The temperature gradient or inversion strength above a canopy is usually much smaller than that over a bare ground surface.

Figure 15.8 gives the hourly mean profiles of wind speed, temperature, water vapor pressure, and carbon dioxide inside and above a barley canopy on a summer day, for which the observed canopy energy budget is given in Figure 2.4. The observed wind and temperature profiles display the same features as discussed above. The daytime profiles of water vapor pressure show the expected decrease with height through much of the canopy and the surface layer above; a slight increase (inversion) in the upper part of the canopy is probably due to increased transpiration from the leaves in that region. Note that both the soil surface and foliage are moisture sources. The nocturnal vapor pressure profiles are more complicated by the processes of dewfall and guttation on leaves, while weak evaporation may continue from the soil surface.

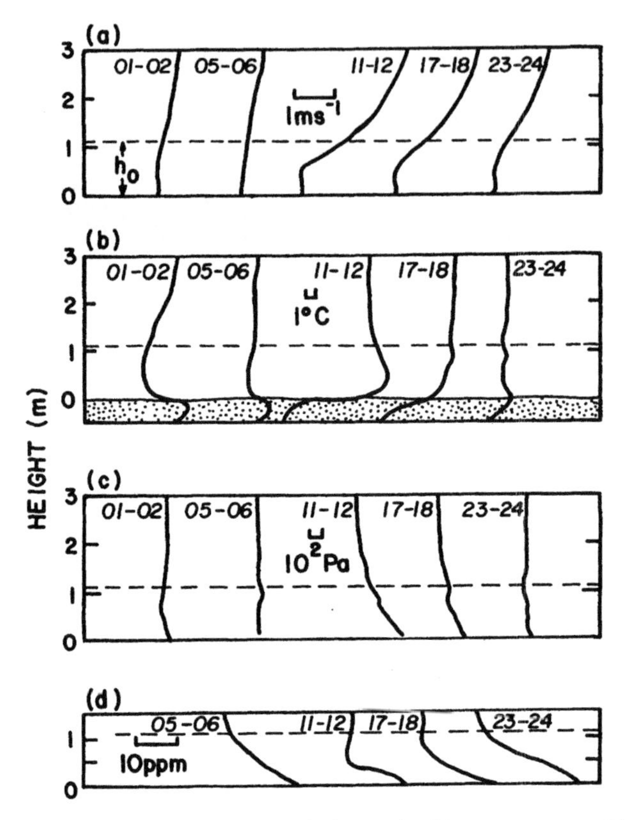

Figure 15.8 Observed profiles of (a) wind speed, (b) temperature, (c) water vapor pressure, and (d) carbon dioxide concentration in and above a barley crop on a summer day at Rothamsted, England. [From Oke (1987) after Long *et al.* (1964).]

Temperature and water vapor pressure profiles in forest canopies are similar to those within tall crops. The main difference is that the gradients are much weaker and diurnal variations are smaller in forest canopies. Even on warm, sunny days, air is cooler and more humid inside forests. Figure 15.9 presents the typical daytime profiles of meteorological variables and foliage area density. Both the temperature and the water vapor pressure profiles show maximum values at the level of maximum foliage density where the sources of heat (radiative absorption) and water vapor (transpiration) are most concentrated. Beneath this there is a temperature inversion, but nearly uniform vapor pressure profile; an inversion in the latter would be expected if the forest floor were dry. At night the temperature profile (not shown) is reversed, with a minimum occurring near the level of maximum foliage density. Due to reduced sky view

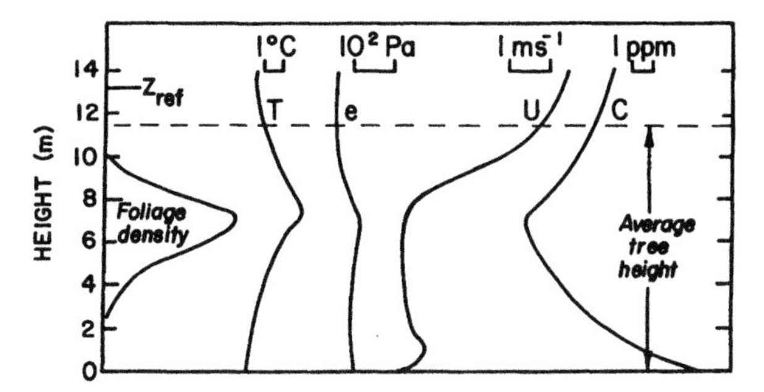

Figure 15.9 Observed mean profiles of foliage density and meteorological variables in a Sitka spruce forest on a sunny summer day near Aberdeen, Scotland. [After Jarvis *et al.* (1976).]

factor and trapping of longwave radiation in the lower part of the canopy, temperatures remain mild there, compared to those in open areas. Dewfall, if any, is largely confined to the upper part of the canopy, just below the crown. In the absence of dewfall, specific humidity or water vapor pressure is a decreasing function of height with weak evaporation from the soil surface as the only source of water vapor in the canopy.

15.5 Turbulence in and above Plant Canopies

The generation of turbulence in a plant canopy is a much more complex process than that over a flat bare surface or in a homogeneous boundary layer above a plant canopy. In addition to the usual sources of turbulence, such as mean wind shear and buoyancy, each canopy element is a source of turbulence due to its interaction with the flow and generation of separated shear layer and wake. The wake-generated turbulence in the canopy has characteristic large-eddy scales comparable to the characteristic sizes of the various canopy elements (e.g., leaves, branches and trunk). These scales are usually much smaller than the large-eddy scales of turbulence ($z - d_0$ and $h - d_0$) above the canopy. However, large-scale turbulence also finds its way through the canopy intermittently during short periods of downsweeps bringing in higher momentum fluid from the overlying boundary layer. It has been observed that the bulk of turbulent exchange between the atmosphere and a canopy may be occurring during these large-scale downsweeps and compensating low-momentum ejections or bursts of high turbulence. Thus, canopy turbulence is distinguished by a multiplicity of scales and generating mechanisms. It is also characterized by high-turbulence intensities, with turbulent velocity fluctuations of the same order or even larger than the mean velocity. These characteristics make canopy turbulence not only extremely difficult to measure, but also preclude a simple representation of it in a similarity form. Still, for convenience, the height above the surface is usually normalized by the canopy height and σ_u, σ_v, σ_w are normalized by the mean wind speed at the measurement height or by the friction velocity u_* based on the momentum flux in the constant flux layer above the canopy.

15.5.1 Turbulence intensities and variances

Limited observations of longitudinal turbulence intensity $i_u = \sigma_u/U$ suggest that, typically, $i_u \cong 0.4$ in agronomic crops, 0.6 in temperate forests and between 0.7 and 1.2 in tropical forests (Raupach and Thom, 1981). In a gross sense, i_u increases with canopy density. The turbulence intensity profiles broadly follow the leaf or plant area density profiles, but are also influenced by mean wind speed and stability within the canopy. The lateral and vertical turbulence intensities (i_v and i_w) are generally proportional to but smaller than i_u. A comparison of turbulence intensity profiles within rice and corn canopies shows that the profile is nearly uniform in the former, with an approximately uniform leaf area density distribution, while the profile in the corn canopy has a maximum near the top of the canopy (Uchijima, 1976). More recent measurements of turbulence in a corn canopy also indicate increasing turbulence activity with height. For example, Figure 15.10 shows the profiles of normalized standard deviations of horizontal and vertical velocity components observed by Wilson *et al.* (1982).

Measurements of turbulence within and above forest canopies have been reported by Bergstrom and Hogstrom (1989) and Leclerc *et al.* (1990). The former study over a pine forest indicates that the turbulence kinetic energy (TKE) remains approximately constant within the height interval ($h_0 < z < 2.4h_0$, where $h_0 = 20$ m is the height of the trees at the location of the measurement tower) studied. The influence of atmospheric stability, as measured by h_0/L within a deciduous forest canopy and $(z - d_0)/L$ above the canopy, on the TKE budget is examined by Leclerc *et al.* (1990). They show that buoyancy effects (production or destruction of TKE) are largest in the upper third of the canopy where the foliage is densest and the radiation load highest. The magnitude of the buoyant production term increases almost linearly with increasing instability in the upper region of the canopy. The onset of stability exerts a strong influence on the shear production of the TKE. Both the shear and buoyancy terms become small in the lower half of the canopy; in strong stratification the former also becomes slightly negative. Within the canopy, turbulence is largely produced in the wakes of individual trees and branches.

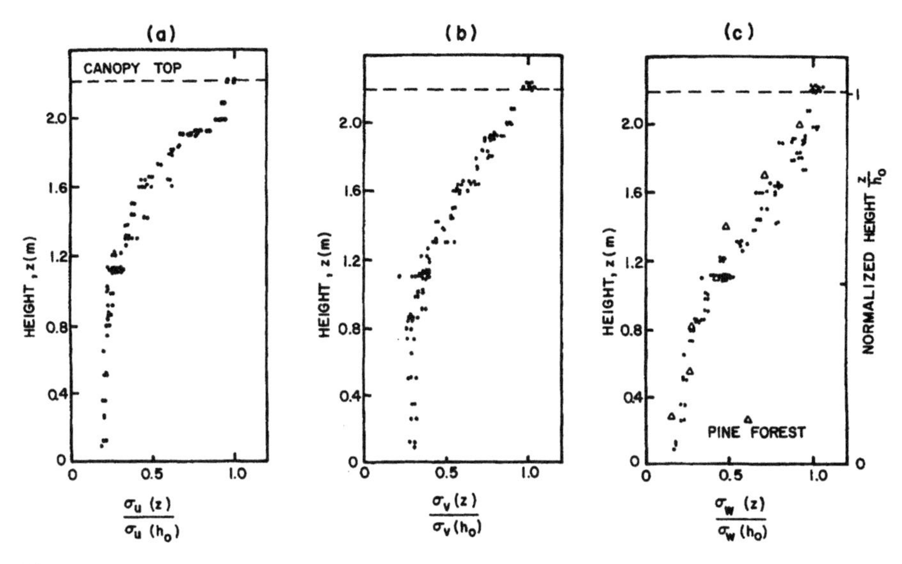

Figure 15.10 Observed profiles of the normalized standard deviations of (a) longitudinal, (b) lateral, and (c) vertical velocity fluctuations in a corn canopy. [After Wilson *et al.* (1982). Copyright © (1982) by D. Reidel Publishing Company. Reprinted by permission.]

15.5.2 Turbulent fluxes

Direct measurements of turbulent fluxes of momentum, heat, and water vapor are extremely difficult to make in plant canopies, and a few attempts have been made only recently. Figures 15.11 and 15.12 show some eddy correlation measurements of momentum and heat fluxes in a corn canopy. In general, fluxes increase with height and reach their maximum values somewhere near the top of the canopy. Their normalized profiles also show some day-to-day variations.

Only a few direct (eddy correlation) measurements of the turbulent fluxes of momentum, sensible heat, and water vapor above forest canopies have been reported in the literature (Verma *et al.*, 1986; Bergstrom and Hogstrom, 1989). Turbulent fluxes are found to remain practically constant (independent of height) in the height interval ($h_0 < z < 2.4h_0$) studied, which includes the roughness sublayer.

More often, turbulent fluxes are estimated from the observed mean profiles using gradient-transport hypotheses (e.g., eddy viscosity or mixing length). This approach suffers from many conceptual and practical limitations, however, and its validity has been seriously questioned by micrometeorologists in recent years (Raupach and Thom, 1981). It has been shown that vertical fluxes in plant

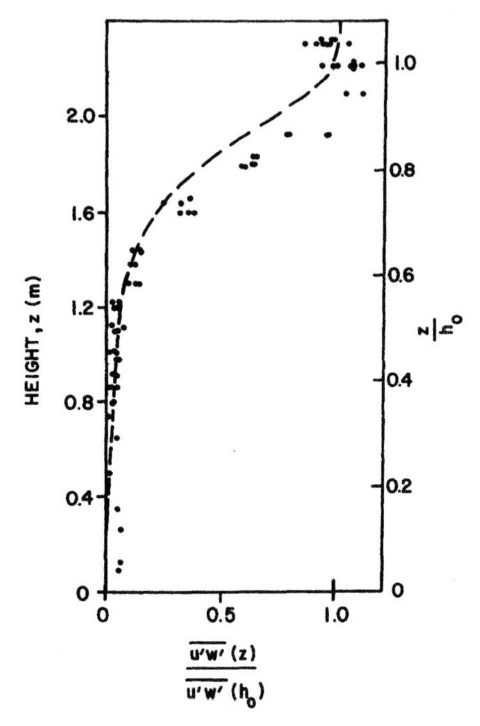

Figure 15.11 Measured and calculated profiles of normalized shear stress in a corn canopy. [After Wilson *et al.* (1982). Copyright © (1982) by D. Reidel Publishing Company. Reprinted by permission.]

canopies are not always down the mean vertical gradients, as implied in simpler gradient-transport hypotheses. For example, the vertical momentum flux $\overline{uw}$ may be driven by the horizontal gradient of mean vertical velocity ($\partial W/\partial x$) and/or by buoyancy, and need not be uniquely related to $\partial U/\partial z$. Even when it can be related to the latter through an eddy viscosity or a mixing-length relationship, the specification of K_m or l_m is not straightforward. A number of mutually contradictory K_m and l_m profiles have been proposed in the literature and there is much confusion regarding the shapes of flux profiles. There are very few direct observations of fluxes and gradients in canopy flows which support the gradient-transport (K) theories. On the contrary, observations show strong evidence for countergradient fluxes. Local-diffusion theories cannot cope with and predict such countergradient fluxes. The main reasons for their failure in canopy flow are three-dimensionality of the mean motion, nonlocal production of turbulence, and the multiplicity of the scales of turbulence.

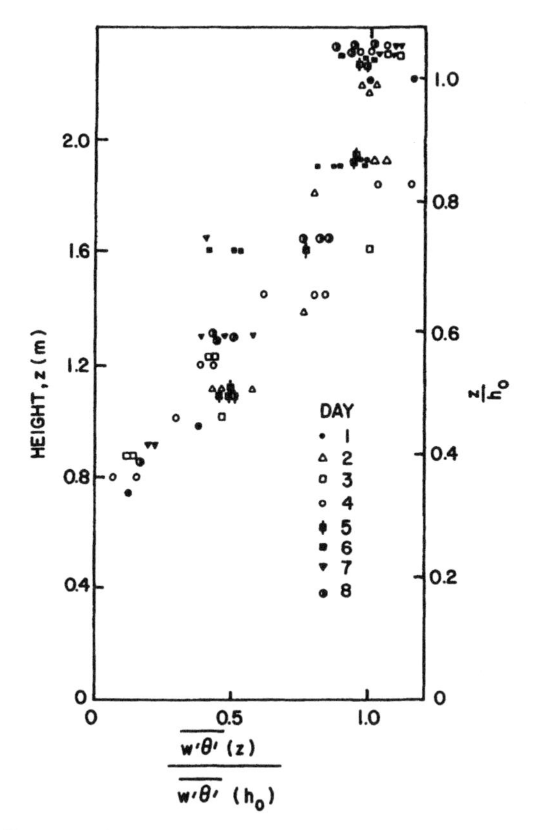

Figure 15.12 Daily normalized sensible heat flux profiles from eddy correlation measurements in a corn canopy. [After Wilson *et al.* (1982). Copyright © (1982) by D. Reidel Publishing Company. Reprinted by permission.]

A far better method of estimating turbulent transport or fluxes from measurements of mean wind, temperature, etc., inside canopies is based on the integration of mean momentum and other conservation equations with respect to height. For example, integration of Equation (15.13) yields

$$\frac{\overline{uw}(z) - \overline{uw}(0)}{\overline{uw}(h_0) - \overline{uw}(0)} = \frac{\int_0^z C_d A U^2 dz}{\int_0^{h_0} C_d A U^2 dz} \tag{15.16}$$

in which the integrals can be numerically evaluated for the given (measured) U profile, the leaf area density $A(z)$, and the average drag coefficient $C_d(z)$. The flux $\overline{uw}(0)$ at the canopy floor can usually be neglected. With this assumption, Equation (15.16) is compared with the observed profile in Figure 15.11. The most uncertain parameter in this indirect estimation of momentum flux is the canopy drag coefficient C_d, which is known only for certain idealized plant shapes such as a circular cylinder.

Similar relationships can be derived for sensible heat and water vapor fluxes, using their appropriate conservation equations for the canopy layer. Figures 15.13 and 15.14 show the observed profiles of soil and air temperatures and water vapor pressure, together with the calculated profiles of sensible and latent heat fluxes in a grass canopy ($h_0 \cong 0.17$ m). Radiative fluxes within this canopy for the same day are given in Figure 15.3. Note that all the fluxes increase with height and attain their maximum values just above the canopy. During the late morning period the sensible heat flux is negative in the shallow surface inversion

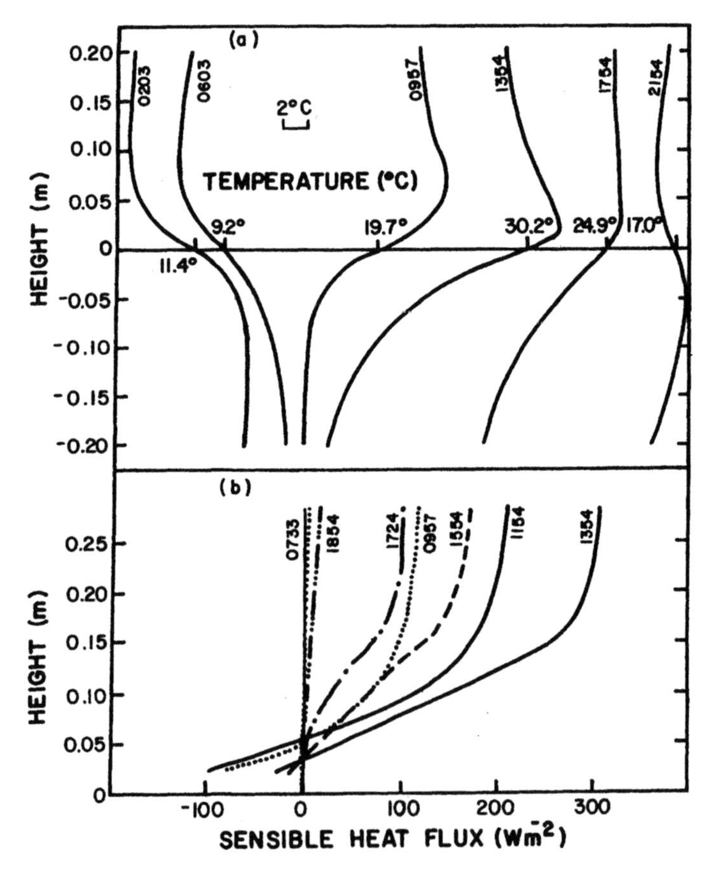

Figure 15.13 Observed profiles of (a) temperature in the canopy and upper layer of soil and (b) calculated profiles of sensible heat flux in a grass canopy at various times during a summer day. [After Ripley and Redmann (1976).]

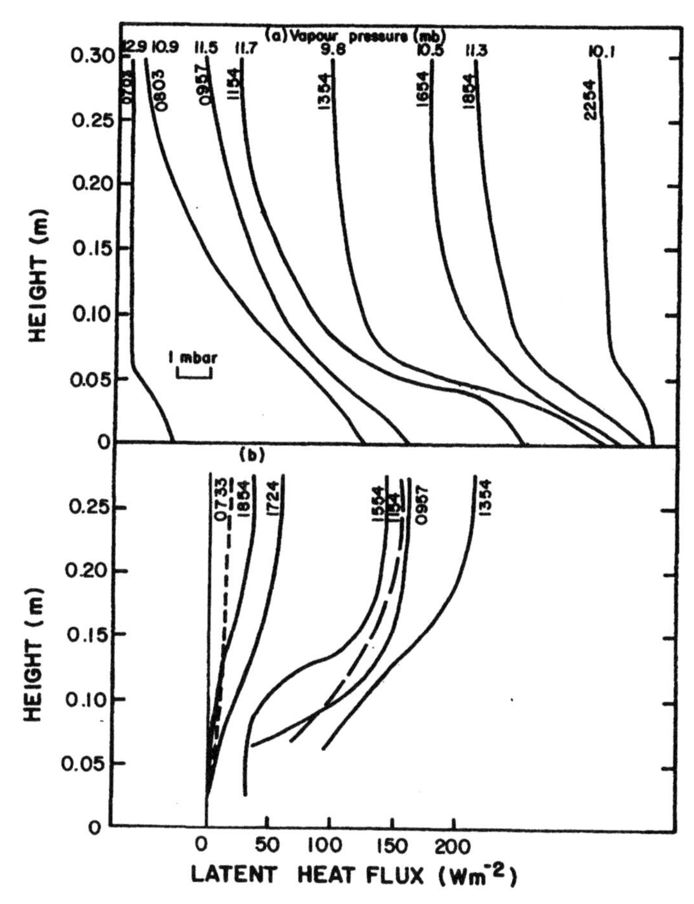

Figure 15.14 Observed profiles of (a) water vapor pressure (in millibars) and (b) calculated profiles of latent heat flux in a grass canopy at various times during a summer day. [After Ripley and Redmann (1976).]

layer that is destroyed later in the afternoon. This inversion in the lower part of the canopy probably produces a build-up of water vapor in this region during the morning period (see Figure 15.14a). The decrease in vapor pressure in early afternoon is a result of more vigorous turbulent mixing, bringing in drier air from higher levels. The highest vapor pressure occurs at the moist ground surface. The computed latent heat flux profiles in Figure 15.14b indicate that the major source region is around $z = d_0 \cong 0.10$ m in the morning and somewhat higher during the afternoon (Ripley and Redman, 1976).

The calculated profiles of eddy diffusivities of heat and momentum based on the above observations in a grass canopy during the morning period are shown

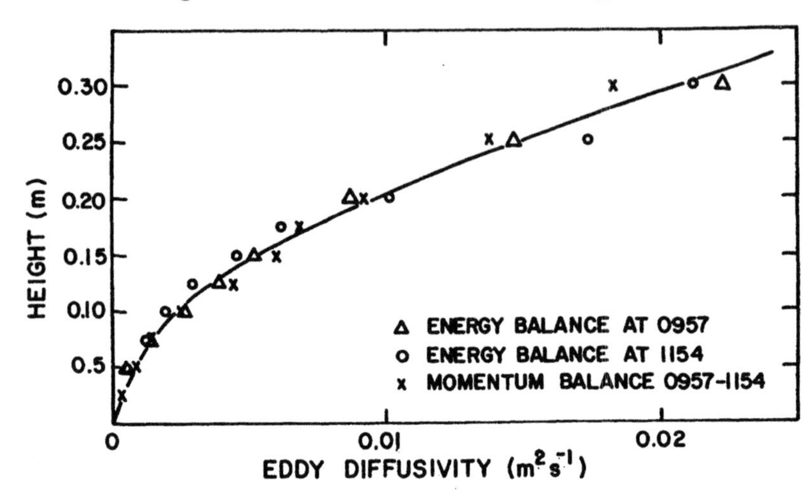

Figure 15.15 Vertical profiles of eddy diffusivity in a grass canopy calculated by the energy-balance and momentum-balance methods. [After Ripley and Redmann (1976).]

in Figure 15.15. Within the canopy the estimated values of K_h and K_m are in close agreement, indicating the dominance of mechanically generated turbulence in the canopy layer. Above the canopy ($z > h_0 \cong 0.17$ m), however, K_h values increase more rapidly with height than K_m values, which might be expected from increasing buoyancy effects on turbulence. These are probably the most consistent estimates of eddy diffusivities in a canopy flow, suggesting the usefulness and validity of gradient-transport relations, at least in a thin grasslike canopy. The observation that other periods of the day did not show such a good agreement cautions against any broad generalization of such results.

15.6 Applications

A primary objective of the various micrometeorological studies involving plant canopies has been to better understand the processes of momentum, heat, and mass exchanges between the atmosphere and the biologically active canopy. These exchanges influence the local weather as well as the microclimate in which plants grow. A basic understanding of exchange mechanisms and subsequent development of practical methods of parameterizing them in terms of more easily and routinely measured parameters is essential for many applications in meteorology, agriculture, forestry, and hydrology. More specific applications of agricultural and forest micrometeorology are as follows:

- Determining the mean wind, temperature, humidity, and carbon dioxide profiles in and above a plant cover.
- Determining the radiative and other energy fluxes in a plant cover.
- Predicting the ground surface and soil temperatures and moisture content.
- Estimating evaporation and transpiration losses in plant canopies.
- Determining photosynthetic activity and carbon dioxide exchange between plants and the atmosphere.
- Protection of vegetation from strong winds and extreme temperature.

Problems and Exercises

1. Explain the concept and physical significance of the zero-plane displacement. How is it related to the height and structure of a plant canopy?

2.

(a) Show that the aerodynamic resistance of a plant canopy in a stably stratified surface layer is given by

$$r_M = r_H = \frac{1}{ku_*}\left[\ln\left(\frac{z - d_0}{z_0}\right) + \beta\left(\frac{z - d_0}{L}\right)\right]$$

(b) Plot r_M as a function of the normalized height for the values of $u_* = 0.25$, 0.50, 0.75, and 1.0 m s^{-1} in neutral conditions.

(c) How does atmospheric stability affect the aerodynamic resistance?

3.

(a) Using the Bear–Bouguer law express the net all-wave radiation (R_N) as a function of height for a constant leaf area density canopy of height h_0.

(b) For the same canopy, plot $R_N(z)/R_N(h_0)$ as a function of z/h_0 for different values of the extinction coefficient $\alpha = 0.5, 0.75$, and 1.0.

(c) How is the net longwave radiation expected to vary with height in such a canopy?

4.

(a) Using Equation (15.13) and the mixing-length relation for $\overline{uw}$ with a constant mixing length l_m, derive the exponential wind profile [Equation (15.14)] for the canopy flow.

(b) Derive an expression for eddy viscosity K_m for the same flow.

(c) Plot $U(z)/U(h_0)$ as a function of z/h_0 for different values of the parameter $n = 1, 2, 3$, and 4.

5.

(a) Using the eddy viscosity relation for $\overline{uw}$ with a constant K_m, derive the canopy wind profile Equation (15.15).

(b) Compare the wind profiles given by Equations (15.14) and (15.15) for $n = m = 3$.

6.

(a) Why do the fluxes of momentum, heat, and water vapor generally increase with height in a plant canopy?

(b) Derive an expression for the momentum flux profile in a uniform plant area density canopy for the given exponential wind profile, Equation (15.14).

References

Alapaty, K., Pleim, J.E., Raman, S., Niyogi, D.S., and Byun, D.W. (1997). Simulation of atmospheric boundary layer processes using local and nonlocal-closure schemes. *J. Appl. Meteorol.*, **36**, 214–233.

Andre, J.C., Moor, G.D., Lacarrere, P., Therry, G., and Vachat, R. (1978). Modeling the 24-hour evolution of the mean and turbulent structures of the planetary boundary layer. *J. Atmos. Sci.*, **35**, 1861–1883.

Andren, A. (1995). The structure of stably stratified atmospheric boundary layers: A large-eddy simulation study. *Q. J. Roy. Meteorol. Soc.*, **121**, 961–985.

Andren, A. Brown, A., Graf, J., Moeng, C.H., Mason, P.J., Nieuwstadt, F.T.M., and Schumann, U. (1994). Large-eddy simulation of neutrally-stratified boundary layer: a comparison of four computer codes. *Q. J. Roy. Meteorol. Soc.*, **120**, 1457–1484.

Anthes, R.A., Seaman, N.L., and Warner, T.T. (1980). Comparisons of numerical simulations of the planetary boundary layer by a mixed-layer and multi-level model. *Monthly Weather Rev.*, **108**, 365–376.

Arritt, R.W. (1987). The effect of water surface temperature on lake breezes and thermal internal boundary layers. *Boundary-Layer Meteorol.*, **40**, 101–125.

Arya, S.P.S. (1972). The critical condition for the maintenance of turbulence in stratified flows. *Q. J. Roy. Meteorol. Soc.*, **98**, 264–273.

Arya, S.P.S. (1975). Geostrophic drag and heat transfer relations for the atmospheric boundary layer. *Q. J. Roy. Meteorol. Soc.*, **101**, 147–161.

Arya, S.P.S. (1977). Suggested revisions to certain boundary layer parameterization schemes used in atmospheric circulation models. *Monthly Weather Rev.*, **105**, 215–227.

Arya, S.P.S. (1978). Comparative effects of stability, baroclinicity and the scale-height ratio on drag laws for the atmospheric boundary layer. *J. Atmos. Sci.*, **35**, 40–46.

Arya, S.P.S. (1981). Parameterizing the height of the stable atmospheric boundary layer. *J. Appl. Meteorol.*, **20**, 1192–1202.

Arya, S.P.S. (1982). Atmospheric boundary layers over homogeneous terrain. In *Engineering Meteorology* (E.J. Plate, ed.), pp. 233–267. Elsevier, New York.

Arya, S.P.S. (1984). Parametric relations for the atmospheric boundary layer. *Boundary-Layer Meteorol.*, **30**, 57–73.

Arya, S.P.S. (1986). The schematics of balance of forces in the planetary boundary layer. *J. Clim. Appl. Meteorol.*, **24**, 1001–1002.

Arya, S.P.S. (1988). *Introduction to Micrometeorology*. Academic Press, San Diego, CA.

Arya, S.P. (1991). Finite-difference errors in estimation of gradients in the atmospheric surface layer. *J. Appl. Meteorol.*, **30**, 251–253.

Arya, S.P. (1999). *Air Pollution Meteorology and Dispersion*. Oxford University Press, New York.

Arya, S.P.S. and Byun, D.W. (1987) Rate equations for the planetary boundary layer depth (urban vs. rural). In *Modeling the Urban Boundary Layer*, pp. 215–251. American Meteorological Society, Boston.

Arya, S.P.S. and Wyngaard, J.C. (1975). Effect of baroclinicity on wind profiles and the geostrophic drag law for the convective planetary boundary layer. *J. Atmos. Sci.*, **32**, 767–778.

Arya, S.P.S., Capuano, M.E., and Fagen, L.C. (1987). Some fluid modeling studies of flow and dispersion over two-dimensional low hills. *Atmos. Environ.*, **21**, 753–764.

Atkinson, B.W. (1981). *Mesoscale Atmospheric Circulation*. Academic Press, New York.

Augstein, E., Schmidt, H., and Ostapoff, F. (1974). The vertical structure of the atmospheric planetary boundary layer in undisturbed trade wind over the Atlantic Ocean. *Boundary-Layer Meteorol.*, **6**, 129–150.

Bains, P.G. (1995). *Topographical Effects in Stratified Flows*. Cambridge University Press, New York.

Batchelor, G.K. (1970). *An Introduction to Fluid Dynamics*. Cambridge University Press, London and New York.

Bergstrom, H. and Hogstrom, U. (1989). Turbulent exchange above a pine forest, II. Organized structures. *Boundary-Layer Meteorol.*, **49**, 231–263.

Bradley, E.F. (1968). A micrometeorological study of velocity profiles and surface drag in the region modified by a change in surface roughness. *Q. J. Roy. Meteorol. Soc.*, **94**, 361–379.

Brost, R.A. and Wyngaard, J.C. (1978). A model study of the stably stratified planetary boundary layer. *J. Atmos. Sci.*, **35**, 1427–1440.

Brown, A.R., Derbyshire, S.H., and Mason, P.J. (1994). Large-eddy simulation of stable atmospheric boundary layers with a revised stochastic subgrid model. *Q. J. Roy. Meteorol. Soc.*, **120**, 1485–1512.

Brutsaert, W.H. (1982). *Evaporation into the Atmosphere*. Reidel, Boston, MA.

Budyko, M.I. (1958). *The Heat Balance of the Earth's Surface*. English translation from the Russian, US Dept. of Commerce, Washington, DC.

Businger, J.A. (1973). Turbulent transfer in the atmospheric surface layer. In *Workshop on Micrometeorology* (D.A. Haugen, ed.), pp. 67–100. American Meteorological Society, Boston, MA.

Businger, J.A. (1975). Aerodynamics of vegetated surfaces. In *Heat and Mass Transfer in the Biosphere, Part I: Transfer Processes in the Plant Environment* (D.A. deVries and N.H. Afgan, eds.), pp. 139–165. Halstead, New York.

Businger, J.A. and Arya, S.P.S. (1974). Height of the mixed layer in a stably stratified planetary boundary layer. *Adv. Geophys.*, **18A**, 73–92.

Businger, J.A., Wyngaard, J.C., Izumi, Y., and Bradley, E.F. (1971). Flux-profile relationships in the atmospheric surface layer. *J. Atmos. Sci.*, **28**, 181–189.

Byun, D.W. (1987). A two-dimensional mesoscale numerical model of St. Louis urban mixed layer. Ph.D. dissertation, North Carolina State University, Raleigh, NC.

Caughey, S.J. and Palmer, S.G. (1979). Some aspects of turbulence structure through the depth of the convective boundary layer. *Q. J. Roy. Meteorol. Soc.*, **105**, 811–827.

Chandrasekhar, S. (1961). *Hydrodynamic and Hydromagnetic Stability*. Oxford University Press, New York.

Charnock, H. (1955). Wind stress on a water surface. *Q. J. Roy. Meteorol. Soc.*, **81**, 639–640.

Ching, J.K.S. (1985). Urban-scale variations of turbulence parameters and fluxes. *Boundary-Layer Meteorol.*, **33**, 335–361.

Cionco, R.M. (1965). A mathematical model for air flow in a vegetative canopy. *J. Appl. Meteorol.*, **4**, 515–522.

Cionco, R.M. (1972). Intensity of turbulence within canopies with simple and complex roughness elements. *Boundary-Layer Meteorol.*, **2**, 435–465.

Clarke, R.H. (1970). Observational studies in the atmospheric boundary layer. *Q. J. Roy. Meteorol. Soc.*, **96**, 91–114.

Clarke, R.H. and Hess, G.D. (1974). Geostrophic departure and the functions of A and B in Rossby-number similarity theory. *Boundary-Layer Meteorol.*, **7**, 267–287.

Clarke, R.H., Dyer, A.J., Brook, R.R., Reid, D.G., and Troup, A.J. (1971). The Wangara Experiment: Boundary-layer data. *Technical Paper No. 19 CSIRO Division of Meteorological Physics, Australia.*

Claussen, M. (1987). The flow in a turbulent boundary layer upstream of a change in surface roughness. *Boundary-Layer Meteorol.*, **40**, 31–86.

Claussen, M. (1988). Models of eddy viscosity for numerical simulation of horizontally inhomogeneous, neutral surface-layer flow. *Boundary-Layer Meteorol.*, **42**, 337–369.

Counihan, J.C. (1975). Adiabatic atmospheric boundary layers: A review and analysis of data from the period 1880–1972. *Atmos. Environ.*, **9**, 871–905.

Crawford, K.C. and Hudson, H.R. (1973). The diurnal wind variation in the lowest 1500 feet in Central Oklahoma: June 1966–May 1967. *J. Appl. Meteorol.*, **12**, 127–132.

Davies, J.A., Robinson, P.J., and Nunez, M. (1970). Radiation measurements over Lake Ontario and the determination of emissivity. First Report, Contract No. H081276, Dept. Geography, McMaster University, Hamilton, Ontario.

Deacon, E.L. (1969). Physical processes near the surface of the earth. In *World Survey of Climatology, Vol. 2: General Climatology* (H. Flohn, ed.). Elsevier, Amsterdam.

Deardorff, J.W. (1968). Dependence of air-sea transfer coefficients on bulk stability. *J. Geophys. Res.*, **73**, 2549–2557.

Deardorff, J.W. (1970a). A three-dimensional numerical investigation of the idealized planetary boundary layer. *Geophys. Fluid Dyn.*, **1**, 377–410.

Deardorff, J.W. (1970b). Convective velocity and temperature scales for the unstable planetary boundary layer and for Raleigh convection. *J. Atmos. Sci.*, **27**, 1211–1213.

Deardorff, J.W. (1972a). Numerical investigation of neutral and unstable planetary boundary layers. *J. Atmos. Sci.*, **29**, 91–115.

Deardorff, J.W. (1972b). Parameterization of the planetary boundary layer for use in general circulation models. *Monthly Weather Rev.*, **100**, 93–106.

Deardorff, J.W. (1973). Three-dimensional numerical modeling of the planetary boundary layer. In *Workshop on Micrometeorology*, pp. 271–311, American Meteorological Society, Boston, MA.

Deardorff, J.W. (1974). Three-dimensional numerical study of the height and mean structure of a heated planetary boundary layer. *Boundary-Layer Meteorol.*, **7**, 81–106.

Deardorff, J.W. (1978). Observed characteristics of the outer layer. In *Short Course on the Planetary Boundary Layer* (A.K. Blackadar, ed.). American Meteorological Society, Boston, MA.

Deardorff, J.W. (1980). Stratocumulus-capped mixed layers derived from a three-dimensional model. *Boundary-Layer Meteorol.*, **18**, 495–527.

Detering, H.W. and Etling, D. (1985). Application of the E–ε turbulence model to the atmospheric boundary layer. *Boundary-Layer Meteorol.*, **33**, 113–133.

Ding, F., Arya, S.P., and Lin, Y.-L. (2001a). Large-eddy simulations of the atmospheric boundary layer using a new subgrid-scale model: Part I. Slighty unstable and neutral cases. *Environ. Fluid Mech.*, (in press).

Ding, F., Arya, S.P., and Lin, Y.-L. (2001b). Large-eddy simulations of the atmospheric boundary layer using a new subgrid-scale model: Part II. Weakly and moderately stable cases. *Environ. Fluid Mech.* (in press).

Doll, D., Ching, J.K.S., and Kaneshiro, J. (1985). Parameterization of subsurface heating for soil and concrete using net radiation data. *Boundary-Layer Meteorol.*, **32**, 351–372.

Drazin, P.G. and Reid, W.H. (1981). *Hydrodynamic Stability*. Cambridge University Press, London.

Driedonks, A.G.M. (1982). Models and observations of the growth of the atmospheric boundary layer. *Boundary-Layer Meteorol.*, **23**, 283–306.

Dutton, J.A. (1976). *The Ceaseless Wind: An Introduction to the Theory of Atmospheric Motion*. McGraw-Hill, New York.

Dycr, A.J. (1967). The turbulent transport of heat and water vapor in unstable atmosphere. *Q. J. Roy. Meteorol. Soc.*, **93**, 501–508.

Dyer, A.J. and Hicks, B.B. (1970). Flux-gradient relationships in the constant flux layer. *Q. J. Roy. Meteorol. Soc.*, **96**, 715–721.

Ekman, V.W. (1905). On the influence of the earth's rotation on ocean currents. *Ark. Mat. Astron. Fys.*, **2**, No. 11, 1–53.

Elliot, W.P. (1958). The growth of the atmospheric internal boundary layer. *Trans. Am. Geophys. Union*, **39**, 1048–1054.

Ferziger, J.H. (1993). Subgrid-scale modeling. In *Large-eddy Simulation of Complex Engineering and Geophysical Flows* (B. Galprin and S.A. Orszag, eds.), pp. 37–54. Cambridge University Press, Cambridge.

Fleagle, R.G., and Businger, J.A. (1980). *An Introduction to Atmospheric Physics*, 2nd edn. Academic Press, New York.

Friehe, C.A., and Schmitt, K.F. (1976). Parameterization of air-sea interface fluxes of sensible heat and moisture by the bulk aerodynamic formulas. *J. Phys. Oceanogr.*, **6**, 801–809.

Fritschen, L.J., Driver, C.H., Avery, C., Buffo, J., Edmonds, R., Kinerson, R., and Schiess, P. (1970). Dispersion of air tracers into and within a forested area: 3. *University of Washington Technical Report, ECOM-68-68-3*, US Army Electronic Command, Ft Hauchuca, Arizona.

Garratt, J.R. (1977). Review of drag coefficients over oceans and continents. *Monthly Weather Rev.*, **105**, 915–929.

Garratt, J.R. (1980). Surface influence on vertical profiles in the atmospheric near-surface layer. *Q. J. Roy. Meteorol. Soc.*, **106**, 803–819.

Garratt, J.R. (1982). Observations in the nocturnal boundary layer. *Boundary-Layer Meteorol.*, **22**, 21–48.

Garratt, J.R. (1992). *The Atmospheric Boundary Layer*. Cambridge University Press, Cambridge.

Garratt, J.R and Brost, R.A. (1981). Radiative cooling effects within and above the nocturnal boundary layer. *J. Atmos. Sci.*, **38**, 2730–2746.

Garratt, J.R. and Hicks, B.B. (1990). Micrometeorological and PBL experiments in Australia. *Boundary-Layer Meteorol.*, **50**, 11–29.

Garratt, J.R., Wyngaard, J.C., and Francey, R.J. (1982). Winds in the atmospheric boundary layer-prediction and observation. *J. Atmos. Sci.*, 39, 1307–1316.

Gates, D.M. (1980). *Biophysical Ecology*. Springer-Verlag, Berlin and New York.

Geiger, R. (1965). *The Climate Near the Ground.* Harvard University Press, Cambridge, MA.

Geiger, R., Aron, R.H., and Todhunter, P. (1995). *The Climate Near the Ground*, 5th edn. Vieweg, Wiesbaden, Germany.

Gray, W.M. and Mendenhall, B.R. (1973). A statistical analysis of factors influencing the wind veering in the planetary boundary layer. In *Climatological Research* (K. Fraedrich, M. Hantel, H.C. Korff, and E. Ruprecht, eds), pp. 167–194. Dummlers, Bonn.

Grimmond, C.S.B. and Oke, T.R. (1995). Comparison of heat fluxes from summertime observations in the suburbs of four North American cities. *J. Appl. Meteorol.*, **34**, 873–889.

Grimmond, C.S.B. and Oke, T.R. (1999). Aerodynamic properties of urban areas derived from analysis of surface form. *J. Appl. Meteorol.*, **38**, 1262–1292.

Halitsky, J. (1968). Gas diffusion near buildings. In *Meteorology and Atomic Energy* (D.H. Slade, ed.), pp. 221–255. Technical Information Center, US Dept. Energy, Oak Ridge, Tennessee.

Haltiner, G.J., and Martin, F.L. (1957). *Dynamical and Physical Meteorology-1968.* McGraw-Hill, New York.

Hanna, S.R. (1982). Applications in air pollution modelling. In *Atmospheric Turbulence and Air Pollution Modelling* (F.T.M. Nieuwstadt and H. van Dop, eds), pp. 275–310. Reidel, Dordrecht.

Hay, D.R. (1980). Fast response humidity sensors. In *Air-Sea Interaction: Instruments and Methods* (F. Dobson, L. Hasse, and R. Davis, eds), pp. 413–432. Plenum Press, New York.

Hess, S.L. (1959). *Introduction to Theoretical Meteorology.* Holt, New York.

Hicks, B.B. (1976). Wind profile relationships from the 'Wangara' experiment. *Q. J. Roy. Meteorol. Soc.*, **102**, 535–551.

Hinze, J.O. (1975). *Turbulence*, 2nd edn. McGraw-Hill, New York.

Hogstrom, U. (1985). Von Karman's constant in the atmospheric boundary layer flow: reevaluated. *J. Atmos. Sci.*, **42**, 263–270.

Hogstrom, U. (1988). Non-dimensional wind and temperature profiles in the atmospheric surface layer: a re-evaluation. *Boundary-Layer Meteorol.*, **42**, 55–78.

Hogstrom, U. (1996). Review of some basic characteristics of the atmospheric surface layer. *Boundary-Layer Meteorol.*, **78**, 215–246.

Hogstrom, U., Bergstrom, H., Smedman, A.-S., Halldin, S., and Lindroth, A. (1989).

Turbulent exchange above a pine forest, I: Fluxes and gradients. *Boundary-Layer Meteorol.*, **49**, 197–217.

Holt, T. and Raman, S. (1988). A review and comprehensive evaluation of multilevel boundary layer parameterizations for first-order and turbulent kinetic energy closure schemes. *Rev. Geophys.*, **26**, 761–780.

Holton, J.R. (1992). *An Introduction to Dynamic Meteorology*, 3rd Edn. Academic Press, San Diego.

Holtslag, A. A. M. (1984). Estimates of diabatic wind speed profiles from near-surface weather observations. *Boundary-Layer Meteorol.*, **29**, 225–250.

Holzworth, G.C. (1964). Estimates of mean maximum mixing depths in the contiguous United States. *Monthly Weather Rev.*, **92**, 235–242.

Hosker, R.P. (1984). Flow and diffusion near obstacles. In *Atmospheric Science and Power Production* (D. Randerson, ed.), pp. 241–326. Technical Information Center, US Dept. Energy, Oak Ridge, Tennessee.

Huang, C.-Y. and Raman, S. (1990). Numerical simulations of cold air advection over the Appalachian mountains and Gulf Stream. *Monthly Weather Rev.*, **118**, 343–362.

Hunt, J.C.R. and Simpson, J.E. (1982). Atmospheric boundary layer over nonhomogeneous terrain. In *Engineering Meteorology* (E.J. Plate, ed.), pp. 269–318. Elsevier, Amsterdam.

Hunt, J.C.R. and Snyder, W.H. (1980). Experiments on stably and neutrally stratified flow over a model three-dimensional hill. *J. Fluid Mech.*, **96**, 671–704.

Hutchinson, B.A., Matt, D.R., McMillen, R.T., Gross, L.J., Tajchman, S.J., and Norman, J.M. (1986). The architecture of a deciduous forest canopy in Eastern Tennessee, U.S.A. *J. Ecol.*, **74**, 635–646.

Inoue, E. (1963). On the turbulent structure of airflow within crop canopies. *J. Meteorol. Soc. Jpn.*, **41**, 317–326.

Izumi, Y. (1971). Kansas 1968 Field Program data report. Environmental Research Paper, No. 379, Air Force Cambridge Research Laboratories, Bedford, MA.

Izumi, Y. and Barad, M.L. (1963). Wind and temperature variations during development of a low-level jet. *J. Appl. Meteorol.*, **2**, 668–673.

Izumi, Y. and Caughey, S.J. (1976). Minnesota 1973 atmospheric boundary layer experiment data report. Environmental Research Paper, No. 547, Air Force Cambridge Research Laboratories, Bedford, MA.

Jarvis, P.,G., James, G.B., and Landsberg, J.J. (1976). Coniferous forest. In *Vegetation and the Atmosphere, Vol. 2, Case Studies* (J.L. Monteith, ed.), pp. 171–140. Academic Press, New York.

Jeram, N., Perkins, R.J., Fung, J.C.H., Davidson, M.J., Belcher, S.E., and Hunt, J.C.R. (1995). Atmospheric flow through groups of buildings and dispersion from localized sources. In *Wind Climate in Cities* (J.E. Cermak, A.G. Davenport, E.J. Plate and D.X. Viegas, eds), pp. 109–130. Kluwer Academic Publishers, Dordrecht.

Johnson, N.K. (1929). A study of the vertical gradient of temperature in the atmosphere near the ground. *Geophys. Memo.*, **46** (Meteorological Office, London).

Kaimal, J.C. and Finnigan, J.J. (1994). *Atmospheric Boundary Layer Flows*. Oxford University Press, New York.

Kaimal, J.C., Wyngaard J.C., Haugen, D.A., Cote, O.R., Izumi, Y., Caughey, S.J., and

Readings, C.J. (1976). Turbulence structure in the convective boundary layer. *J. Atmos. Sci.* **33**, 2152–2169.

Kazanski, A.B. and Monin, A.S. (1961). On the dynamical interaction between the atmosphere and the earth's surface. *Izv. Akad. Nauk. SSSR, Geophys. Ser.* **5**, 514–515.

Kitaigorodski, S.A. (1970). *The Physics of Air-Sea Interaction.* Gidromet. Izdatel'stvo, Leningrad (Israel Prog. Sci. Transl., Jerusalem).

Kolmogorov, A.N. (1941). Local structure of turbulence in an incompressible fluid at very high Reynolds numbers. *Dokl. Akad. Nauk SSSR*, **30**, 299–303.

Kondo, J. (1976). Heat balance of the East China Sea during the Air Mass Transformation Experiment. *J. Meteorol. Soc. Jpn.*, **54**, 382–398.

Kondratyev, K.Y. (1969). *Radiation in the Atmosphere.* Academic Press, New York.

Koracin, D. and Berkowicz, R. (1988). Nocturnal boundary-layer height: Observations by acoustic sounders and predictions in terms of surface-layer parameters. *Boundary-Layer Meteorol.*, **43**, 65–83.

Kosovic, B. and Curry, J.A. (2000). A large-eddy simulation study of a quasi-steady, stably stratified atmospheric boundary layer. *J. Atmos. Sci.*, **57**, 1052–1068.

Kraus, E.B. (1972). *Atmosphere-Ocean Interaction.* Oxford University Press, London and New York.

Kundu, P.K. (1990). *Fluid Mechanics.* Academic Press, San Diego, CA.

Kutzbach, J.E. (1961). Investigations of the modifications of wind profiles by artificially controlled surface roughness. Annual Report, Contract DA-36-039-SC-80282, University of Wisconsin, pp. 71–114.

Lacser, A. and Arya, S.P.S. (1986). A comparative assessment of mixing-length parameterizations in the stably stratified nocturnal boundary layer (NBL). *Boundary-Layer Meteorol.*, **36**, 53–70.

Lamb, H. (1932). *Hydrodynamics*, 6th edn. Cambridge University Press, London and New York.

Landsberg, H.E. (1981). *The Urban Climate.* Academic Press, New York.

Leclerc, M.Y., Beissner, K.C., Shaw, R.H., Hartog, G.D., and Neumann, H.H. (1990). The influence of atmospheric stability on the budgets of the Reynolds stress and turbulent kinetic energy within and above a deciduous forest. *J. Appl. Meteorol.*, **29**, 916–933.

Lenschow, D.H. (1973). Two examples of planetary boundary layer modification over the Great Lakes. *J. Atmos. Sci.*, **30**, 568–581.

Lenschow, D.H. (1986). *Probing the Atmospheric Boundary Layer.* American Meteorological Society, Boston, MA.

Lenschow, D.H., Li, X.S., Zhu, C.J., and Stankov, B.B. (1988). The stably stratified boundary layer over the Great Plains. I: Mean and turbulence structure. *Boundary-Layer Meteorol.*, **42**, 95–121.

Lettau, H.H. and Davidson, B. (1957). *Exploring the Atmosphere's First Mile.* Pergamon, Oxford.

Lettau, H., Riordan, A., and Kuhn, M. (1977). Air temperature and two-dimensional wind profiles in the lowest 32 meters as a function of bulk stability. *Antarct. Res. Ser.*, **25**, 77–91.

Liou, K.N. (1980). *An Introduction to Atmospheric Radiation.* Academic Press, New York.

Long, I.F., Monteith, J.L., Penman, H.L., and Szeicz, G. (1964). The plant and its environment. *Meteorol. Rundsch.*, **17**, 97–101.

Lowry, W.P. (1967). The climate of cities. *Sci. Am.*, **217**, 15–23.

Lowry, W.P. (1970). *Weather and Life: An Introduction to Biometeorology*. Academic Press, New York.

Lumley, J.L. (1980). Second-order modeling of turbulent flows. In *Prediction Methods for Turbulent Flows* (W. Kollmann, ed.), pp. 1–31. Hemisphere, London.

Mahrt, L. (1981). The early evening boundary layer transition. *Q. J. Roy. Meteorol. Soc.*, **107**, 329–343.

Mahrt, L., Heald, R.C., Lenschow, D.H., and Stankov, B.B. (1979). An observational study of the structure of the nocturnal boundary layer. *Boundary-Layer Meteorol.*, **17**, 247–264.

Mason, P.J. (1994). Large-eddy simulation: A critical review. *Q. J. Roy. Meteorol. Soc.*, **120**, 1–26.

Mason, P.J. and Derbyshire, S.H. (1990). Large-eddy simulation of the stably stratified atmospheric boundary layer. *Boundary-Layer Meteorol.*, **53**, 117–162.

Mason, P.J. and King, J.C. (1984). Atmospheric boundary layer over a succession of two-dimensional ridges and valleys. *Q. J. Roy. Meteorol. Soc.*, **110**, 821–845.

Mason, P.J. and Sykes, R.I. (1979). Flow over an isolated hill of moderate slope. *Q. J. Roy. Meteorol. Soc.*, **105**, 383–395.

Mason P.J. and Thompson D.J. (1992). Stochastic backscatter in large-eddy simulations of boundary layers. *J. Fluid Mech.*, **242**, 51–78.

McComb, W.D. (1990). *The Physics of Fluid Turbulence*. Clarendon Press, Oxford.

McDonald, R., Griffiths, R., and Hall, D. (1998). An improved method for the estimation of surface roughness of obstacle arrays. *Atmos. Environ.*, **32**, 1857–1864.

McNaughton, K. and Black, T.A. (1973). A study of evapotranspiration from a Douglas fir forest using the energy balance approach. *Water Resource Res.*, **9**, 1579–1590.

Mellor, G.L. and Yamada, T. (1974). A hierarchy of turbulence closure models for planetary boundary layers. *J. Atmos. Sci.*, **31**, 1791–1806.

Mellor, G.L. and Yamada, T. (1982). Development of a turbulence closure model for geophysical fluid problems. *Rev. Geophys.*, **20**, 851–875.

Moeng, C.-H. (1984). A large-eddy simulation model for the study of planetary boundary-layer turbulence. *J. Atmos. Sci.*, **41**, 2052–2062.

Moeng, C.-H. and Wyngaard, J.C. (1988). Spectral analysis of large-eddy simulations of the convective boundary layer. *J. Atmos. Sci.*, **45**, 3575–3587.

Monin, A.S. and Obukhov, A.M. (1954). Basic turbulent mixing laws in the atmospheric surface layer. *Tr. Geofiz. Inst. Akad. Nauk. SSSR*, **24**(151), 163–187.

Monin, A.S. and Yaglom, A.M. (1971). *Statistical Fluid Mechanics: Mechanics of Turbulence*, Vol. 1. MIT Press Cambridge, MA.

Monin, A.S. and Yaglom, A.M. (1975). *Statistical Fluid Mechanics: Mechanics of Turbulence*, Vol. 2. MIT Press, Cambridge, MA.

Monji, N. (1972). Budgets of turbulent energy and temperature variance in the transition zone from forced to free convection. Ph.D. thesis, University of Washington, Seattle.

Monteith, J.L. (1973). *Principles of Environmental Physics*. Edward Arnold, London.

Monteith, J.L. (1975). *Vegetation and the Atmosphere, Vol. 1: Principles*. Academic Press, New York.

Monteith, J.L. (1976). *Vegetation and the Atmosphere, Vol. 2: Case Studies*. Academic Press, New York.

Monteith, J.L. and Unsworth, M.H. (1990). *Principles of Environmental Physics*, 2nd edn. Edward Arnold, London.

Munn, R.E. (1966). *Descriptive Micrometeorology*. Academic Press, New York.

Nicholls, S. (1985). Aircraft observations of the Ekman layer during the Joint Air-Sea Interaction Experiment. *Q. J. Roy. Meteorol. Soc.*, **111**, 391–426.

Nicholls, S. and LeMone, M.A. (1980). The fair-weather boundary layer in GATE: The relationship of subcloud fluxes and structure to the distribution and enhancement of cumulus clouds. *J. Atmos. Sci.*, **37**, 2051–2067.

Nieuwstadt, F.T.M. (1984). Turbulence structure of the stable nocturnal boundary layer. *J. Atmos. Sci.*, **41**, 2202–2216.

Nieuwstadt, F.T.M., Mason, P.J., Moeng, C.-H., and Schumann, U. (1992). Large-eddy simulation of the convective boundary layer: A comparison of four computer codes. In *Turbulent Shear Flows* (F. Durst *et al.*, eds), **8**, pp. 343–367. Springer-Verlag, Berlin.

Obukhov, A.M. (1946). Turbulence in an atmosphere with a non-uniform temperature. *Tr. Inst. Teor. Geofiz., Akad. Nauk. SSSR*, **1**, 95–115 (English Translation in *Boundary-Layer Meteorol.*, **2**, 7–29, 1971).

Oke, T.R. (1970). Turbulent transport near the ground in stable conditions. *J. Appl. Meteorol.*, **9**, 778–786.

Oke, T.R. (1974). Review of urban climatology 1968–1973. WMO Technical Note No. 134, World Meteorological Organization, Geneva.

Oke, T.R. (1978). *Boundary Layer Climates*. Halsted, New York.

Oke, T.R. (1987). *Boundary Layer Climates*, 2nd edn, Halsted, New York.

Oke, T.R. (1988). The urban energy balance. *Progr. in Phys. Geog.*, **12**, 471–508.

Oke, T.R. (1995). The heat island of the urban boundary layer: Characteristics, causes and effects. In *Wind Climate in Cities* (J.E. Cermak, A.G. Davenport, E.J. Plate, and D.X. Viegas, eds), pp. 81–107. Kluwer Academic Publishers, Dordrecht.

Oke, T.R. and East, C. (1971). The urban boundary layer in Montreal. *Boundary-Layer Meteorol.*, **1**, 411–437.

Panofsky, H.A. and Dutton, J.A. (1984). *Atmospheric Turbulence*. Wiley (Interscience), New York.

Panofsky, H.A. and Townsend, A.A. (1964). Change of terrain roughness and the wind profile. *Q. J. Roy. Meteorol. Soc.*, **90**, 147–155.

Panofsky, H.A., Tennekes, H., Lenschow, D.H., and Wyngaard, J.C. (1977). The characteristics of turbulent velocity components in the surface layer under convective conditions. *Boundary-Layer Meteorol.*, **11**, 355–361.

Paulson, C.A. (1967). Profiles of wind speed, temperature and humidity over the sea. Ph.D. dissertation, University of Washington, Seattle.

Paulson, C.A., Leavitt, E., and Fleagle, R.G. (1972). Air-sea transfer of momentum, heat and water determined from profile measurements during BOMEX. *J. Phys. Oceanogr.*, **2**, 487–497.

Pedlosky, J. (1979). *Geophysical Fluid Dynamics*. Springer-Verlag, Berlin and New York.

Pendergrass, W. and Arya, S.P.S. (1984). Dispersion in neutral boundary layer over a step change in surface roughness – I. Mean flow and turbulence structure. *Atmos. Environ.*, **18**, 1267–1279.

Penman, H.L. (1948). Natural evaporation from open water, bare soil and grass. *Proc. R. Soc. London, A*, **193**, 120–145.

Pennell, W.T. and LeMone, M.A. (1974). An experimental study of turbulence structure in the fair weather trade wind boundary layer. *J. Atmos. Sci.*, **31**, 1308–1323.

Pielke, R.A. (1984). *Mesoscale Numerical Modeling*. Academic Press, New York.

Plate, E.J. (1971). Aerodynamic characteristics of atmospheric boundary layers. *AEC Crit. Rev. Ser. TID-15465*, Technical Information Center, US Department of Energy.

Pleim, J.E. and Chang, J.S. (1992). A non-local closure model for vertical mixing in the convective boundary layer. *Atmos. Environ.*, **26A**, 965–981.

Pond, S. (1975). The exchanges of momentum, heat and moisture at the ocean–atmosphere interface. *Numerical Models of Ocean Circulation*. National Academy of Science, Washington, DC.

Prandtl, L. (1905). *Verh. Int. Math. Kongr., 3rd,* Heidelburg, **1904**, pp. 484–491.

Rao, K.S. and Nappo, C.J. (1998). Turbulence and dispersion in the stable atmospheric boundary layer. In *Dynamics of Atmospheric Flows: Atmospheric Transport and Diffusion Processes* (M.P. Singh and S. Raman, eds), pp. 39–91. Computation Mechanics Publishers, Southampton.

Rao, K.S., Wyngaard, J.C., and Cote, O.R. (1974). The structure of two-dimensional internal boundary layer over a sudden change of surface roughness *J. Atmos. Sci.*, **31**, 738–746.

Raupach, M.R. (1979). Anomalies in flux-gradient relationships over forest. *Boundary-Layer Meteorol.*, **16**, 467–486.

Raupach, I. and Thom, A.S. (1981). Turbulence in and above plant canopies. *Ann. Rev. Fluid Mech.*, **13**, 97–129.

Raynor, G.S., Michael, P., Brown, R.A., and SethuRaman, S. (1975). Studies of atmospheric diffusion from a nearshore oceanic site. *J. Appl. Meteorol.*, **14**, 1080–1094.

Raynor, G.S., SethuRaman, S., and Brown, R.M. (1979). Formation and characteristics of coastal internal boundary layers during onshore flows. *Boundary-Layer Meteorol.*, **16**, 487–514.

Reynolds, O. (1894). On the dynamical theory of incompressible viscous fluids and the determination of the criterion. *Phil. Trans. Roy. Soc. London*, **186**, 123–161.

Richardson, L.F. (1920). The supply of energy from and to atmospheric eddies. *Proc. R. Soc. London*, **A97**, 354–373.

Rider, N.E., Philip, J.R., and Bradley, E.F. (1963). The horizontal transport of heat and moisture–a micrometeorological study. *Q. J. Roy. Meteorol. Soc.*, **89**, 506–531.

Ripley, E.A. and Redmann, R.E. (1976). Grassland. In *Vegetation and the Atmosphere, Vol. 2: Case Studies* (J.L. Monteith, ed.), pp. 351–398. Academic Press, New York.

Rosenberg, N.J., Blad, B.L., and Verma, S.B. (1983). *Microclimate: The Biological Environment*, 2nd edn. Wiley (Interscience), New York.

Ross, J. (1975). Radiative transfer in plant communities. In *Vegetation and the Atmosphere, Vol. 1: Principles* (J.L. Monteith, ed.), pp. 13–55. Academic Press, New York.

Roth, M. (1991). Turbulent transfer characteristics over a suburban surface. Ph.D Thesis, University of British Columbia, Vancouver, Canada.

Royal Aeronautical Society (1972). Characteristics of wind speed in the lower layers of the atmosphere near the ground: Strong winds (neutral atmosphere). Engineering Science Data Unit. No. 72026, London.

Schlichting, H. (1960). *Boundary-Layer Theory*, 4th edn. McGraw-Hill, New York.

Sellers, W.D. (1965). *Physical Climatology*. University of Chicago Press, Chicago, IL.

SethuRaman, S. (1978). Influence of mean wind direction on sea surface wave development. *J. Phys. Oceanogr.*, **8**, 926–929.

Sheppard, P.A. (1956). Airflow over mountains. *Q. J. Roy. Meteorol. Soc.*, **82**, 528–529.

Slatyer, R.O. and McIlroy, I.C. (1961). *Practical Micrometeorology*. CSIRO Division of Meteorological Physics, Melbourne.

Smagorinski, J. (1963). General circulation experiments with the primitive equations, Part I: The basic experiment. *Monthly Weather Rev.*, **91**, 99–152.

Smedman, A.-S. (1991). Some turbulence characteristics in stable atmospheric boundary layer flow. *J. Atmos. Sci.*, **48**, 856–868.

Smedman, A.-S. and Hogstrom, U. (1983). Turbulent characteristics of a shallow convective internal boundary layer. *Boundary-Layer Meteorol.*, 25, 271–287.

Smith, S.D. (1980). Wind stress and heat flux over the ocean in gale force winds. *J. Phys. Oceanogr.*, **10**, 709–726.

Smith, S.D. and Banke, E.G. (1975). Variation of the sea surface drag coefficient with wind speed. *Q. J. Roy. Meteorol. Soc.*, **101**, 665–673.

Smith, S.D., Fairall, C.W., Geernaert, G.L., and Hasse L. (1996). Air–sea fluxes: 25 years of progress. *Boundary-Layer Meteorol.*, **78**, 247–290.

Snyder, W.H., Thompson, R.S., Eskridge, R.E., Lawson, R.E., Castro, I.P., Lee, J.T., Hunt, J.C.R., and Ogawa, Y. (1985). The structure of strongly stratified flow over hills: Dividing-streamline concept. *J. Fluid Mech.*, **152**, 249–288.

Sorbjan, Z. (1989). *Structure of Atmospheric Boundary Layer*. Prentice-Hall, Englewood Cliffs, NJ.

Spangler, T.C. and Dirks, R.A. (1974). Mesoscale variations of the urban mixing height. *Boundary-Layer Meteorol.*, **6**, 423–441.

Stanhill, G. (1969). A simple instrument for field measurement of turbulent diffusion flux. *J. Appl. Meteorol.*, **8**, 509–513.

Stearns, C.R., and Lettau, H.H. (1963). Report on two wind profile modification experiments in air flow over the ice of Lake Mendota. Annual Report, Contract DA-36-039-AMC-00878, pp. 115–138. University of Wisconsin, Madison, Wisconsin.

Stull, R.B. (1988). *An Introduction to Boundary Layer Meteorology*. Kluwer Academic, Dordrecht.

Stull, R.B. (1991). Static stability–an update. *Bull. Amer. Meteorol. Soc.*, **72**, 1521–1529.

Stull, R.B. (1993). Review of transilient turbulence theory and nonlocal mixing. *Boundary-Layer Meteorol.*, **62**, 21–96.

Stunder, M. and SethuRaman, S. (1985). A comparative evaluation of the coastal internal boundary-layer height equations. *Boundary-Layer Meteorol.*, **32**, 177–204.

Sullivan, P.P., McWilliams, J.C., and Moeng, C.-H. (1994). A subrid-scale model for large-eddy simulation of planetary boundary-layer flows. *Boundary-Layer Meteorol.*, **71**, 247–276.

Sutton, O.G. (1953). *Micrometeorology*. McGraw-Hill, New York.

Taylor, G.I. (1915). Eddy motion in the atmosphere. *Philos. Trans. Roy. Soc. London, A*, **CCXV**, 1–26.

Taylor, G.I. (1938). The spectrum of turbulence. *Proc. Roy. Soc. London*, **A164**, 476–490.

Taylor, P.A., Mason, P.J., and Bradley, E.F. (1987). Boundary-layer flow over low hills (a review). *Boundary-Layer Meteorol.*, **39**, 107–132.

Tennekes, H. (1982). Similarity relations, scaling laws and spectral dynamics. In *Atmospheric Turbulence and Air Pollution Modeling* (F.T.M. Nieuwstadt and H. van Dop, eds.), pp. 37–68. Reidel, Dordrecht.

Tennekes, H. and Lumley, J.L. (1972). *A First Course in Turbulence*. MIT Press, Cambridge, MA.

Therry, G., and Lacarrere, P. (1983). Improving the eddy kinetic energy model for planetary boundary layer description. *Boundary-Layer Meteorol.*, **25**, 63–88.

Thom, A.S. (1975). Momentum, mass and heat exchange of plant communities. In *Vegetation and the Atmosphere, Vol. 1: Principles* (J.L. Monteith, ed.), pp. 57–109. Academic Press, New York.

Townsend, A.A. (1965). Self-preserving flow inside a turbulent boundary layer. *J. Fluid Mech.*, **22**, 773–797.

Townsend, A.A. (1976). *The Structure of Turbulent Shear Flow*, 2nd edn. Cambridge University Press, London and New York.

Tritton, D.J. (1988). *Physical Fluid Dynamics*, 2nd edn. Oxford University Press, New York.

Tunick, A. (1999). A review of previous works on observing the atmospheric boundary layer through meteorological measurements. Report No. ARL-MR-448, Army Research Laboratory, Adelphi, MD.

Uchijima, Z. (1976). Maize and rice. In *Vegetation and the Atmosphere, Vol. 2: Case Studies* (J.L. Monteith, ed.), pp. 33–64. Academic Press, New York.

Vehrencamp, J.E. (1953). Experimental investigation of heat transfer at an air–earth interface. *Trans. Am. Geophys. Union*, **34**, 22–30.

Venkatram, A. (1977). A model of internal boundary-layer development. *Boundary-Layer Meteorol.*, **11**, 419–437.

Verma, S.B., Rosenberg, N.J., and Blad, B.L. (1978). Turbulent exchange coefficients for sensible heat and water vapor under advective conditions. *J. Appl. Meteorol.*, **17**, 330–338.

Verma, S.B., Baldocchi, D.D., Anderson, D.E., Matt, D.R., and Clement, R.J. (1986). Eddy fluxes of CO_2, water vapor, and sensible heat over a deciduous forest. *Boundary-Layer Meteorol.*, **36**, 71–91.

Vugts, H.F. and Businger, J.A. (1977). Air modification due to a step change in surface temperature. *Boundary-Layer Meteorol.*, **11**, 295–305.

Wayland, R.J. and Raman, S. (1989). Mean and turbulent structure of a baroclinic marine boundary layer during the 28 January 1986 cold-air outbreak (GALE 86). *Boundary-Layer Meteorol.*, **48**, 227–254.

Webb, E.K. (1970). Profile relationships: the log-linear range, and extension to strong stability. *Q. J. Roy. Meteorol. Soc.*, **96**, 67–90.

Welch, R.M., Cox, S.K., and Davis, J.M. (1980). Solar radiation and clouds. Meteorological Monographs No. 39.

West, E.S. (1952). A study of annual soil temperature wave. *Aust. J. Sci. Res. Ser. A*, **5**, 303–314.

Wilson, J.D., Ward, D.P., Thurtell, G.W., and Kidd, G.E. (1982). Statistics of atmospheric turbulence within and above a corn canopy. *Boundary-Layer Meteorol.*, **24**, 495–519.

Woo, H.G.C., Peterka, J.A., and Cermak, J.E. (1977). Wind-tunnel measurements in the wakes of structures. *NASA Contract. Report No. 2806.*

Wu, J. (1980). Wind stress coefficients over sea surface near neutral conditions–a revisit. *J. Phys. Oceanogr.*, **10**, 727–740.

Wylie, D.P. and Young, J.A. (1979). Boundary-layer observations of warm air modification over Lake Michigan using a tethered balloon. *Boundary-Layer Meteorol.*, **17**, 279–291.

Wyngaard, J.C. (1973). On surface-layer turbulence. In *Workshop on Micrometeorology* (D. Haugen, ed.), pp. 101–149. American Meteorological Society, Boston, MA.

Wyngaard, J.C. and Brost, R.A. (1984). Top-down and bottom-up diffusion of a scalar in the convective boundary layer. *J. Atmos. Sci.*, **41**, 102–112.

Wyngaard, J.C. and Moeng, C.-H. (1993). Large eddy simulation in geophysical turbulence parameterization: An overview. In *Large-Eddy Simulation of Complex Engineering and Geophysical Flows* (B. Galperin and S.A. Orszag, eds), pp. 349–398. Cambridge University Press, New York.

Wyngaard, J.C., Cote, O.R., and Rao, K.S. (1974). Modeling the atmospheric boundary layer. *Adv. Geophys.*, **18A**, 193–211.

Wyngaard, J.C., Pennell, W.T., Lenschow, D.H., and LeMone, M.A. (1978). The temperature–humidity covariance budget in the convective boundary layer. *J. Atmos. Sci.*, **35**, 47–58.

Yaglom, A.M. (1977). Comments on wind and temperature flux-profile relationships. *Boundary-Layer Meteorol.*, **11**, 89–102.

Yamada, T. (1976). On the similarity functions A, B and C of the planetary boundary layer. *J. Atmos. Sci.*, **33**, 781–793.

Zilitinkevich, S.S. (1972). On the determination of the height of the Ekman boundary layer. *Boundary-Layer Meteorol.*, **3**, 141–145.

Zilitinkevich, S.S. and Deardorff, J.W. (1974). Similarity theory for the planetary boundary layer of time-dependent height. *J. Atmos. Sci.*, **31**, 1449–1452.

Zilitinkevich, S.S., Laikhtman, D.L., and Monin, A.S. (1976). Dynamics of the atmospheric boundary layer. *Izv. Atmos. Ocean. Phys.*, **3**, 170–191.

Index

Page numbers in **bold** refer to main discussion and those in *italic* to illustrations and tables.

International Geophysics Series

EDITED BY

RENATA DMOWSKA
Division of Engineering and Applied Science
Harvard University
Cambridge, MA 02138

JAMES R. HOLTON
Department of Atmospheric Sciences
University of Washington
Seattle, Washington

H. THOMAS ROSSBY
University of Rhode Island
Kingston, Rhode Island

Volume 1 BENO GUTENBERG. Physics of the Earth's Interior. 1959*

Volume 2 JOSEPH W. CHAMBERLAIN. Physics of the Aurora and Airglow. 1961*

Volume 3 S. K. RUNCORN (ed.) Continental Drift. 1962*

Volume 4 C. E. JUNGE. Air Chemistry and Radioactivity. 1963*

Volume 5 ROBERT G. FLEAGLE AND JOOST A. BUSINGER. An Introduction to Atmospheric Physics. 1963*

Volume 6 L. DUFOUR AND R. DEFAY. Thermodynamics of Clouds. 1963*

Volume 7 H. U. ROLL. Physics of the Marine Atmosphere. 1965*

Volume 8 RICHARD A. CRAIG. The Upper Atmosphere: Meteorology and Physics. 1965*

Volume 9 WILLIS L. WEBB. Structure of the Stratosphere and Mesosphere. 1966*

Volume 10 MICHELE CAPUTO. The Gravity Field of the Earth from Classical and Modern Methods. 1967*

* Out of Print.

Volume 11 S. MATSUSHITA AND WALLACE H. CAMPBELL (eds.) Physics of Geomagnetic Phenomena. (In two volumes.) 1967*

Volume 12 K. YA KONDRATYEV. Radiation in the Atmosphere. 1969*

Volume 13 E. PALMÁN AND C. W. NEWTON. Atmospheric Circulation Systems: Their Structure and Physical Interpretation. 1969*

Volume 14 HENRY RISHBETH AND OWEN K. GARRIOTT. Introduction to Ionospheric Physics. 1969*

Volume 15 C. S. RAMAGE. Monsoon Meteorology. 1971*

Volume 16 JAMES R. HOLTON. An Introduction to Dynamic Meteorology. 1972*

Volume 17 K. C. YEH AND C. H. LIU. Theory of Ionospheric Waves. 1972*

Volume 18 M. I. BUDYKO. Climate and Life. 1974*

Volume 19 MELVIN E. STERN. Ocean Circulation Physics. 1975*

Volume 20 J. A. JACOBS. The Earth's Core. 1975*

Volume 21 DAVID H. MILLER. Water at the Surface of the Earth: An Introduction to Ecosystem Hydrodynamics. 1977

Volume 22 JOSEPH W. CHAMBERLAIN. Theory of Planetary Atmospheres: An Introduction to Their Physics and Chemistry. 1978*

Volume 23 JAMES R. HOLTON. An Introduction to Dynamic Meteorology, Second Edition. 1979*

Volume 24 ARNETT S. DENNIS. Weather Modification by Cloud Seeding. 1980*

Volume 25 ROBERT G. FLEAGLE AND JOOST A. BUSINGER. An Introduction to Atmospheric Physics, Second Edition. 1980

Volume 26 KUG-NAN LIOU. An Introduction to Atmospheric Radiation. 1980*

Volume 27 DAVID H. MILLER. Energy at the Surface of the Earth: An Introduction to the Energetics of Ecosystems. 1981*

Volume 28 HELMUT G. LANDSBERG. The Urban Climate. 1981

Volume 29 M. I. BUDYKO. The Earth's Climate: Past and Future. 1982*

Volume 30 ADRIAN E. GILL. Atmosphere–Ocean Dynamics. 1982

Volume 31 PAOLO LANZANO. Deformations of an Elastic Earth. 1982*

Volume 32 RONALD T. MERRILL AND MICHAEL W. MCELHINNY. The Earth's Magnetic Field. Its History, Origin, and Planetary Perspective. 1983*

Volume 33 JOHN S. LEWIS AND RONALD G. PRINN. Planets and Their Atmospheres: Origin and Evolution. 1983

Volume 34 ROLF MEISSNER. The Continental Crust: A Geophysical Approach. 1986

Volume 35 M. U. SAGITOV, B. BODKI, V. S. NAZARENKO, AND KH. G. TADZHIDINOV. Lunar Gravimetry. 1986*

* Out of Print.

Volume 36 JOSEPH W. CHAMBERLAIN AND DONALD M. HUNTEN. Theory of Planetary Atmospheres: An Introduction to Their Physics and Chemistry, Second Edition. 1987

Volume 37 J. A. JACOBS. The Earth's Core, Second Edition. 1987*

Volume 38 J. R. APEL. Principles of Ocean Physics. 1987

Volume 39 MARTIN A. UMAN. The Lightning Discharge. 1987*

Volume 40 DAVID G. ANDREWS, JAMES R. HOLTON AND CONWAY B. LEOVY. Middle Atmosphere Dynamics. 1987

Volume 41 PETER WARNECK. Chemistry of the Natural Atmosphere. 1988

Volume 42 S. PAL ARYA. Introduction to Micrometeorology. 1988

Volume 43 MICHAEL C. KELLEY. The Earth's Ionosphere. 1989*

Volume 44 WILLIAM R. COTTON AND RICHARD A. ANTHES. Storm and Cloud Dynamics. 1989

Volume 45 WILLIAM MENKE. Geophysical Data Analysis: Discrete Inverse Theory, Revised Edition. 1989

Volume 46 S. GEORGE PHILANDER. El Niño, La Niña, and the Southern Oscillation. 1990

Volume 47 ROBERT A. BROWN. Fluid Mechanics of the Atmosphere. 1991

Volume 48 JAMES R. HOLTON. An Introduction to Dynamic Meteorology, Third Edition. 1992

Volume 49 ALEXANDER A. KAUFMAN. Geophysical Field Theory and Method.
Part A: Gravitational, Electric, and Magnetic Fields. 1992
Part B: Electromagnetic Fields I. 1994
Part C. Electromagnetic Fields II. 1994

Volume 50 SAMUEL S. BUTCHER, GORDON H. ORIANS, ROBERT J. CHARLSON, AND GORDON V. WOLFE. Global Biogeochemical Cycles. 1992*

Volume 51 BRIAN EVANS AND TENG-FONG WONG. Fault Mechanics and Transport Properties of Rocks. 1992

Volume 52 ROBERT E. HUFFMAN. Atmospheric Ultraviolet Remote Sensing. 1992

Volume 53 ROBERT A. HOUZE JR. Cloud Dynamics. 1993

Volume 54 PETER V. HOBBS. Aerosol–Cloud–Climate Interactions. 1993

Volume 55 S. J. GIBOWICZ AND A. KIJKO. An Introduction to Mining Seismology. 1993

Volume 56 DENNIS L. HARTMANN. Global Physical Climatology. 1994

Volume 57 MICHAEL P. RYAN. Magmatic Systems. 1994

Volume 58 THORNE LAY AND TERRY C. WALLACE. Modern Global Seismology. 1995

Volume 59 DANIEL S. WILKS. Statistical Methods in the Atmospheric Sciences. 1995

Volume 60 FREDERIK NEBEKER. Calculating the Weather. 1995

* Out of Print.

Volume 61 MURRY L. SALBY. Fundamentals of Atmospheric Physics. 1996

Volume 62 JAMES P. MCCALPIN. Paleoseismology. 1996

Volume 63 RONALD T. MERRILL, MICHAEL W. MCELHINNY, AND PHILLIP L. MCFADDEN. The Magnetic Field of the Earth: Paleomagnetism, the Core, and the Deep Mantle. 1996

Volume 64 NEIL D. OPDYKE AND JAMES E. T. CHANNELL. Magnetic Stratigraphy. 1996

Volume 65 JUDITH A. CURRY AND PETER J. WEBSTER. Thermodynamics of Atmospheres and Oceans. 1998

Volume 66 LAKSHMI H. KANTHA AND CAROL ANNE CLAYSON. Numerical Models of Oceans and Oceanic Processes. 2000

Volume 67 LAKSHMI H. KANTHA AND CAROL ANNE CLAYSON. Small Scale Processes in Geophysical Fluid Flows. 2000

Volume 68 RAYMOND S. BRADLEY. Paleoclimatology, Second Edition. 1999

Volume 69 LEE-LUENG FU AND ANNY CAZANAVE. Satellite Altimetry. 2000

Volume 70 DAVID A. RANDALL. General Circulation Model Development. 2000

Volume 71 PETER WARNECK. Chemistry of the Natural Atmosphere, Second Edition. 2000

Volume 72 MICHALE C. JACOBSON, ROBERT J. CHARLSON, HENNING RODHE, AND GORDON H. ORIANS. Earth System Science: From Biogeochemical Cycles to Global Change. 2000

Volume 73 MICHAEL W. MCELHINNY AND PHILLIP L. MCFADDEN. Paleomagnetism: Continents and Oceans. 2000

Volume 74 ANDREW E. DESSLER. The Chemistry and Physics of Stratospheric Ozone. 2000

Volume 75 BRUCE DOUGLAS, MICHAEL KEARNEY AND STEPHEN LEATHERMAN. Sea Level Rise: History and Consequences. 2000

Volume 76 ROMAN TEISSEYRE AND EUGENIUSZ MAJEWSKI. Earthquake Thermodynamics and Phase Transformations in the Interior. 2001

Volume 77 GEROLD SIEDLER, JOHN CHURCH AND JOHN GOULD. Ocean Circulation and Climate: Observing and Modelling The Global Ocean. 2001

Volume 78 ROGER PIELKE. Mesoscale Meteorological Modeling, 2nd Edition. 2001

Printed and bound by CPI Group (UK) Ltd, Croydon, CR0 4YY

08/05/2025

01864786-0003